Mechanical Behaviour of Engineering Materials

Mechanical Behaviour of Engineering Materials

Volume 1: Static and Quasi-Static Loading

by

Yehia M. Haddad

Faculty of Engineering,
Mechanical Engineering,
University of Ottawa,
Canada

SPRINGER-SCIENCE+BUSINESS MEDIA, B.V.

A C.I.P. Catalogue record for this book is available from the Library of Congress.

ISBN 978-1-4020-0349-3   ISBN 978-94-010-9500-6 (eBook)
DOI 10.1007/978-94-010-9500-6

Printed on acid-free paper

To My Family.

Table of Contents

Preface

I wish to express my full indebtedness to all researchers in the field. Without their outstanding contribution to knowledge, this book would not have been written.

The author wishes to express his sincere thanks and gratitude to Professors M. F. Ashby (University of Cambridge), D. R. Axelrad (McGill University), N. D. Cristescu (University of Florida), N. Davids (The Pennsylvania State University), H. F. Frost (Dartmouth College), A. W. Hendry (University of Edinburgh), F. A. Leckie (University of California, Santa Barbara), A. K. Mukherjee (University of California, Davis), T. Nojima (Kyoto University), J. T. Pindera (University of Waterloo), J. W. Provan (University of Victoria), K. Tanaka (Kyoto University), Y. Tomita (Kobe University) and G. A. Webster (Imperial College), and to Dr. H. J. Sutherland (Sandia National Laboratories).

Permission granted to the author for the reproduction of figures and/or data by the following scientific societies, publishers and journals is gratefully acknowledged: ASME International, ASTM, Academic Press, Inc., Addison Wesley Longman (Pearson Education), American Chemical Society, American Institute of Physics, Archives of Mechanics / Engineering Transactions (*archiwum mechaniki stosowanej / rozprawy inzynierskie, Warsaw, Poland*), British Textile Technology Group, Butterworth-Heinemann Ltd. (USA), Chapman & Hall Ltd. (International Thomson Publishing Services Ltd.), Elsevier Science-NL (The Netherlands), Elsevier Science Limited (U.K.), Elsevier Sequoia S. A. (Switzerland), John Wiley & Sons, Inc., IOP Publishing Limited (UK), Kluwer Academic Publishers (The Netherlands), Les Editions de Physique Les Ulis (France), Pergamon Press Ltd. (U.S.A), Society for Experimental Mechanics, Inc. (USA), Springer-Verlag (Heidelberg, Germany), Steinkopff Verlag Darmstadt, Tappi, Technomic Publishing Co., Inc. (U.S.A.) and The Institute of Physics (UK).

I wish to express my deep appreciation to Drs. Karel Nederveen and Sabine Freisem (Publishing Editors), the sub-editorial staff at *Kluwer academic publishers*, and to *Kluwer* for the reviewing, editing, and the efficient production and distribution of the book.

I like to extend my thanks to past and present graduate students: Mrs. Y. Ping, and Messrs J. Feng, S. Iyer, P. Mirfakhraei and G. Molina, for their conscientious assistance during the preparation of the text. I am, also, grateful to my family for their understanding, patience and support.

I hope that the work presented in this book will provide guidance to science and engineering students, educators and researchers who are working in the field. Also, it is hoped that the book will be of value to scientists and engineers who are involved in the production, processing, and application of engineering materials.

Y. M. Haddad
University of Ottawa, Ottawa, Canada

List of Symbols

A	Energy per unit mass
A_0	Mean free energy
ds	Line element
C(t)	Creep function
$D(\cdot)/Dt$	Material derivative
e_k (k=1,2,3)	Unit vectors associated with an external Cartesian frame of reference
E	Elastic modulus
$E_1(\omega)$	Storage modulus (frequency-dependent)
$E_2(\omega)$	Loss modulus (frequency-dependent)
$E^*(i\omega)$	Dynamic complex modulus
E(t)	Nonlinear strain measure
$f(\cdot)$	Constitutive functional
F	Deformation measure
g	Deformation gradient
I	Identity matrix
I_1, I_2, I_3	Stress invariants
II_1, II_2, II_3	Strain invariants
$J^*(i\omega)$	Dynamic creep compliance
K	Bulk modulus
L	Velocity gradient
$L(\cdot)$	Constitutive function
n	Unit normal vector

$\overset{1}{n}, \overset{2}{n}, \overset{3}{n}$	Unit vectors associated with the stress tensor principal axes
$N(s), N'(s)$	Frequency distribution (creep, relaxation)
$s = 1/\lambda$	Frequency
t, τ	Time parameter
T	Absolute temperature
T	Stress-traction vector
u	Displacement vector
v	Velocity
v_{ij}	Rate of deformation tensor
γ	Specific rate (per unit mass) of entropy production
Δ	Dilatation
$\delta_{ij}, \delta_{\alpha\beta}$	Kronecker delta
∇^2	Laplace operator
$\epsilon_{ijk}, \epsilon_{\alpha\beta}$	Alternating tensor
ϵ_{ij}	Infinitesimal strain tensor
ϵ_{ij}'	Deviatoric strain tensor
η	Viscosity modulus
θ	Empirical temperature
υ	Poisson's ratio
ρ, ρ_0	Mass density (current, reference configuration)
$\sigma(t), \sigma_{ij}(t)$	Cauchy stress tensor (time-dependent)
σ_{ij}'	Stress deviator
$\bar{\sigma}$	Mean stress
$\sigma_1, \sigma_2, \sigma_3$	Principal components of stress
Σ	Piola-Kirchhoff stress tensor
χ	Body force

INTRODUCTION

During the past four decades, the research efforts of investigating the mechanical response behaviour of engineering materials, under various types of loading, have been ultimately significant. The interpretations and applications of mechanical response data have simulated powerful advances in research interest and in engineering practice. In this context, widespread research work on the subject has established well-profound concepts, principles and results.

The purpose of this monograph is to introduce the principles of the mechanical response of various classes of engineering materials, the identification and interpretation of the mechanical response data, properties evaluation, and, whenever possible, application of the data to structure-properties relationships. The monograph deals with the subject matter in two volumes. Volume I, contains eight chapters and three appendices, concerns itself with the basic concepts as pertain to the entire monograph, together with the response behaviour of engineering materials under static and quasi-static loading, Thus, Volume I is dedicated to the introduction, the basic concepts and principles of the mechanical response of engineering materials, together with the pertaining analysis of elastic, elastic-plastic, and viscoelastic behaviour. Volume II, consists of ten chapters and one appendix, concerns itself with the mechanical behaviour of various classes of materials under dynamic loading, together with the effects of local and microstructural phenomena on the response behaviour of the material. Volume II contains also selected topics concerning intelligent material systems and pattern recognition and classification methodology for the characterization of material response states. In the majority of the presentation, the two volumes of the monograph treat the considered subjects in a generalized three-dimensional fashion.

Static loading?

In the case of static loading, one has, at any particular instant of time, a condition of static equilibrium. A conventional static tensile test of a material specimen within its linear elastic range would be a typical example of this situation.

Quasi-static Loading?

A quasi-static deformation process, although it is, in general, time-dependent, is in reality a sequence of states of static equilibrium. Typical illustrations of a quasi-static deformation process are the quasi-static creep and relaxation processes of engineering materials.

Dynamic Loading?

The deformation process that occurs in the material under dynamic loading differs to a large extent from those due to static or quasi-static loading. When the material is subjected to dynamic loading, e.g., a very high rate impulse, the portion of the body that contains the point of impact is stressed instantaneously, while the other portions may not have yet experienced the effect of the imposed impact. This is due to the fact that the imposed dynamic effect will require time to travel, i.e., to propagate, through the body. Such propagation of the dynamic effect through the body occurs with a particular velocity of propagation which would depend on the specific characteristics of the material and the boundary conditions at the instant of time considered. This phenomenon is referred to as "*wave propagation*". That is, the dynamic deformation of materials, under dynamic loading, involves stress wave propagation, whereby the inertia and inner kinetics of the material specimen play important roles.

At strain rates of the order 10^{-6} to 10^{-5} s^{-1}, the creep behaviour of the material is the primary consideration and creep laws are used to describe the mechanical behaviour. At higher rates, e.g., in the range of 10^{-4} to 10^{-3} s^{-1}, a uniaxial test, or a quasi-static stress-strain curve obtained from a constant strain-rate test is used to describe the material behaviour. Although the quasi-static stress-strain curve is often treated as an inherent property of the material, it is a valid description of the material only at the strain rate at which the test was conducted. At higher strain rates, the mechanical response of the material may change, and alternate testing techniques have to be used. The range of strain rates from 10^{-1} to 10^{2} s^{-1} is generally referred to as the intermediate or medium strain-rate regime. Within this regime, strain-rate effects become a consideration in most materials (e.g., metals), although the magnitude of such effects may be quite small. Strain rates of 10^{3} s^{-1} or higher are generally treated as the range of high strain-rate response. It is within the high strain-rate range (10^{3} s^{-1} or higher) that inertia and wave-propagation effects become important in interpreting experimental data. At these high rates of strain, care must be taken to distinguish between average and local values of stress and that may be the result of one, or more, high-intensity stress wave propagating through the material. At strain rates of 10^{5} s^{-1} or higher, we are generally dealing with "*shock waves*" propagating through the material. At these high rates, there exists a transition from nominally isothermal conditions to adiabatic conditions.

In the mechanics of deformable media, we deal with physical events, e.g. deformation and flow, that occur and evolve, in both space and time, independent of any particular coordinate system that may be used to observe them. In a proper mathematical description, such events and their governing laws are expressed in terms of tensorial quantities. The invariance of tensors under coordinate transformation highlights a principal reason for employing tensor calculus in the study of the mechanics of deformable media. When transformation is carried out from one homogeneous (rectangular) coordinate system to another, the resulting tensors are identified as "*Cartesian tensors*". However, in dealing with tensor transformation between general "*curvilinear*" coordinate systems, the pertaining tensorial quantities are referred to as "*Curvilinear*" or "*General*" tensors. In *Chapter 1*, the reader is introduced to Cartesian tensors. Curvilinear tensors, however, are considered in *Appendix A (Volume I)*.

Two mechanical approaches are generally considered in the study of phenomena and problems concerning the mechanics of deformable media, i.e., the "*microstructural*" approach and the "*continuum mechanics*" approach.

In the "*microstructural*" approach, the macroscopic medium is considered to consist of a large number of structural elements. Such elements are assumed to be in continuous interaction with each other, and, hence their individual responses are seen to be mutually inter-dependent. The behavior of a statistical ensemble of such elements may be studied using, for instance, statistical or stochastic mechanics.

Conventionally, however, the description of material behavior is based on "*continuum mechanics*" models that generally refer to homogeneous media. In the "*continuum mechanics*" approach, the actual microstructure of the medium is disregarded and the medium is pictured as a "*continuum*" without gaps or empty spaces. Hence, the configuration of the assumed continuous medium would be described by a continuous mathematical model whose geometrical points are identified with material particles of the actual physical medium. The aim of *Chapter 2* is to provide the reader with a concise introduction of the basic assumptions and principles of *Continuum Mechanics* with an emphasis on those specifically used in the remainder of the book.

As mentioned earlier, engineering materials, when subjected to external loading, experience deformation and flow that evolve in space and time. Thus, in *Chapter 3*, we first consider the kinematics of involved deformation in the continuous material body and the determination of the pertaining strain by adopting a number of conventional measures. Second, we analyze the relationships between the sequential configurations that the parts of a "*continuous*" material body may acquire with the passage of time. Subsequently, in *Chapter 4*, we attempt to study the restrictions that classical thermodynamics impose on the theory of deformation of solids, and to seek information concerning the thermodynamics of continuous media.

Different materials of the same geometry may respond differently under identical external effects. Such difference in response is often attributed to the inherent constitution of the material. Consequently, the response behaviour of a particular material, or of a class of such material, is described mathematically by so-called "*constitutive relations*". These constitutive equations define the response behaviour of idealized media within a specific range of external effects. Accordingly, they only approximate the response characteristics of real materials, within a specified domain of actual service conditions. Constitutive relations establish, under certain physical and thermodynamical restrictions, the connection between the stimuli acting on the material specimen and the evolution of the occurring response. Thus, *Chapter 5* attempts to guide the reader throughout a transition between the general concepts and principles, which are presented in Chapters 1 to 4, and the task of establishing the response behaviour of engineering materials, as presented in Chapters 6 to 15. In this, the elastic response behaviour of the material is dealt with first.

Elastic behaviour of an engineering material depends only upon the stress level in the material, meanwhile, it is not strain- or time-history dependent. Further, an elastic deformation process is described, from a thermodynamical point of view, as dealt with in *Chapter 5*, to be a *reversible* process. Thus, upon the removal of the load, a complete recovery to the undeformed configuration would take place. An elastic response of an engineering material is formulated within the realm of *"classical elasticity"*. Such an elastic response could be linear or nonlinear pending on the form of the constitutive law that is used in its description. In this context, *Chapter 6* deals first with the general nonlinear elastic behavior, then it introduces the required assumptions and postulates in order to reduce such response to the idealized case of linear (perfect) elastic behavior.

Two ways in which the behaviour of real solids deviates from a perfect elastic one:

First, the stress-strain relationship may be nonlinear and may also depend on the loading path. Further, the pertaining stress-strain curve may show hysteresis loops. Thus, the resulting stress-strain relationships may not be *"uni-valued"*.

Second, the stress-strain relationship may be time-dependent. Thus, phenomena such as creep and stress-relaxation could become of importance, in determining the mechanical response of the real solid.

In general, *"inelastic"* solids show the above mentioned two types of deviation from a perfect (linear) elastic behavior. That is, the stress-strain relation is both time-dependent and nonlinear. Thus, inelastic deformation depends, in general, as dealt with in *Chapter 7*, on the stress level and both the strain- and time-history of the material. A transition to the important subject of creep and stress-relaxation of metallic systems is dealt with at the end of Chapter 7.

With the recent advances in material science and the parallel extensive industrial demands on advanced industrial materials such as high polymers and polymeric base composite systems, the identification of the viscoelastic response of engineering materials has gained recently a strong momentum in the realms of industrial techniques and applications.

High polymeric materials are organic substances of high molecular weight, the technical importance of which depends on their particular microstructure. This class of materials may include, for example, rubber in its various forms, synthetic rubber-like materials, commercial plastics, and natural and synthetic textile fibres. Other few examples of a viscoelastic material would include a wide range of inorganic polymeric systems such as silicones and glass resins, constituents of polymeric base systems, natural fibres such as wood and the by-products of such fibres as, for instance, paper and board, building materials such as concrete, and a large class of biomaterials, among others. These materials are *"time-dependent"* in response and possess a *"time-memory"*. Attempts to characterize the behaviour of such materials under the action of external loading, consequently, gave rise to the science of *"rheology"* within which the phenomena now labelled *"viscoelasticity"* is well defined and intended to convey mechanical behaviour combining response characteristics of both an elastic solid and a viscous fluid. A viscoelastic material is, thus, characterized by a certain level of rigidity of an elastic solid body, but, at the same time, it flows and dissipates energy by frictional losses as a viscous fluid. *Chapter 8* treats the subject of viscoelasticity of engineering materials in a quite comprehensive manner. The important subject of thermoviscoelasticity is also dealt with in Chapter 8.

The significant importance of the subject of the dynamic response of engineering materials has, also, gained in recent years a strong momentum in a wide scope of engineering practice. Dynamic properties of materials appear to be receiving more attention at present as a result of such current applications as space structures, machine components, advanced aircraft, and nondestructive evaluation of engineering materials and structures. Familiar applications of the study of dynamic deformation of engineering materials may include, for instance,

- identification, modelling, and prediction of the response behaviour of different classes of engineering materials under the effect of rapidly changing loads.
- development of new materials that can perform favourably from a design point of view when subjected to dynamic loading.
- study of the dynamic response of engineering members and structures with the inclusion of the dynamic behaviour of the pertaining materials.
- identification of the response of materials during dynamic fabrication processes, e.g. metal forming operations under rapidly changing loads, explosive welding and compaction operations.
- development of nondestructive evaluation techniques that are based on dynamic-effect phenomena, e.g., acoustic emission, ultrasonics and acousto-ultrasonics.
- shock synthesis to produce new elements or compounds.
- study of crash worthiness.
- development of anti-collision shielding for space vehicles.
- traditional and novel armour and anti-armour concepts for military applications.

In *Chapter 9*, we introduce the subject of the response of metallic materials to dynamic loading. In this, the distinction of higher rates from lower rates is made not on the basis of time-dependence of the material behavior, as we dealt with, for instance, in Chapter 8, but rather on the necessity of including inertia forces in the pertaining dynamic analysis.

Chapter 10 deals with the subject of plastic instability and localization effects in engineering materials. In this context, a decrease in stiffness due to geometrical change and/or material softening caused by deformation is responsible for the occurrence of instability phenomena in engineering materials within the plastic range; i.e., beyond the yield point. Such phenomena manifest themselves in various ways; e.g., buckling, bulging, necking and shear banding. Once such instabilities are started, they tend to persist and the stiffness of the specific cross-sectional area of the specimen decreases; therefore deformation intensifies locally and eventually leads to final collapse and/or failure. Because the occurrence of such instabilities is an important precursor to collapse or failure, computational prediction of the onset and of the augmentation of these instabilities is essential and indispensable in understanding the ultimate strength of engineering materials, and in predicting and improving, for instance, the formability of ductile solids.

In rigid body dynamics, it is assumed that, when an external force is applied to any one point of the body, the resulting effect sets every other point of the body instantaneously in motion, and the applied force can be considered as producing a linear acceleration of the whole body, together with an angular acceleration about its center of gravity. In the theory of deformable media, however, the body is considered to be in equilibrium under the action of the external applied forces, and the occurring deformations are assumed to have reached their equilibrium static values. This assumption could be sufficiently accurate for problems in which the time between the application of the force and the setting up of effective equilibrium is short compared with the time in which the observation is made. Meanwhile, If the external force is applied for only a short period of time, or it is changing rapidly, the resulting effect must be considered from the point of view of *"stress wave"* motion. Thus, when a localized disturbance is applied suddenly into a medium, it will propagate to other parts of this medium. The local excitation is not detected at other positions of the medium instantaneously, as some time would be necessary for the disturbance to propagate from its source to other parts of the medium. This simple fact constitutes a general basis for the interesting subject of *"wave propagation"*. In the particular case, when the suddenly applied disturbance is mechanical, e.g., an impact force, the resulting waves in the medium are due to mechanical stress effects and, thus, these waves are referred to as *"mechanical stress waves"*, or simply *"stress waves"*.

The propagation of stress waves in solids can be divided into two categories, *"elastic"* and *"inelastic"*. When loading conditions result in stresses below the yield point, solids behave elastically and obey Hook's Law, and consequently stress waves are *"elastic"*. As the intensity of applied loading is increased, the response of the material is driven out of the elastic range to a possible inelastic behavior. The behavior here may involve large deformation, internal heat generation, and often failure of the solid through a variety of mechanisms. In this context, *"plastic"* waves can be propagated in a material, such as a metal, which exhibits the phenomenon of yielding, when stressed beyond its proportional limit. The subject of elastic wave propagation in engineering materials is dealt with in *Chapter 11*. Meanwhile, in *Chapter 12*, we consider the plastic response of engineering materials under dynamic loading, whereby a rate-effect phenomenon might be occurring in the material and the inertia forces would be included in the equation of motion.

Chapter 13 deals with the identification problem of the linear viscoelastic response behaviour of an engineering material using dynamic experimental measurements. In this context, a linear viscoelastic material is considered as a dynamic system, whereby, a dynamic system identification method is developed for the determination of the relaxation or creep function of the material.

In most of viscoelastic material components, the presence of mechanical dissipation can effectively change the nature of wave motion in them. In addition to the significant mechanical dissipation that can occur in viscoelastic materials, it is well-recognized that these materials are "*dispersive*". In view of the latter property, phase velocity of a wave propagating in a viscoelastic material will depend on wave frequency. More specifically, waves of high frequency will propagate in viscoelastic materials with a greater phase velocity than if these waves have a low frequency. Consequently, a mechanical disturbance would continually change in shape during its motion in a viscoelastic medium. Further, the attenuation of high frequency waves in viscoelastic materials is greater than that of waves of low frequency. In this context, *Chapter 14* concerns itself with the phenomenon of wave propagation in a viscoelastic solid and the associated with boundary value problem.

The current technology of the design and manufacturing of laminated and fibre-reinforced composites is faced with problems essentially related to the inherent nature of the mechanical response of the different constituents of the microstructure, the formation of interfaces between such constituents and the evolution of the associated deformation processes under loading. Optimal design of such material systems is becoming a very progressive and challenging domain in both applied mechanics and material science. Thus, the increasing use of such materials is inciting new developments to be made within the context of macro- and micro-mechanical constitutive modelling, applications of such materials under variable boundary conditions, experimental testing methods, computational methods of analysis, and optimization. A new dimension of optimal design is being realized by building new composite systems through direct tailoring of the microstructure, e.g., by judicious reinforcement and mixing (hybridization) of the constituents of the microstructure within a specific topological frame of reference and to satisfy the boundary conditions involved. In this context, theoretical and experimental studies of the dynamic stress-strain relations of hybrid composites have become significantly important. The increased interest in the subject matter has been motivated recently by the increasing number of engineering applications and, as well, by the contributions provided by such studies to a better understanding of the mechanisms of deformation of such material systems when subjected to a dynamic loading environment. In this, *Chapter 15* reviews recent research efforts pertaining to the micromechanics of polymeric fibre-composite systems, in general, and the optimization of the microstructure in the case of short-fibre composite systems.

Chapter 16 deals with the microstructural or microscopic effects on the response behaviour of structured material systems. In this, the material system is considered as a heterogenous medium of actual microstructural elements. These elements are seen to exhibit random geometric and physical characteristics. Due to the discrete nature of the microstructure , the pertaining deformation process and its space- and time-evolutions are considered to be stochastic in character. Thus, the overall response behaviour of the material is formulated by the use of probabilistic concepts and statistical theory. An important feature of the presented analysis is the introduction of a so called "*Material Operator*" of the structured material system that contains in its argument the significant response characteristics of the microstructure. These concepts are, first, utilized to formulate the outlines of a stochastic micromechanical model for the deformation of a heterogenous elastoplastic system. Then, the presented approach is extended to include the analysis of probable internal damage mechanisms in this class of material.

Engineering materials are used either for their inherent structural strength or for their functional properties. Often a feed back control loop is designed so that the mechanical response of the material is monitored and the environment that is causing such a response can be controlled. The evolution of a new kind of material termed "*Intelligent*", "*Smart*", or "*Adaptive*" witnesses a significant development in materials science whereby the referred-to smart material adapts itself to suit the environment rather than necessitating to control the same. In this context, development in the area of materials research aims at incorporating intelligence into engineering materials, enabling them to sense the external stimuli and alter their own properties to adapt to the changes in the environment.
Chapter 17 presents "*an overview*" of possible forms of intelligence that may be incorporated in these materials. Three basic mechanisms of intelligent materials, namely, the sensor, processor and actuator functions are described. Implementation of these in the microstructure of various materials, as well as associated algorithms and techniques are illustrated. Different models, control algorithms and analyses are reviewed and their potential applications in engineering materials are presented.

Chapter 18 deals with the design procedure of a computer-based expert system, in conjunction with a non-destructive quantitative examination technique, e.g., acousto-ultrasonics, for the identification of material response states.

Acousto-ultrasonics (AU) is a relatively new quantitative non-destructive examination technique that combines aspects of conventional "*Ultrasonic*" and "*Acoustic Emission*" practices. It has been proven to be a suitable approach to quantify microstructural and morphological states of materials and the related mechanical properties.

In the AU practice, the multi-interactions of the ultrasonic-wave with the material microstructure usually result in complicated waveforms that are quite difficult to analyse. A relatively new approach to the analysis of AU signals is the use of "*Pattern-recognition and Classification Methodologies*". In this approach, acousto-ultrasonic waveforms are identified as belonging to a number of classes, where each class represents one of different states of the tested material-property. For this purpose, each waveform is mathematically treated as a multi-parametric entity, which is called a "*pattern vector*". Each component of such a pattern vector represents a value of a parameter, called "*feature*", which is used for the identification of the AU signal. In the pattern-recognition practice, a computer-based pattern-recognition system, labelled "*Pattern-recognition Classifier*", is designed on the basis of AU signals pertaining to known material states of a particular tested response or material property. Two case studies are being dealt with in Chapter 18, i.e., characterization of the stress-relaxation response of a class of polymeric system, and the identification of residual impact properties of such a system.

Throughout the text, generalized tensorial notations are used. For simplification, however, the presentation has been limited, as much as possible, to Cartesian tensors. *Appendix A* (Volume I), however, introduces to the reader the basics of *"Curvilinear or General tensors"*. This will prove to be particularly useful when reading Chapter 10. *Appendix B* (Volume I) presents the definition and a summary of the properties of both the delta and step functions. These functions are used frequently throughout the text. *Meantime*, the important subject of integral transformation is dealt with in *Appendix C* (Volume I). *Appendix D* (Volume II) deals with the definition and basic properties of z-transform. The latter is employed throughout Chapter 13.

In the presentation, vectors and unindexed tensor quantities are indicated in general by bold. The author has used majuscules to identify the undeformed configuration or the original state of the material and minuscules to designate the corresponding deformed state. Equations, figures and tables are numbered within the chapter; for example, Fig. 2.1 identifies Fig. 1 of Chapter 2.

Further Reading

Atkins, A. G. (1969) *The Rheology of Solids: High Speed Testing*, Vol. II, Wiley Interscience, New York.
Campbell, J. D. (1970) *Dynamic Plasticity of Metals*, Course held at the Department of Mechanics of Deformable Bodies, Udine, July 1970, Springer-Verlag, New York.
Cristescu, N. (1967) *Dynamic Plasticity*, North Holland, Amsterdam.
Davison, L. and Graham, R. A. (1979) Shock compression of solids, *Phys. Rep.* 55 (4), 255-379.
Duvall, G. F. (1962) Shock waves in the study of solids, *Appl. Mech. Rev.* 15, 849-54.
Edington, J. W. (1968) Effect of strain rate on the dislocation substructure in deformed Niobium single crystals, In *Mechanical Behavior of Materials under Dynamic Loads*, edited by U. S. Lindholm, Symposium held in San Antonio, Texas, Sept. 6-8, 1967, Springer-Verlag, New York, pp. 191-240.
Greszczuk, L. B. (1982) Damage in composite materials due to low velocity impact, in *Impact Dynamics*, Zukas, J. A., Nicholas, T., Swift, H. F., Greszczuk, L.B. and Curran (eds.), John Wiley & Sons, New York, pp. 55-94.
Haddad, Y. M. (1995) *Viscoelasticity of Engineering Materials*, Kluwer, Dordrecht, The Netherlands.
Haddad, Y. M., editor (1998) *Advanced Multilayered and Fibre-Reinforced Composites*, Kluwer, Dordrecht, The Netherlands.
Hahn, G. T., Kanninen, M. F., Mukherjee, A. K. and Rosenfield, A. R. (1968) The speed of ductile-crack propagation and the dynamics flow in metals. In: *Mechanical Behavior of Materials under Dynamic Loads*, edited by U. S. Lindholm, Symposium held in San Antonio, Texas, Sept. 6-8, 1967, Springer-Verlag, New York, pp. 96-133.
Hopkins, H. G. (1968) *Engineering Plasticity*, University Press, Cambridge,

Huffington, N. J., Jr. , ed. (1965) *Behavior of Materials Under Dynamic Loading*, ASME, New York.

Johnson, W. (1979) Application: Processes involving high strain rates. In *Mechanical Properties at High Rates of Strain*, 1979, Proceedings of the Second Conference on the Mechanical Properties of Materials at High Rates of Strain, J. Harding (editor), Oxford, 1979, The Institute of Physics, Conference Series Number 47, pp. 337-59.

Kolsky, H.(1963) *Stress Waves in Solids*, Dover, New York.

Kornhauser, M. (1964) *Structural Effects of Impact*, Cleaver Hume, London.

Kumar, S. (1968) Introduction- Mechanics/Materials, Aspects of Dynamic Loading. In: *M e c h a n i c a l Behavior of Materials under Dynamic Loads*, edited by U. S. Lindholm, Symposium held in San Antonio, Texas, Sept. 6-8, 1967, Springer-Verlag Inc., New York, pp. ix-xv.

Lindholm, U. S. (1965) Dynamic deformation of metals, in *Behavior of Materials Under Dynamic Loading*, Huffington, N. J., Jr. , (ed.), ASME, New York, pp. 42-61..

Lindholm, U. S. (1968) Some experiments in dynamic plasticity under combined stress, In: *Mechanical Behavior of Materials under Dynamic Loads*, edited by U. S. Lindholm, Symposium held in San Antonio, Texas, Sept. 6-8, 1967, Springer-Verlag, New York, pp. 77-95.

Meyers, M. A. (1994) *Dynamic Behaviour of Materials*, John Wiley & Sons, New York.

Nicholas, T. (1982) Material behavior at high strain rates, in *Impact Dynamics*, Zukas, J. A., Nicholas, T., Swift, H. F., Greszczuk, L.B. and Curran (eds.), John Wiley & Sons, New York, pp. 277-332.

Perzyna, P. (1968) On thermodynamic foundations of viscoplasticity, In *Mechanical Behavior of Materials under Dynamic Loads*, edited by U. S. Lindholm, Symposium held in San Antonio, Texas, Sept. 6-8, 1967, Springer-Verlag, New York, pp. 61-76.

Ripperger, E. A. and Watson, H., Jr. (1968) The relationship between the constitutive equation and one-dimensional wave propagation, In: *Mechanical Behavior of Materials under Dynamic Loads*, edited by U. S. Lindholm, Symposium held in San Antonio, Texas, Sept. 6-8, 1967, Springer-Verlag, New York, pp. 294-313.

CHAPTER 1

CARTESIAN TENSORS

1.1. Introduction

In the mechanics of deformable media, we deal with physical events, e.g. deformation and flow, that occur and evolve, in both space and time, independent of any particular coordinate system that may be used to observe them. That is if two coordinate systems, fixed at different locations in space, with a particular relative orientation to each other, are used to observe and, subsequently, to formulate a governing law of an occurring physical event, the resulting physical law should hold valid in the two coordinate systems and also in any other coordinate system in the same reference frame.

In a proper mathematical description, such events and their governing laws are expressed in terms of tensorial quantities. As a mathematical entity, a *"tensor"* has an existence independent of the coordinate system. That is, the components of a tensor in one coordinate system would determine the corresponding components of the same tensor in any other coordinate system that is not in motion relative to the first one. Further, the formulated tensorial equations that govern specific physical events would be also invariant under coordinate transformation. This invariance of tensors under coordinate transformation highlights a principal reason for employing tensor calculus in the study of the mechanics of deformable media. Invariance of the form of a physical law when referred to two coordinate systems not in the same frame of reference (i.e., in general motion relative to each other) would require an alternative approach within a general relative theory. In this case, a four dimensional frame of reference that would incorporate space and time coordinates would be required. In some cases, however, the so-called *"principal of material indifference"*, is used (e.g., Truesdell and Toupin, 1960, Truesdell and Noll, 1965, and Malvern, 1969).

When transformation is carried out from one homogeneous (rectangular) coordinate system to another, the resulting tensors are identified as *"Cartesian tensors"*. However, in dealing with tensor transformation between general *"curvilinear"* coordinate systems, the pertaining tensorial quantities are referred to as *"Curvilinear or General tensors"*. In this Chapter, the reader is introduced to Cartesian tensors. Curvilinear tensors, however, are dealt with in Appendix A of the book.

1.2. Indicial Notation

In writing a set of N quantities, say $a_1, a_2, ..., a_N$, the notation $a_i (i = 1, 2, ..., N)$ is generally used. In this notation, the subscript i is referred to as the *"index"* and it implies that it may take on any integer value from the set $\{1, 2, ..., N\}$. The number N is called the *"range"* of the index. Hence, the quantity a_i represents any element of the set $(a_1, a_2, ..., a_N)$ and simultaneously identifies the entire set.

In tensorial notation, a tensorial quantity which depends on *"m"* indices is conventionally referred to as an *" m-th order system"*. This is provided that any of the m indices associated with the tensorial quantity does not repeat itself. When an index appears unrepeated in a system, this index is referred to as *"free"* or *"alive"* index. The number and location of the free indices reveal directly the exact tensorial character of the system being dealt with. *Thus, the tensorial order of a given system is conventionally considered to be equal to the number of free indices attached to the system.* On the other hand, indices that appear repeated in a system are often referred to as *"dummy"*. A letter representing a dummy index can be replaced by any other letter (not appearing as a free index in the same system) without changing the meaning of the system or term to which it is attached.

EXAMPLE 1.1

Given below are examples of tensorial systems with *free* and *dummy* indices.

Free indices: $a_i b_j, c_{kl}, d_{ijkl}$
Dummy indices: $a_i b_i = a_j b_j = a_k b_k, \delta_{kk} = \delta_{ii}, C_{ijkk} = C_{ijss}$

If we denote the range of the index by the letter N, then, the number of components belonging to an m-th order tensorial system is given by N^m.

QUIZ 1.1

In Example 1.1, above, let the range N of the index to be equal 3, determine the number of components belonging to each tensorial system.

Following Example 1.1, if one adopts the convention that a Latin index has the range of 3, then, the coordinates of a considered point are identified, in a three-dimensional coordinate system, as x_i : x_1, x_2, x_3.

On the other hand, if a point is to be identified in a two-dimensional space, we may select to use a Greek index with the convention that the latter has a range of 2. Thus, the coordinates of the point are identified, in a two-dimensional coordinate system, as x_α: x_1, x_2.

As mentioned earlier, a tensorial quantity which depends on *"m" free* indices is conventionally referred to as an *"m-th order system"*. As illustrative examples, we mention the following systems:

First-order tensor. It is denoted by a kernel letter(s) bearing one free index only irrespective of the number of dummy indices that may be present. Examples of a first order tensor are

$$a_i,\ a^j,\ a_{ij}\,b_j,\ c_{jkk}$$
$$a_\alpha,\ a^\beta,\ a_{\alpha\beta}\,b_\beta,\ c_{\alpha\beta\beta}$$

A first-order tensor has three components in a three-dimensional coordinate system. Meantime, it has two components in a two-dimensional coordinate system. Thus, the components of a first order tensor represent the components of a vector, e.g. the deformation vector components u_k (or u_α), and the velocity vector components v_i (or v_α).

Second-order tensor. It is identified by two free indices. Thus, a second-order system has nine components in a three-dimensional coordinate system. An example of a second-order tensor is the system

$$a_{ij} = \begin{bmatrix} a_{11} & a_{12} & a_{13} \\ a_{21} & a_{22} & a_{23} \\ a_{31} & a_{32} & a_{33} \end{bmatrix}$$

Other designations of a second-order system in a three-dimensional space may take forms such as,

$$a^{ij},\ a_{ijkk},\ a_{.jll}^{i},\ a_i b_j$$

The dot, which appears in the third term above, indicates that the index j is in the second position. Meantime, in the last term, the second-order tensor $a_i b_j$ is expressed in a combination of two first-order tensors a_i and b_j.

Similarly, a second-order system in a two-dimensional space may take forms such as,

$$a_{\alpha\beta}, \, a_\alpha b_\beta, \, a_{\alpha\beta\gamma\gamma}, \, a^{\alpha\beta}$$

A second-order tensor has four components in any two-dimensional coordinate system. For instance, the second-order system

$$a_{\alpha\beta} = \begin{bmatrix} a_{11} & a_{12} \\ a_{21} & a_{22} \end{bmatrix}$$

Two familiar examples of a second-order system are, for instance, the stress tensor σ_{kl} (or $\sigma_{\gamma\delta}$) and the strain tensor ϵ_{ij} (or $\epsilon_{\alpha\beta}$).

Third-order tensor. It has three free indices which correspond to either 27 components in a three-dimensional space or 8 components in a two-dimensional space. Examples of a third-order tensor are

$$a_{ijk}, \, a^{ijk}, \, a_{.jk}^{i}, \, a_{ijkll}$$

$$a_{\alpha\beta\gamma}, \, a^{\alpha\beta\gamma}, \, a_{\alpha\beta\gamma\delta\delta}$$

A third-order tensor may be also constructed through a combination of first-order and second-order tensors. For example,

$$a_i \, b_{jk}, \, a_i \, b_j \, c_k, \, A_{ij} \, C_k$$

$$a_\alpha \, b_{\beta\gamma}, \, a_\alpha \, b_\beta \, c_\gamma, \, A_{\alpha\beta} \, C_\gamma$$

Fourth-order tensor. It has four free indices which correspond to 81 components in a three-dimensional space and 16 components in a two-dimensional space. Examples of a fourth-order tensor are

$$a_{ijkl}, \, a^{ijkm}, \, a_i \, b_{jkl}, \, a_{ij} \, b_{kl}$$

$$a_{\alpha\beta\gamma\delta}, \, a^{\alpha\beta\gamma\delta}, \, a_\alpha \, b_{\beta\gamma\delta}, \, a_{\alpha\beta} \, b_{\gamma\delta}$$

A generalized system of the m-th order. It has *m* free indices which correspond to 3^m components in a three-dimensional space, e.g.,

$$a_{ijkl...u} , a_{ijk} b_{lm...u}$$

Alternatively, a generalized system of the χ-order will have χ free indices attached to it. This corresponds to 2^χ components in a two-dimensional space. For instance,

$$a_{\alpha\beta\gamma...\eta} , a_{\alpha\beta} b_{\gamma\delta...\eta}$$

Meantime, a tensorial system with no index is referred to as *"zero-order tensor system"*. It has one component only, thus it is a scalar. Examples of zero-order tensors are, for instance, constants a, b, c; time parameter t; temperature T, ..., etc.

1.3. Coordinate Transformation

We consider the N-dimensional space Ω^N of real numbers. The coordinates of a point *P* in this space is presented by the set of real numbers, say x_i (i = 1, 2, ..., N), with the correspondence

$$P \sim x_i (i = 1, 2, ..., N) \tag{1.1}$$

For the same point *P*, another correspondence in an alternative coordinate system, in the same space Ω^N, may exist, i.e.,

$$P \sim x_i' (i = 1, 2, ... N) \tag{1.2}$$

Comparing (1.1) and (1.2), it is clear that

$$x_i \sim x_i' \quad \text{with} \quad i = 1, 2, ..., N \tag{1.3}$$

Meantime, the correspondence (1.3) expresses a relationship in the form

$$x_i' = f_i(x_i) \quad ; \quad (i = 1, 2, ..., N) \tag{1.4}$$

In view of (1.3), the inverse of (1.4) exists and is uniquely determined in the form

$$x_i = f_i'(x_i') \quad ; \quad (i = 1, 2, ..., N) \tag{1.5}$$

if and only if the *"Jacobean"* determinants $\left|\dfrac{\partial x_i'}{\partial x_j}\right|$ and $\left|\dfrac{\partial x_i}{\partial x_j'}\right|$, $(i, j = 1, 2, ..., N)$, are every where different from zero.

We consider now the set of reals $(u_1, u_2, ..., u_N)$ as components of a point relative to the coordinate system x_i $(i = 1, 2, ..., N)$ and another set of reals $(u_1', u_2', ..., u_N')$ as components of the same point relative to the coordinate system x_i' $(i = 1, 2, ..., N)$. Thus, provided that the *"Jacobean"* determinants between the two coordinate systems do not vanish, one can write that

$$u_i' = u_j \frac{\partial x_i'}{\partial x_j} \tag{1.6}$$

Now, as an extension of (1.6), we may consider the set of reals $\sum_{ij}$ with components defined relative to the x_i $(i = 1, 2, ..., N)$ coordinate system and the corresponding set of reals $\sum_{ij}'$ with components defined in the x_i' $(i = 1, 2, ..., N)$ coordinate system. Then, as an extension to (1.6), one writes that

$$\sum_{ij}' = \sum_{kl} \frac{\partial x_i'}{\partial x_k} \frac{\partial x_j'}{\partial x_l} \tag{1.7}$$

Further, in a general manner, one can write that

$$\sum_{ij...u}' = \sum_{kl...p} \frac{\partial x_i'}{\partial x_k} \frac{\partial x_j'}{\partial x_l} ... \frac{\partial x_u'}{\partial x_p} \tag{1.8}$$

Equations (1.7) and (1.8) are valid provided that the pertaining Jacobean determinants do not vanish.

Accordingly, one may define a tensor as a set of reals in the N-dimensional space Ω^N which transforms according to (1.8).

A tensor with indices written as subscripts is called a *"covariant tensor"*, e.g., the covariant tensor $\sum_{ij...n}$. Meantime, a tensor with indices designated as superscripts is referred to as a *"contravariant tensor"*, e.g., the contravariant tensor $\sum^{ij...n}$. A system may be also defined as one with both superscript- and subscript-indices at the same time, e.g., the ***mixed tensor*** $\sum_{ij...n}^{pq...u}$. The transformation law for a mixed tensor is written as combination of the transformation laws of both covariant and contravariant tensors. For instance, the transformation law for the mixed tensor $\sum_{ij...n}^{pq...u}$ is written as

$$\sum_{ij...n}^{pq...u} = \sum_{ab...f}^{gh...z} \frac{\partial x_p'}{\partial x_g} \cdots \frac{\partial x_u'}{\partial x_z} \frac{\partial x_a}{\partial x_i'} \cdots \frac{\partial x_f}{\partial x_n'} \tag{1.9}$$

The subject of covariant, contravariant and mixed tensors is further elaborated upon in *Appendix A* within the context of *"Curvilinear tensors"*.

1.4. Tensor Algebra

(i)   *Addition*: Addition of two tensorial systems $A_{ij...n}$ and $B_{pq...u}$ of the same order and range has the following properties:

-       Addition is commutative
        $A_{ij...n} + B_{pq...u} = B_{pq...u} + A_{ij...n}$
-       Addition is associative
        $A_{ij...n} + (B_{pfq...u} + C_{ab...f}) = (A_{ij...n} + B_{pq...u}) + C_{ab...f}$
-       There exists a unique system, 0, such that
        $A_{ij...n} + 0 = A_{ij...n}$
-       To every tensorial system $A_{ij...n}$, there corresponds a unique system $(-A_{ij...n})$ such that
        $A_{ij...n} + (-A_{ij...n}) = 0$

(ii)   *Multiplication*: An m-th order tensorial system may be multiplied with an n-th order system of the same range to produce an (m + n)th order system with the following properties:

-       Multiplication is commutative
        $A_i B_{jk} = B_{jk} A_i$
-       Multiplication is associative
        $A_i (B_{jk} C_{mn}) = (A_i B_{jk}) C_{mn}$
-       Multiplication is associative with respect to addition
        $A_{ij} (B_k + C_k) = A_{ij} B_k + A_{ij} C_k$
-       There exists a unique scalar of magnitude 1 (unity) such that,
        $1 . A_{ij...n} = A_{ij...n}$

(iii)   *Summation over repeated indices*: If x_i is a set of N variables and b_i is a set of N constants, a linear form may be written as
As shown above, the summation is carried out over repeated indices and, thus, the

$$\sum_{i=1}^{N} b_i x_i = b_1 x_1 + b_2 x_2 + \dots + b_N x_N$$

summation sign could be omitted.   Hence, the following convention may be adopted:

> *"The repetition of an index implies the summation over the range of that index in the absence of an explicit statement to the contrary"*.

As an example, consider the first-order tensor $a_{\alpha\beta}\, b_\beta$ in a two-dimensional coordinate system:

$$a_{\alpha\beta}\, b_\beta = a_{\alpha 1}\, b_1 + a_{\alpha 2}\, b_2$$
$$= (a_{11}\, b_1 + a_{12}\, b_2,\ a_{21}\, b_1 + a_{22}\, b_2)$$

(iv)   *Symmetry*: A tensor is described to be **symmetric**, if its matrix of components is symmetric. In a system of two or more indices, if the values of the components do not change by interchanging two indices, the system is said to be symmetric with respect to these two indices,   More generally, if the system is symmetric with respect to all indices, the system is referred to as *completely symmetric*. Examples of symmetric tensorial systems:

Symmetric in i & j: $A_{ijk} = A_{jik}$
Completely symmetric: $A_{ij} = A_{ji}$
$$A_{ijk} = A_{kij} = A_{jki} = A_{kij} = A_{ikj} = A_{jik}$$

The adjective *"symmetric"* is a tensorial property in a real sense. For instance, if the matrix of a tensor is symmetric in one Cartesian coordinate system, it would be symmetric in all such systems within the same frame of reference. However, the product of two symmetric tensors might not be necessarily symmetric.

(v)   *Skew-symmetry*: The definitions of skew-symmetry in tensors follow those for symmetry except that an interchange of a pair of indices would change the sign of the tensor. For example,

Skew-symmetric in i & j: $b_{ijk} = - b_{jik}$

Completely skew-symmetric: $b_{ij} = - b_{ji}$

$$b_{ijk} = b_{kij} = b_{jki}$$
$$= - b_{kji} = - b_{ikj} = - b_{jik}$$

In this context, it should be noted that:

- If, in a component of a skew-symmetric system, any two indices are indistinct, the value of such component must be zero. For example, if b_{ijk} is skew-symmetric in j and k, then $b_{i11} = - b_{i11} = 0$. Thus, in the skew-symmetric system $b_{ij} = - b_{ji}$, the diagonal elements a_{11}, a_{22}, a_{33} are all zero.
- If the order of a completely skew-symmetric system is equal to the range of its indices, then, the non-vanishing components of the system have only one distinct absolute numerical value. For example, if b_{ijk} is completely skew-symmetric, then, the only distinct term is $b_{1\,2\,3}$. The student is encouraged to illustrate the validity of this argument.

The property of "*skew-symmetry*" is sometimes referred-to as "*anti-symmetry*"

1.5. Special Tensors

The following tensors, in rectangular Cartesian coordinate systems, are frequently used. They are termed "*special tensors*" due to their particular characteristics as illustrated below.

1.5.1. THE KRONECKER DELTA

It is defined, in a three-dimensional coordinate-system, as

$$\delta_{ij} = \begin{cases} 1 & \text{if } i = j \\ 0 & \text{if } i \neq j \end{cases} = \begin{vmatrix} 1 & 0 & 0 \\ 0 & 1 & 0 \\ 0 & 0 & 1 \end{vmatrix} \qquad (1.10)$$

In a two-dimensional space, the Kronecker delta $\delta_{\alpha\beta}$ is defined as

$$\delta_{\alpha\beta} = \begin{cases} 1 & \text{if } \alpha = \beta \\ 0 & \text{if } \alpha \neq \beta \end{cases} = \begin{vmatrix} 1 & 0 \\ 0 & 1 \end{vmatrix} \qquad (1.11)$$

Consider the transformation of the Kronecker delta δ_{ij}, one can write

$$\delta_{ij} = \frac{\partial x_i}{\partial x_k'} \frac{\partial x_k'}{\partial x_j} = \delta_{km} \frac{\partial x_i'}{\partial x_k} \frac{\partial x_m}{\partial x_j'} \tag{1.12}$$

It can be seen from (1.12) above that δ_{ij} transforms as a second-order tensor.

QUIZ 1.2

Prove the substitution property of the Kronecker delta:

$$a_i \, \delta_{ij} = a_j, \, a_{ij} = a_{ik}\delta_{kj} \text{ and } a_\alpha \, \delta_{\alpha\beta} = a_\beta$$

QUIZ 1.3

Show that the Kronecker delta δ_{ij} is completely symmetric.

1.5.2. THE "ALTERNATING" OR "PERMUTATION" TENSOR

It exists for any completely skew-symmetric tensorial system in which the number of indices is equal to the range of index. The most important are the second and third-order alternating tensors defined, respectively, by

$$\varepsilon_{\alpha\beta} = \begin{cases} +1 & \text{if } \alpha, \beta \text{ is an even permutation of } 1, 2 \\ -1 & \text{if } \alpha, \beta \text{ is an odd permutation of } 1, 2 \\ 0 & \text{if } \alpha, \beta \text{ are indistinct} \end{cases} \tag{1.13}$$

That is

$$\varepsilon_{\alpha\beta} = \begin{vmatrix} 0 & 1 \\ -1 & 0 \end{vmatrix}$$

and

$$\varepsilon_{ijk} = \begin{vmatrix} 0 & 0 & 0 & & 0 & 0 & -1 & & 0 & 1 & 0 \\ 0 & 0 & 1 & & 0 & 0 & 0 & & -1 & 0 & 0 \\ 0 & -1 & 0 & & 1 & 0 & 0 & & 0 & 0 & 0 \end{vmatrix}$$

That is

$$\varepsilon_{ijk} = \begin{cases} +1 & \text{if } i, j, k \text{ is an even permutation of } 1, 2, 3 \\ -1 & \text{if } i, j, k \text{ is an odd permutation of } 1, 2, 3 \\ 0 & \text{if any two of } i, j, k \text{ are indistinct} \end{cases} \qquad (1.14)$$

In view of the definition of the alternating tensor ε_{ijk}, one can prove the following expansion of the determinant $[a_{ij}]$,

$$[a_{ij}] = \varepsilon_{ijk} \, a_{i1} \, a_{j2} \, a_{3k} \qquad (1.15)$$

Further, if one multiplies both sides of (1.15) by another alternating tensor, say ε_{lmn}, one obtains

$$\varepsilon_{lmn} [a_{ij}] = \varepsilon_{lmn} \, \varepsilon_{ijk} \, a_{il} \, a_{jm} \, a_{kn}$$

i.e.,

$$\varepsilon_{lmn} = [a_{ij}]^{-1} \, \varepsilon_{ijk} \, a_{il} \, a_{jm} \, a_{kn} \qquad (1.16)$$

Relation (1.16) above shows that the alternating tensor ε_{ijk} transforms as a third-order tensor. Because the transformation of the alternating tensor ε_{ijk} incorporates the system $[a_{ij}]^{-1}$, ε_{ijk} is often referred to as a relative tensor of weight -1. The alternating tensor ε_{ijk}

complies with Cartesian tensor transformation law for third order tensors provided that the transformation is a proper one (i.e., det. $a_{ij} = 1$ in Eqn.(1.16)) above.

From the definition of the alternating tensor ε_{ijk}, the cross product $a \times b = c$ is expressed in indicial notations by $\varepsilon_{ijk} a_i b_j = c_k$ Using this relationship, the treble product $a \times b \cdot c = \lambda$ is written as $\lambda = \varepsilon_{ijk} a_i b_j c_k$. Meantime, the "*dual vector*" of an arbitrary second-order Cartesian tensor T_{ij} is defined by:

$$v_i = \varepsilon_{ijk} T_{jk}$$

QUIZ 1.4

Show that the completely skew-symmetric systems $b_{\alpha\beta}$ and b_{ijk} can be written, respectively, as

$$b_{\alpha\beta} = \varepsilon_{\alpha\beta} b_{12} \quad \text{and} \quad b_{ijk} = \varepsilon_{ijk} b_{123}$$

where $\varepsilon_{\alpha\beta}$ and ε_{ijk} are alternating tensors.

1.6. Some Applications of Cartesian Tensors

1.6.1. ISOTROPY

We consider the transformation of a second-order tensor with components B_{ij} in the x_i coordinate system to corresponding components B'_{ij} in the x'_i coordinate-system, within the same frame of reference. If this transformation is orthogonal and if further the components B_{ij} are found to be invariant with respect to this transformation, then, the tensor is called "*isotropic*". Considering the different ranks (orders) of a tensor, it can be noted that:

- All tensors of zero-order (scalars) are isotropic.
- There is no isotropic first-order tensor, since all first-order tensors are vectors which are, of course, directional.
- The Kronecker delta δ_{ij} (or $\delta_{\alpha\beta}$) and scalar multiples of it are the only nontrivial

second-order isotropic tensors.

- The only nontrivial isotropic third-order tensors are those representing rectangular Cartesian components that are given by the alternating tensor ε_{ijk} and scalar multiples of it.
- The most general fourth-order isotropic tensor E_{ijkl} has rectangular Cartesian components expressed by

$$E_{ijkl} = \lambda\, \delta_{ij}\, \delta_{kl} + \mu\, (\delta_{ik}\, \delta_{jl} + \delta_{il}\, \delta_{jk}) + \upsilon\, (\delta_{ik}\, \delta_{jl} - \delta_{il}\, \delta_{jk})$$

where λ, μ and υ are parameters that are invariants with respect to coordinate transformation (see, e. g., Aris, 1962 and Malvern, 1969).

1.6.2. AN IMPORTANT PROPERTY OF A SECOND-ORDER TENSOR

Any second-order system may be represented as the sum of a symmetric and a skew-symmetric part. For instance, the second order system a_{ij} may be expressed by

$$a_{ij} = a_{(ij)} + a_{[ij]} \tag{1.17}$$

in which the first term on the right hand side is symmetric while the second is skew-symmetric. These two terms are expressed, respectively, by

$$a_{(ij)} = \frac{1}{2}\, (a_{ij} + a_{ji}) \; ; \; \text{symmetric} \tag{1.17a}$$

$$a_{[ij]} = \frac{1}{2}\, (a_{ij} - a_{ji}) \; ; \; \text{skew symmetric} \tag{1.17b}$$

EXAMPLE 1.2

$$\text{If } a_{\alpha\beta} = \begin{vmatrix} \cos\theta & -\sin\theta \\ \sin\theta & \cos\theta \end{vmatrix}$$

The symmetric part of the above system is expressed by

$$a_{(\alpha\beta)} = \tfrac{1}{2}(a_{\alpha\beta} + a_{\beta\alpha}) = \begin{vmatrix} \cos\theta & 0 \\ 0 & \cos\theta \end{vmatrix}$$

and the skew-symmetric part is

$$a_{[\alpha\beta]} = \tfrac{1}{2}(a_{\alpha\beta} - a_{\alpha\beta}) = \begin{vmatrix} 0 & -\sin\theta \\ \sin\theta & 0 \end{vmatrix}$$

Accordingly, it is clear that

$$a_{\alpha\beta} = a_{(\alpha\beta)} + a_{[\alpha\beta]}$$

EXAMPLE PROBLEM 1.1

If δ_{ij} denotes the Kronecker delta and ε_{ijk} designates the permutation tensor, show that:

(a) $\delta_{ii} = 3$
(b) $\delta_{ij}\,\delta_{jk} = \delta_{ik}$
(c) $\delta_{ik}\,\varepsilon_{ikl} = 0$

Solution:

(a)
$$\delta_{ii} = \delta_{11} + \delta_{22} + \delta_{33}$$
$$= 1 + 1 + 1 = 3$$

(b)
$$\delta_{ij}\,\delta_{jk} = \delta_{il}\,\delta_{lk}$$
$$= \delta_{ik}$$

(c)
$$\delta_{ik}\,\varepsilon_{ikl} = \sum_{i=1}^{3} \sum_{k=1}^{3} \left(\delta_{ik}\,\varepsilon_{ikl} \right)$$
$$= \sum_{i=1}^{3} \left[\left(\delta_{ik}\,\varepsilon_{ikl} \right)_{k=i} + \sum_{k=1}^{3} \left(\delta_{ik}\,\varepsilon_{ikl} \right)_{k \neq i} \right]$$
$$= 0$$

QUIZ 1.5

Prove the ε - δ relationship: $\varepsilon_{\alpha\beta}\,\varepsilon_{\delta\gamma} = \delta_{\alpha\delta}\,\delta_{\beta\gamma} - \delta_{\alpha\gamma}\,\delta_{\beta\delta}$

EXAMPLE PROBLEM 1.2

Using tensor notation, show that

$$A \times (B \times C) = (A . C)\,B - (A . B)\,C$$

Solution:

$$A \times (B \times C) = A_i\, e_i \times (B_j\, C_l\, \varepsilon_{jlk}\, e_k)$$
$$= \varepsilon_{jlk}\, A_i\, B_j\, C_l\, \varepsilon_{ikm}\, e_m$$
$$= -\varepsilon_{kjl}\, \varepsilon_{kim}\, A_i\, B_j\, C_l\, e_m$$
$$= -(\delta_{ji}\, \delta_{lm} - \delta_{jm}\, \delta_{li})\, A_i\, B_j\, C_l\, e_m$$
$$= A_i\, C_l\, B_j\, e_j - A_i\, B_i\, C_l\, e_l$$
$$= (A . C)\, B - (A . B)\, C$$

1.7. Principle Values and Principle Directions of Symmetric Second-Order Tensors

Consider the symmetric second-order tensor $\sum_{ij}$ to be defined at some point in space with reference to a three-dimensional Cartesian frame of reference. Then, there is associated with each direction, specified by a unit normal n_j, at that point, a vector, say T_i, given by the relation

$$T_i = \sum_{ij} n_j \tag{1.18}$$

In equation (1.18) above, $\sum_{ij}$ may be seen as an operator on the unit normal n_j to produce its conjugate T_i. If the direction in question is one for which the vector T_i is parallel to the unit normal n_j, then, the inner product (1.18) may be expressed as a scalar multiple of n_j, i.e.,

$$\sum_{ij} n_j = \lambda\, n_i \tag{1.19}$$

whereby the direction of n_j defines a principal direction or principal axis of the tensor $\sum_{ij}$. Combining (1.19) with the identity $n_i = \delta_{ij}\, n_j$, it follows that

$$\left(\sum_{ij} - \lambda\, \delta_{ij}\right) n_j = 0 \tag{1.20}$$

For a non-trivial solution of (1.20), the determinant of coefficients, in this equation, must be zero. That is

$$\left| \Sigma_{ij} - \lambda \delta_{ij} \right| = 0 \qquad (1.21)$$

The expansion of the determinant (1.21) produces the following cubic polynomial in λ

$$\lambda^3 - I\lambda^2 + II\lambda - III = 0 \qquad (1.22)$$

where the scalar coefficients I, II and III are given by

$$I = \Sigma_{ij} = \Sigma_{11} + \Sigma_{22} + \Sigma_{33}$$
$$= tr \, \Sigma_{ij}$$
$$II = \tfrac{1}{2} \left(\Sigma_{ii} \, \Sigma_{jj} - \Sigma_{ij} \, \Sigma_{ij} \right) \qquad (1.23)$$
$$III = \left| \Sigma_{ij} \right| = \frac{1}{6} \varepsilon_{ijk} \, \varepsilon_{pqr} \, \Sigma_{ip} \, \Sigma_{jq} \, \Sigma_{kr}$$

The above scalar coefficients are referred-to as *first, second* and *third invariants*, respectively, of the second-order tensor Σ_{ij}. Meantime, the three roots $\lambda_{(1)}$, $\lambda_{(2)}$ and $\lambda_{(3)}$ of the cubic equation (1.22) are referred-to as the *"principal values"* of the tensor Σ_{ij} (see *Chapter 2*).

1.8. Tensor Fields

Consider the first-order tensor v_i which may represent, for instance, the velocity vector at a point in a three-dimensional space. If the components of this tensor depend in some manner upon the coordinates of the point, say

$$v_i(\mathbf{x}) = v_i(x_1, x_2, x_3)$$

these components are said to be *"point functions"* and the tensor quantity is referred to as *"tensor field"*; or simply, in the example cited, a velocity vector field.

Further, a tensor field could be time-dependent if its components, in addition to being dependent upon the coordinates, are, time-dependent. For instance, the time-dependent tensor field:

$$v_i(\mathbf{x}, t) = v_i(x_1, x_2, x_3, t)$$

In the expression of the tensor field above, it is understood that the position vector $\mathbf{x} : (x_1, x_2, x_3)$ varies over a specified domain in space and the time parameter t changes within a particular extent of time. Other examples of tensor fields which are dependent on the position and time are

- the temperature field $T = T(\mathbf{x}, t)$. It is a scalar field.
- the displacement field $u_i = u_i(\mathbf{x}, t)$. It is a first-order tensor field (i.e., a vector field).
- the strain tensor field $\epsilon_{ij}(\mathbf{x}, t)$, and, for instance, the stress tensor field $\sigma_{kl}(\mathbf{x}, t)$. Both are tensor fields of second-order.
- the mth-order tensor field $A_{ij\ldots u}(\mathbf{x}, t) = A_{ij\ldots u}(x_1, x_2, x_3, t)$
- if the components of a tensor field are dependent on the position vector x only, this tensor field is identified as being *steady*, e.g. the isothermal temperature tensor field $T(\mathbf{x})$ and the elastic strain tensor field $\epsilon_{ij}^E(\mathbf{x})$ where the superscript E refers to the strain as being elastic. Further, a tensor field is described as continuous (or differentiable) if its components are continuous (or differentiable) functions of its arguments. In continuum mechanics, one is usually concerned with tensor fields the components of which, together with all partial derivatives, with respect to the coordinates are continuous functions (see *Chapter 2*).

An important property of tensor fields is that if all components of a tensor field vanish in one coordinate system, they vanish likewise in all coordinate systems which can be obtained by admissible transformations.

1.9. Divergence and Gradient Operations

1.9.1. GRADIENT OF A SCALAR

A time-dependent scalar field $\phi = \phi(x_i, t)$, where x_i represent the components of a position vector x, may be partially differentiated with respect to x_i. This partial differentiation is conventionally denoted by a comma, i.e.

$$\frac{\partial \phi}{\partial x_i} = \phi_{,i}$$

and the associated vector is represented by

$$\nabla \phi = \phi_{,i} \, e_i$$

where e_i are the unit vectors associated with the pertaining Cartesian coordinate system.

In the above expression, the vector field $\nabla\phi$ is referred to as the gradient of the scalar field ϕ. In this context, the symbol ∇ may be regarded to be an operator such that

$$\nabla \equiv {}_{,i} \, e_i$$

1.9.2. GRADIENT OF A VECTOR

Consider a vector $A = A_i \, e_i$ with components A_i being a time-dependent vector field: $A_i = A_i \, (x_i, t)$. The partial derivative of the components with respect to the coordinates is expressed as

$$\frac{\partial A_i}{\partial x_j} = A_{i,j} = \begin{vmatrix} A_{1,1} & A_{1,2} & A_{1,3} \\ A_{2,1} & A_{2,2} & A_{2,3} \\ A_{3,1} & A_{3,2} & A_{3,3} \end{vmatrix}$$

These nine quantities $A_{i,j}$ are the components of the gradient of the vector A.

1.9.3. DIVERGENCE OF A VECTOR

The quantity

$$A_{i,i} = A_{1,1} + A_{2,2} + A_{3,3}$$

is defined as the divergence of the vector **A** and is expressed by

$$\text{div } \mathbf{A} = A_{i,i}$$

QUIZ 1.6

Show that $\text{div } \mathbf{A} = \nabla \cdot \mathbf{A} = A_{i,i}$

1.9.4. LAPLACIAN OPERATOR

Following the definitions of the operations above, one may consider the divergence of the gradient of a scalar as

$$\nabla \cdot \nabla \phi = \phi_{,ii}$$
$$= \nabla^2 \phi$$

where the operator ∇^2 is the Laplacian operator.

1.9.5. CURL OF A VECTOR

The vector $\nabla \times \mathbf{A}$ is referred to as the curl of the vector A, and it may be represented as follows

$$\nabla \times \mathbf{A} = {}_{,i}\, e_i \times A_j\, e_j$$
$$= A_{j,i}\, e_i \times e_j$$
$$= A_{j,i}\, \varepsilon_{ijk}\, e_k$$
$$= \varepsilon_{ijk}\, A_{j,i}\, e_k$$

1.10. Common Tensor Operations

– *Inner Product (Composition)*

$$A_{ij} = B_{ik} C_{kj}$$

$$(\alpha A_{ij} + \beta B_{kl}) R_{mn} = \alpha A_{ij} R_{mn} + \beta B_{kl} R_{mn}$$

$$A_{ij}^2 = A_{ij} A_{ij}^T$$

– *Transpose*

$$(A_{ij} + B_{kl})^T = A_{ij}^T + B_{k\ell}^T$$

$$B_{ij}^{-1} B_{ij} = B_{ij} B_{ij}^{-1} = 1$$

$$(A_{ij} B_{kl})^{-1} = B_{kl}^{-1} A_{ij}^{-1}$$

$$\det(A_{ij}^{-1}) = (\det A_{ij})^{-1}$$

$$(A_{ij}^T)^{-1} = (A_{ij}^{-1})^T$$

– *Trace*

$$\alpha = \operatorname{tr} A_{ij} = A_{ii}$$

$$\operatorname{tr}(A_{ij} + B_{kl}) = \operatorname{tr} A_{ij} + \operatorname{tr} B_{kl}$$

$$= A_{ii} + B_{kk}$$

$$\operatorname{tr}(A_{ij} B_{kl}) = \operatorname{tr}(B_{kl} A_{ij}) = B_{kk} A_{ii}$$

$$= A_{ii} B_{kk}$$

$$\operatorname{tr}(A_{ij}^T) = \operatorname{tr}(A_{ij}) = A_{ii}$$

–   *Scalar Product*

$$\alpha = A_{ij} \cdot B_{kl} = A_{ij} B_{kl}$$
$$A_{ij} \cdot B_{kl} = B_{kl} \cdot A_{ij}$$

–   *Magnitude*

$$\alpha = |A_{ij}| = \sqrt{A_{ij} \cdot A_{ij}} = \sqrt{A_{ij} A_{ij}}$$

–   *Determinant*

$$\beta = \det A_{ij}$$
$$= \frac{1}{3!} \, \varepsilon_{ijk} \, \varepsilon_{rst} \, A_{ir} \, A_{js} \, A_{kt}$$

where ε_{ijk} is the previously defined alternating (permutation) tensor.

$$\det (a \, A_{ij}) = a^3 \det (A_{ij})$$
$$\det (A_{ij} \, B_{kl}) = \det (B_{kl} \, A_{ij}) = \det A_{ij} \ \det B_{kl}$$
$$\det (-A_{ij}) = -\det (A_{ij})$$
$$\det (A_{ij}^{T}) = \det (A_{ij})$$

–   *Inverse*

If $A_{ri} = B_{st}^{-1}$

then,

$$A_{ri} = \frac{1}{2 \det B_{st}} \ \varepsilon_{ijk} \, \varepsilon_{rst} \, B_{js} \, B_{kt}$$

where ε_{ijk} is the alternating (permutation) tensor.

1.11. Problems

1. Determine the order of the following tensorial systems:

$$\delta_{ij}, \ a_i, \ \varepsilon_{ijk}, \ b, \ a_{ij} \, b_{ij}, \ \varepsilon_{\alpha\beta}, \ \delta_{ij} \, \delta_{ik} \, \delta_{jk}$$

2. If $a_i \sim (2,4,8)$ and $b_i \sim (-2,4,-3)$, find the components of $c_i = a_i + b_i$.

3. If δ_{ij} denotes the Kronecker delta and ε_{ijk} is the alternating tensor, show that:

   (i) $\delta_{ij} \, \delta_{ij} = 3$
   (ii) $\varepsilon_{ijk} \, \varepsilon_{jki} = 6$
   (iii) $\delta_{ij} = \delta_{ik}$

4. In the skew-symmetric system $a_{\alpha\beta}$, identify the values of the diagonal elements of the corresponding determinant.

5. Display the elements of the skew-symmetric system d_{ijk}

6. (a) If a_{ij} is symmetric and b_{ij} is skew-symmetric, show that
   $$a_{ij} \, b_{ij} = 0$$

   (b) Furthermore, if d_{ijk} is skew-symmetric in the indices i and j, show that
   $$a_{ij} \, d_{ijk} = 0$$

7. In Cartesian tensor coordinates, show that the Kronecker delta δ_{ij} transforms as a second-order tensor, and that the alternating tensor ε_{ijk} transforms as a third-order tensor.

8. If $\alpha b_{ij} + \beta b_{ji} = 0$ show that either

   i) $\alpha + \beta = 0$ and b_{ij} is symmetric
   or ii) $\alpha - \beta = 0$ and b_{ij} is skew-symmetric.

9. (i) Show that $a_{ij} \, \delta_{ik} \, \delta_{jl} = a_{kl}$

   (ii) Express both the cross product $\mathbf{a} \times \mathbf{b} = \mathbf{c}$ and the treble product

$\mathbf{a} \times \mathbf{b} \cdot \mathbf{c} = \lambda$ in indicial notation using the alternating tensor ε_{ijk}

10. Prove that $\delta_{ij} \delta_{ik} \delta_{jk} = 3$, where δ_{ij} is the Kronecker delta.

11. Show that:

a) $\varepsilon_{\alpha\beta} \varepsilon_{\gamma\beta} = \delta_{\alpha\gamma}$
b) $\varepsilon_{\alpha\beta} \varepsilon_{\alpha\beta} = 2$

12. (a) Using the indicial notation, prove the following vector identities

(i) $\nabla \times \nabla\phi = 0$ \qquad\qquad (ii) $\nabla \cdot \nabla \times \mathbf{a} = 0$

(b) If A_{tm} is a second order tensor, show that its derivative with respect to X_n, i.e., $\partial A_{tm}/\partial X_n$, is a Cartesian tensor of the third order.

13. (a) For arbitrary vectors $\mathbf{P}$ and $\mathbf{Q}$, both of order unity, show that:

$$\mu = (\mathbf{P} \times \mathbf{Q}) \cdot (\mathbf{P} \times \mathbf{Q}) + (\mathbf{P} \cdot \mathbf{Q})^2 = P^2 Q^2$$

(b) If A_{ij} is a symmetric tensor and B_{ij} a skew-symmetric, show that:

$$A_{ij} B_{ij} = 0$$

14. (a) If $b = \text{Det. } b_{ij}$, verify that $b = \varepsilon_{ijk} b_{1i} b_{2j} b_{3k}$

(b) Show that:

$$\text{div. } \mathbf{T} = \nabla \cdot \mathbf{T} = T_{i,i}$$

15. Show that:

(a) $\varepsilon_{ijm} \varepsilon_{ijk} = 2 \delta_{mk}$

(b) $\varepsilon_{ijm} \varepsilon_{ijm} = 6$

16.   Given the components of the transformation

$$a_{ij} = \begin{vmatrix} \dfrac{1}{2} & -\dfrac{1}{2} & 0 \\[2mm] \dfrac{1}{2} & \dfrac{1}{2} & 0 \\[2mm] 0 & 0 & 1 \end{vmatrix}$$

and that the components of the vector **A** in the unprimed coordinate system are

$$A_i - (1,2,3),$$

find A_i'.

17.   Show that $\varepsilon_{ijk}\, a_j b_k\, c_i = \varepsilon_{ijk}\, a_i b_j c_k$

18.   Given the vectors **Q**, **V** and **U**, show, using Cartesion tensor methods, that

(a) $\nabla \cdot (Q\,V) = \nabla Q \cdot V + Q \nabla \cdot V$ (b) $\nabla X (Q\,V) = \nabla Q\, X\, V + Q \nabla\, X\, V$

(c) $\nabla \cdot (U\, X\, U) = (\nabla\, X\, U)$

19.   Show that:

(i)

$$A_{ij} \cdot (B_{kl}\, C_{mn}) = (A_{ij}\, C_{mn}^{T}) \cdot B_{kl}$$
$$= (B_{kl}^{T}\, A_{ij}) \cdot C_{mn}$$

(ii)

If $a = \det A_{ij}$

then $a = \dfrac{1}{3!}\, \varepsilon_{ijk}\, \varepsilon_{rst}\, A_{ir}\, A_{js}\, A_{kl}$

where ε_{ijk} is the alternating (permutation) tensor.

20.   Prove that:

(i) $\det(a\,A_{ij}) = a^3 \det(A_{ij})$

where a is a scalar

(ii) $\det(-A_{ij}) = -\det(A_{ij})$

(iii) $\det(A_{ij}^T) = \det(A_{ij})$

(iv) $\det(A_{ij}\,B_{kl}) = \det(B_{kl}\,A_{ij}) = \det A_{ij} \det B_{kl}$

21.   Illustrate that:

$$A_{kl}^{ij} = A_{kl}^{(ij)} + A_{kl}^{[ij]}$$

1.12. References

Aris, R. (1962) *Vectors, Tensors and the Basic Equations of Fluid Mechanics*, Prentice-Hall, Englewood Cliffs, NJ.

Malvern, L.E. (1969) *Introduction to the Mechanics of a Continuous Medium*, Prentice Hall, Englewood Cliffs, NJ.

Truesdell, C. and Noll, W. (1965) The Nonlinear field theories of mechanics, in *Encyclopedia of Physics*, Vol. III/3 (ed. S. Flügge), Springer, Berlin.

Truesdell, C. and Toupin, R.A. (1960) Classical field theories, in *Handbuch der Physic*, Vol. III/1 (ed. S. Flügge), Springer, Berlin, pp. 226-90.

1.13. Further Reading

Bishop, R.L. and Goldberg. S.I. (1968) *Tensor Analysis and Manifolds*, Dover Publications, New York.

Borg, S.F. (1963) *Matrix-Tensor Methods in Continuum Mechanics*, Van Nostrand, New York.

Brillouin, L. (1946) *Les Tenseurs en Mécanique et en Élasticité*, Dover Publications, New York.

Coburn, N. (1955) *Vector and Tensor Analysis*, Macmillan, New York.

Cotter, B.A. and Rivlin, R.S. (1955) Tensors associated with time-dependent stress, *Quart. Appl. Math.* **13**, 177-82

Eisenhardt, L.P. (1926) *An Introduction to Differential Geometry*, Princeton University Press, Princeton, NJ.

Eisenhart, L.P. (1949) *Riemannian Geometry*, Princeton University Press, Princeton, NJ.

Ericksen, J.L. (1960) Tensor fields, in *Encyclopedia of Physics*, Vol. 3/1 (ed. S. Flügge), Springer, Berlin, pp. 794-858.

Eringen, A.C. (1962) *Nonlinear Mechanics of Continua*, McGraw-Hill, New York.

Eringen, A. C. (1967) *Mechanics of Continua*, Wiley, New York.

Eringen, A. C. (1971) *Continuum Physics*, Vol. 1, Academic Press, New York.

Fung, Y.C. (1965) *Foundations of Solid Mechanics*, Prentice Hall, Englewood Cliffs, NJ.

Hay, G.E. (1953) *Vector and Tensor Analysis*, Dover Publications, New York.

Jefferys, H. (1931) *Cartesian Tensors*, Cambridge University Press, Cambridge.

Landau, L. and Lifshitz, E. (1951) *The Classical Theory of Fields* (translated from Russian by H. Hammermesh), Addison-Wesley, Reading, MA.

Lass, H. (1950) *Vector and Tensor Analysis*, McGraw-Hill, New York.

Levi-Civita, T. (1927) *The Absolute Differential Calculus* (translated from Italian by M. Long), Blackie, London.

Lodge, A.S. (1951) On the use of convected coordinate systems in the mechanics of continuous media, *Proc. Cambridge Phil. Soc.* **47**, 575-84.

McConnell, A.J. (1946) *Applications of the Absolute Differential Calculus*, Blackie, London.

Michal, A.D. (1947) *Matrix and Tensor Calculus with Applications to Mechanics, Elasticity and Aeronautics*, Wiley, New York.

Naghdi, P.M. and Wainwright, W.L. (1961) On the time derivative of tensors in mechanics of continua, *Quart. Appl. Math.* **19**, Number 1, 95-119.

Nakada, O. (1960) Theory of nonlinear responses, *J. Phys. Soc. Jpn.* **15**, 2280-8.

Ricci, G. and Levi-Civita, T. (1901) Methodes du Calcul différentiel absolu et leurs applications, *Math. Ann.* **54**, 125-201.

Schouten, J. (1951) *Tensor Analysis for Physicists*, Oxford University Press, New York.

Sokolnikoff, I. (1964) *Tensor Calculus*, Wiley, New York.

Spain, B. (1953) *Tensor Calculus*, Interscience, New York.

Spiegel, M.R. (1959) *Theory and Problems of Vector Analysis and an Introduction to Tensor Analysis*, Schaum, New York.

Synge, J. and Schild, A. (1949) *Tensor Calculus*, University of Toronto Press, Toronto.

Thomas, T.Y. (1955) Kinematically Preferred Coordinate System, *Proc. Natl. Acad. Sci.* **41**, pp.762-70.

Truesdell, C. (1953) The physical components of vectors and tensors, *Zeis Angew. Math. U. Mech.* **33**, 345-56.

CHAPTER 2

ANALYSIS OF STRESS

2.1. Introduction

Two mechanical approaches are generally considered in the study of phenomena and problems concerning the mechanics of deformable media, i.e., the *microstructural* approach and the *continuum mechanics* approach.

In the microstructural approach, the macroscopic medium is considered to consist of a large number of structural elements. Such elements are often seen to be in continuous interaction with each other, and, hence their individual responses are considered to be mutually inter-dependent. The behaviour of a statistical ensemble of such elements may be studied using, for instance, statistical or stochastic mechanics (e.g., Axelrad and Haddad, 1998, and Haddad, 1998).

Conventionally, however, the description of material behaviour is based on continuum mechanics models that generally refer to homogeneous media.

In the continuum mechanics approach, the actual microstructure of the medium is disregarded and the medium is pictured as a "*continuum*" without gaps or empty spaces. Hence, the configuration of the assumed continuous medium is described by a continuous mathematical model whose geometrical points are identified with material particles of the actual physical medium. Further, when such a continuum changes its configuration under some boundary conditions, such change is assumed to be continuous, i.e., neighbourhoods evolve into neighbourhoods. Thus, the mathematical functions entering the analysis of a deformation process, are assumed to be continuous functions with continuous derivatives. Any creation of new boundary surfaces, such as those developed by internal fracture may, then, be seen as extraordinary events that might require alternative formulations outside the realm of continuum mechanics.

The aim of this chapter is to provide the reader with a concise introduction of the basic assumptions and principles of Continuum Mechanics with an emphasis on those specifically used in the remainder of the book. Section 2.2 deals with the definition of the "*continuous*" medium, Section 2.3 introduces to the reader a number of fundamental principles of the continuum mechanics theory. Section 2.4 treats the analysis of stress from a continuum mechanics point of view. Throughout the Chapter, Cartesian tensor notations are used. A reader who is not familiar with Cartesian tensor operations is advised to consult

Chapter 1 concurrently with this chapter.

2.2. The "Continuous" Medium

As mentioned in the foregoing, problems of the response of deformable substances are often mathematically analyzed by introducing the concept of a *"continuum"* or *"continuous"* medium. In this idealization, it is assumed that properties, averaged over a very small element, e. g., the mean mass density, the mean displacement, the mean interaction force, etc., vary continuously with position in the medium, so that we may speak about the mass density, the displacement and the stress, as functions of position. Such functions become time-dependent if we are dealing with a time-dependent deformation or flow process.

2.2.1. THE CONCEPT OF "STRESS AT A POINT"

A basic concept in continuum mechanics is that of the definition of *"stress at a point"* (Cauchy, 1827a&b, 1828). Reference is being made here to a geometric point that has no volume and may be associated with a mathematical limit of an elemental region of the continuum when the volume of such region shrinks down to zero. This is, in essence, similar to the definition of the derivative in differential calculus and follows immediately from the postulate of continuity of the medium. Through its connection with the definition of the derivative, the concept of the stress at a point makes the powerful methods of calculus available for the analysis of the deformation process or flow in continuous media.

2.2.2. HOMOGENEOUS MATERIAL

A continuous material specimen is homogeneous if it has identical material properties at all points. Hence, under this assumption, the material specimen is uniform and its material properties are independent of the position.

2.2.3. ISOTROPIC MATERIAL

A continuous material specimen is isotropic with respect to a certain material property if such property has the same value in all directions, i.e., independent of the orientation of any reference coordinate system that may be chosen to measure the property. Hence, such a property remains constant in any plane that passes through a point in the material specimen.

2.3. Fundamental Principles of Continuum Mechanics

2.3.1. CONSERVATION OF MASS

From a classical mechanics point of view, mass is assumed to be conserved. Hence, the mass of a material body is considered to be a time-independent property. In continuum mechanics, it is further postulated that mass is an absolutely continuous function of volume. Hence, it is assumed that a positive quantity ρ, referred to as the "*density*", can be defined at every point in the body by (see, e.g., Fung, 1965)

$$\rho(x) = \underset{i \to \infty}{\text{Limit}} \frac{\text{mass of } \Omega_i}{\text{volume of } \Omega_i} \tag{2.1}$$

where Ω_i is a suitably chosen infinite sequence of particle-sets shrinking down upon the point of the medium. This point is identified by the current position vector x referred to a particular coordinate frame of reference. The counterpoint of the latter, in the undeformed configuration, is given the majuscule symbol X. The mapping between the two positions x and X is dealt with in Chapter 3 where we discuss the deformation kinematics, and the measures of strain and strain-rate in a continuous medium.

The conservation of mass is expressed by

$$\int \rho(x)dx_i = \int \rho(X)dX_I \tag{2.2}$$

where the integrals extend over the same sets of particles. Hence,

$$\rho(X) = \rho(x) \left| \frac{\partial x_i}{\partial X_I} \right| \quad \text{and} \quad \rho(x) = \rho(X) \left| \frac{\partial X_I}{\partial x_i} \right| \tag{2.3}$$

Equation (2.2), or (2.3), relates, then, the densities of the body for different configurations of the coordinate-system.

2.3.2. MATERIAL DERIVATIVE OF A FIELD FUNCTION

A field function is a function that is attributable to the motion of a point in a continuum. It is often expressed in terms of the position vector x and the time t, e.g., the field function $\phi(x_i, t)$. The derivative of a field function is expressed by the so-called "*material derivative*". The latter is denoted by an over-dot or, alternatively, by the derivative symbol D/DT. It can be shown that the material derivative of a field function $\phi(x,t)$ takes the form

$$\dot{\phi}(x,t) = \frac{D\phi(x,t)}{Dt}$$
$$= \left(\frac{\partial\phi}{\partial t}\right)_{x=constant} + v_i \frac{\partial\phi}{\partial x_i}$$

$$(2.4)$$

where v_i is the velocity associated with the current position x_i. The first term on the right-hand side of (2.4) is due to the time-dependence of the function ϕ and is interpreted as the "*local part*" of the field function. Meantime, the second term, on the right-hand side of (2.4), is contributed by the motion of the particle in the current field of the function ϕ and is referred to as the "*convective part*" of the field function. The field function $\phi(x,t)$ may take the form of a tensor field of any order (see *Chapter 1*).

2.3.3. CONTINUITY OF MASS EQUATION

In the derivation of the continuity of mass equation, one may consider a constant volume while taking into account the variation in mass between entry and exit. Thus, the rate of mass increase in an arbitrary fixed volume V is equal to the influx of matter through its surface S, i.e.,

$$\int_V \frac{\partial\rho}{\partial t}\, dV = -\int_S \rho v_i\, n_i\, dS \qquad (2.5)$$

In the above equation, the negative sign accounts for the influx being opposite to the direction of the outward unit normal n_i and v_i is the velocity of the material entering the control volume.

Recalling Gauss's divergence theorem (e.g., Flügge,1972), then, Eqn. (2.5) can be written as

$$\int_V \left[\frac{\partial \rho}{\partial t} + (\rho v_i)_{,i} \right] dV = 0$$

whereby the *"continuity of mass equation"* is expressed as

$$\frac{\partial \rho}{\partial t} + (\rho v_i)_{,i} = 0 \tag{2.6}$$

An alternative expression may be found by carrying through the indicated partial differentiation in (2.6) to obtain

$$\frac{\partial \rho}{\partial t} + \rho_{,i} v_i + \rho v_{i,i} = 0 \tag{2.7}$$

Employing, the derivative operator $D(\cdot)/Dt$ introduced earlier by Eqn.(2.4), then, in terms of this operator, the *continuity of mass equation*, (2.7), is expressed as

$$\frac{D\rho}{Dt} + \rho v_{i,i} = 0 \tag{2.8}$$

2.3.4. CONTINUITY OF MOMENTUM

Following the approach adopted above in deriving the continuity of mass equation, we consider an arbitrary fixed volume V whereby the total rate of change of linear momentum has two components; one is associated with the change of mass within the volume V and the other is associated with the influx of the mass through the bounding surface S. Thus,

$$\frac{d}{dt} \int_V \rho v_i dV = \int_V \frac{\partial(\rho v_i)}{\partial t} dV + \int_S (\rho v_i) v_j n_j dS \tag{2.9}$$

which can be written by using Gauss's divergence theorem, introduced earlier, as

$$\frac{d}{dt} \int_V \rho v_i dV = \int_V \left[\frac{\partial \rho}{\partial t} v_i + \rho \frac{\partial v_i}{\partial t} + (\rho v_i v_j)_{,j} \right] dV$$

$$= \int_V \left\{ \left[\frac{\partial \rho}{\partial t} + \rho_{,j} v_j + \rho v_{j,j} \right] v_i \right. \tag{2.10}$$

$$\left. + \rho \frac{\partial v_i}{\partial t} + \rho v_{i,j} v_j \right\} dV$$

In view of the *continuity of mass equation* (2.7), expression (2.10) reduces to

$$\frac{d}{dt} \int_V \rho v_i dV = \int_V \rho \left[\frac{\partial v_i}{\partial t} + v_{i,j} v_j \right] dV \tag{2.11}$$

Expression (2.11) is referred to as the *"continuity of momentum equation"*.

2.4. Analysis of Stress

2.4.1. BODY AND SURFACE FORCES

The external forces acting at any time on a free body are classified, from a continuum mechanics point of view, in two categories, namely *"body forces"* and *"surface forces"*.

Body Forces
Such forces act on the elements of mass or volume inside the body. Hence, in continuum mechanics, they are expressed as forces per unit mass or forces per unit volume. Examples of body forces are those due to gravity, magnetic effects and inertia. In this book, *unless otherwise stated*, we regard the body forces acting on a free body as being expressed per unit volume. Hence, the term *"volumetric forces"* may be used as a reference to these body forces throughout the text.

We shall denote the body force per unit volume acting on an infinitesimal volume element dV of the body by the vector χ. If the resultant of body forces acting on an elemental volume ΔV is designated by ΔB, then, the body force is defined by

$$\chi = \lim_{\Delta V \to 0} \frac{\Delta B}{\Delta V} \tag{2.12}$$

In general, the vector χ, at any given instant of time, varies from a point to another in the free body and it may also vary with time at any particular point of the body. It can be also dependent on other state variables such as the temperature. The vector sum of all body forces acting on a free body of a finite volume V, at any particular instant of time, is then given by the space integral over this volume, i.e., $\int_V \chi \, dV$.

However, in many applications, the body forces are likely to be uniform, e.g., gravity forces, or, otherwise, they may be small enough so that they could be assumed to be negligible.

Surface forces

These forces are exerted on the bounding surface of the free body, by another body or due to other external effects. Hence, they are usually expressed per unit area of the surface upon which they are acting.

The limit of the surface force per unit area, as the unit area tends to zero, is often referred to as *"stress vector"*, or, alternatively, *"traction vector"*. It is denoted, in this book, by the symbol **T**. Accordingly, the force acting on an elemental area dS of the bounding surface is **T** dS and the vector sum of all surface forces acting on a finite region S of this surface is given by the corresponding vector surface integral, that is $\int_S \mathbf{T} \, dS$.

Generally, when applied to a given solid body, the definitions of body and surface forces are taken with reference to the current deformed configuration of the body. However, in many applications of the theory of elasticity, the occurring deformations might be so small that the definitions of such forces could be expressed with reference to the undeformed configuration of the body without a significant error. Alternatively, in applications concerning fluid flow, one may use as a reference, a given fixed volume of space, through which different substance or fluid passes at different times. In this context, the concept of the *"imaginary control surface"* in a fluid is well recognized in the studies of fluid mechanics (e.g., Malvern, 1969).

2.4.2. STRESS VECTOR

We consider, in Fig. 2.1, below, a continuous free body being acted upon by an externally applied equilibrium force system $F_1, F_2, \ldots F_N$. Suppose, now, that the body be divided, through a bounding surface S, into two parts R_1 and R_2. If the part of the body R_1, for example, is to be in equilibrium, forces must be exerted on S by the part R_2. These forces would be in equilibrium with the externally applied forces on R_1. We consider a point p on S with unit normal n surrounded by an elemental surface area ΔS of S. The forces acting on ΔS are statically equivalent to a force and a couple. At the limit, as ΔS shrinks down to a point, the couple (per unit area) produced, at this point by the continuous distribution of the internal forces, may be taken to be zero. This does not, of course, exclude the possibility of the existence of a "*couple-stress*" whose value may be different from zero. Such couple stress has been proposed in "higher order" continuum mechanics theories; see, e.g., Malvern (1969). However, for our purpose, we shall assume, following classical continuum mechanics, that there is no couple stress acting on ΔS and that the action of one body on another across an infinitesimal surface area can be presented solely by the *stress vector*. The latter is defined, at the point p of the elemental surface area ΔS with normal n, (Fig. 2.1), by

$$T(n) = \underset{\Delta S \to 0}{\text{Lim}} \frac{\Delta F}{\Delta S} \tag{2.13}$$

Thus, $T(n)$ is a stress vector function defined on the elemental surface area ΔS and is dependent on the unit normal n defining ΔS .

The stress state at any point in a continuum can be determined in terms of the three stress vectors acting on three mutually perpendicular planes intersecting at this point. Hence, the stress vectors acting on planes perpendicular to the axes of a rectangular coordinate system originating at a point which is embedded in a continuum are considered to be of particular interest. Figure 2.2 shows three such stress vectors acting on the center points of three faces of an infinitesimal cube surrounding a chosen particular point of interest "O". The point "O" represents, in turn, the center of the cube and, at the same time, the origin of the rectangular coordinate system. In Fig. 2.2, $T_i (n_j)$ is the stress vector acting on a plane whose normal n_j is pointing in the positive x_j-direction.

With reference to Fig. 2.2, the three stress vectors $T_i (n_j)$ are considered to be the average traction vectors on the respective faces of the cube. As the cube shrinks down towards its centre point O, then, the limit approached by the stress vector on a face of the cube is taken, from continuum mechanics point of view, as a stress vector acting at the point O on this particular face which, in effect, represents a plane perpendicular to one of the coordinate axes.

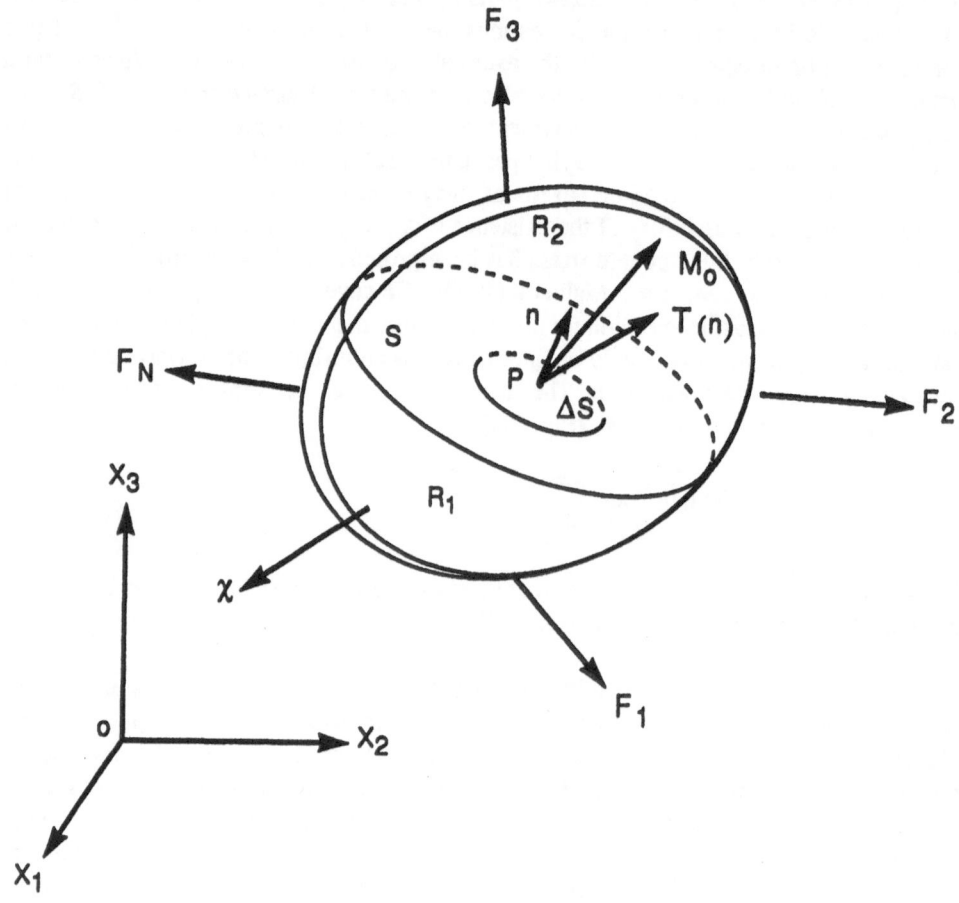

Figure 2.1. Body and surface forces.

Hence, the three stress vectors may be seen to represent the local stress state at the point Ω. This approach illustrates a basic concept in continuum mechanics, i.e., the *"concept of stress at a point"*, mentioned earlier in the introduction to this chapter. This concept refers to a geometric point in space visualized as a mathematical limit in a manner similar to the definition of the derivative in differential calculus. *The concept of "stress at a point" has been, in fact, a key to the development of continuum mechanics as it immediately makes the powerful methods of calculus available for the study of the deformation and flow in a physical continuum.*

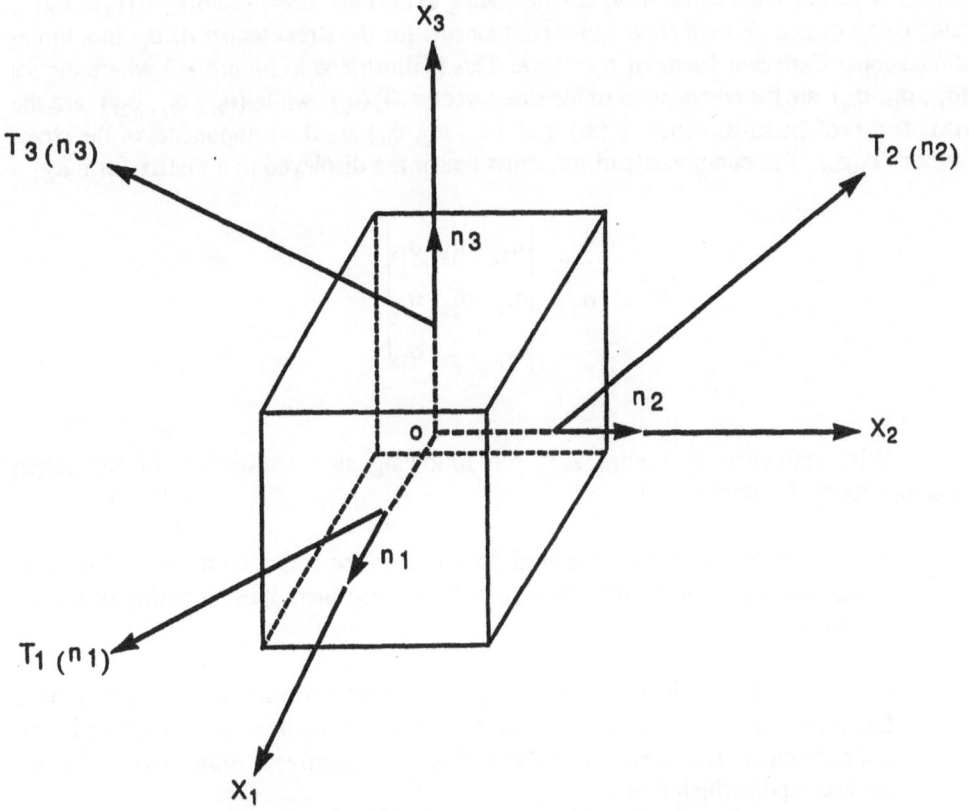

Figure 2.2. Traction vectors acting on three planes perpendicular
to the rectangular coordinate axes (*with origin al at O*).

2.4.3. STRESS TENSOR

As mentioned earlier, the set of the three stress vectors $T_1(n_1)$, $T_2(n_2)$ and $T_3(n_3)$ are
taken as a representation of the stress state at the particular point under consideration. This
translates into a corresponding stress tensor defined as the linear vector function which
associates with each unit normal vector n, on a face of the cube, a pertaining traction
vector $T(n)$. Denoting the referred-to stress tensor by σ, then, it follows that

$$T(n) = n\sigma \tag{2.14}$$

The three sets of stress components corresponding to the three stress vectors $T_1(n_1)$, $T_2(n_2)$ and $T_3(n_3)$ constitute nine rectangular components for the stress tensor σ: σ_{ij}, in a three-dimensional Cartesian frame of reference. This is illustrated in Figure 2.3 where the set $(\sigma_{11}, \sigma_{12}, \sigma_{13})$ are the components of the stress vector $T_1(n_1)$, while $(\sigma_{21}, \sigma_{22}, \sigma_{23})$ are the components of the stress vector $T_2(n_2)$ and $(\sigma_{31}, \sigma_{32}, \sigma_{33})$ are the components of the stress vector $T_3(n_3)$. The components of the stress tensor are displayed in a matrix form as

$$\sigma_{ij} = \begin{vmatrix} \sigma_{11} & \sigma_{12} & \sigma_{13} \\ \sigma_{21} & \sigma_{22} & \sigma_{23} \\ \sigma_{31} & \sigma_{32} & \sigma_{33} \end{vmatrix}$$

With reference to Figure 2.3, the following sign convention of the stress components σ_{ij} is adopted.

(a) σ_{ij} $(i \neq j)$ is the shear component of the stress tensor acting on a face of the cube whose normal is in the ith direction and the component itself is acting in the jth direction.

(b) σ_{ii} (no sum over i) is the normal component of the stress tensor. It is acting on a face whose normal is in the ith direction and the component itself is acting in the same direction. The normal component of stress is positive if drawn outward from the face upon which it acts.

(c)   If the outward normal n to a surface is in the positive direction of the associated-with coordinate axis, then, the shear component of stress acting on this surface is positive if it is in the positive direction of the axis with which the shear component being associated. Alternatively, if the outward normal to a surface is in the negative direction of the associated-with coordinate axis, then, the shear component is positive if it is in the negative direction of the axis with which the shear component being associated. With reference to Figure 2.3, positive σ_{23}, for instance, represents an upward-acting stress component on the right side, or a down-acting component on the left side of the cube. Negative σ_{23}, on the other hand, acts downward on the right side and upward on the left side of the cube. Further, to comply with the condition of static equilibrium of the cube, the components acting on the negative sides of the cube will have senses opposite to those acting on the positive sides.

(d)   When the normal component of stress, σ_{ii} (no sum over i), is positive, it represents a tensile stress, but if it is negative, it designates compression. The algebraic sign of a tangential shear component, however, has not an intrinsic physical significance. Hence, positive and negative shear components represent the same kind of loading, but, in different directions.

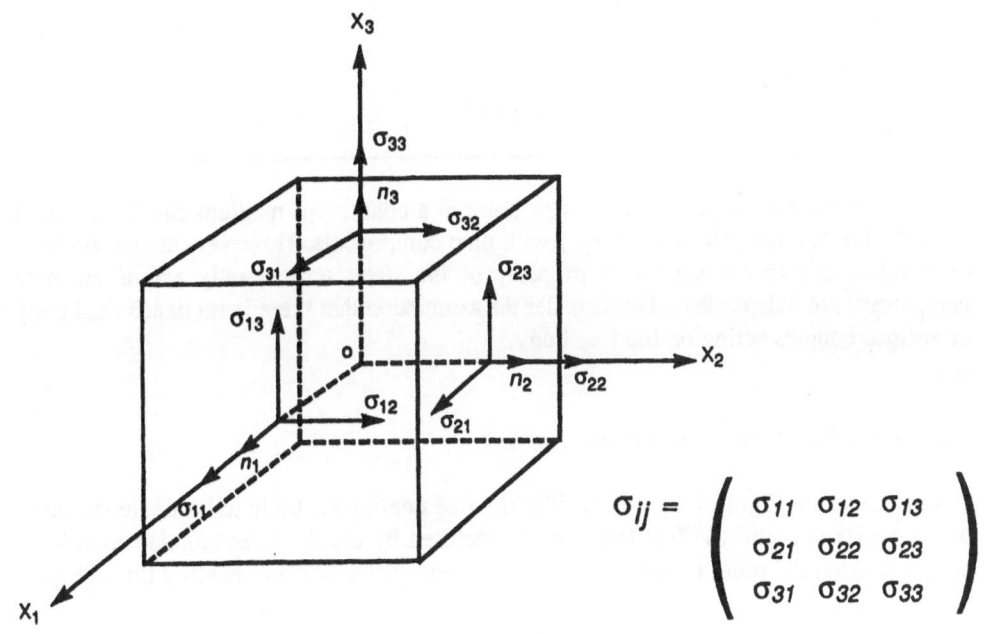

$$\sigma_{ij} = \begin{pmatrix} \sigma_{11} & \sigma_{12} & \sigma_{13} \\ \sigma_{21} & \sigma_{22} & \sigma_{23} \\ \sigma_{31} & \sigma_{32} & \sigma_{33} \end{pmatrix}$$

Figure 2.3. Stress tensor components.

QUIZ 2.1

Label the stress components shown in the figure below.

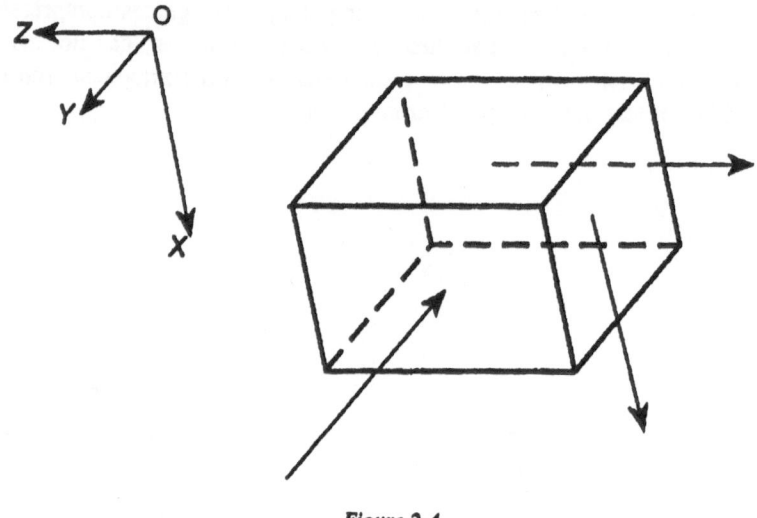

Figure 2.4

The state of stress at a particular point in a continuous medium can be specified fully by the second-order tensor σ_{ij}, with nine components. However, as will be dealt with later, due to the symmetry property of the stress tensor, only six of the nine components are independent. This is under the assumption that there is no distributed body or surface couples acting on the free body.

2.5. Stress Boundary Conditions

With reference to Figure 2.5, we consider the free continuous body to be the tetrahedron or triangular pyramid *OABC*. The latter is enclosed by the three rectangular coordinate planes through the point *O* (the origin) and a fourth plane *ABC* not passing through *O*.

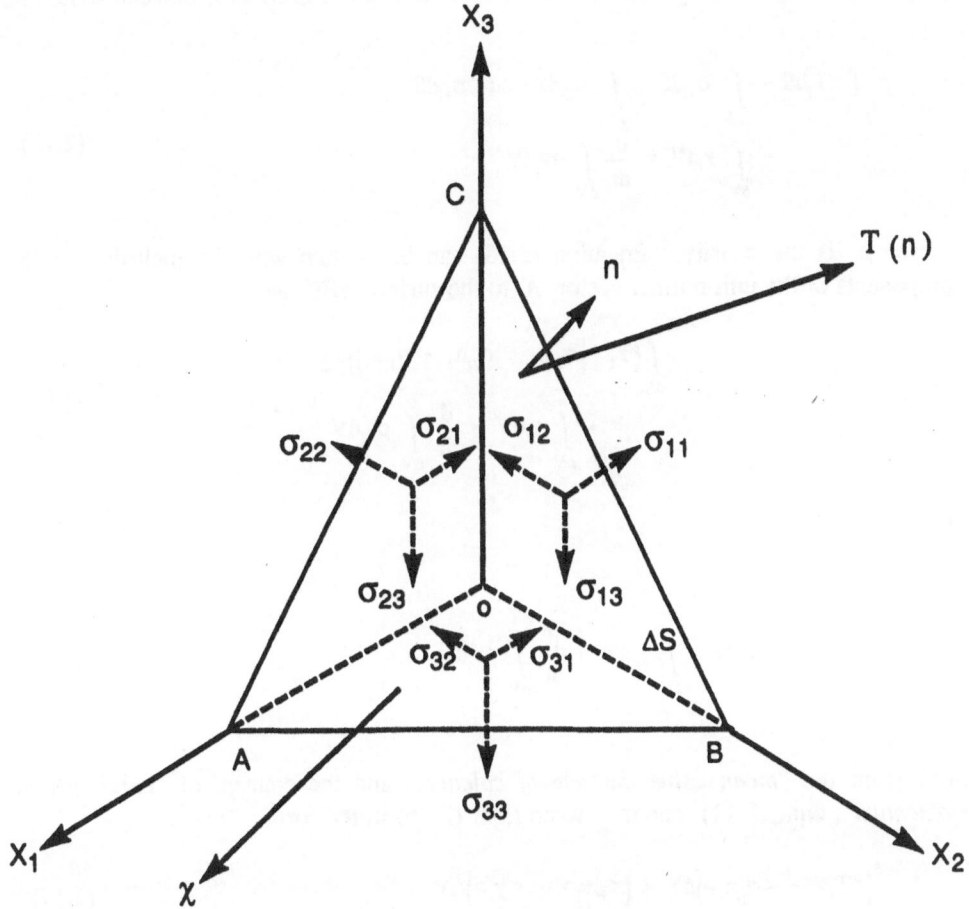

Figure 2.5. Cauchy's tetrahedron.

Let **T(n)** be the stress vector at a point of the oblique surface *ABC* whose normal is **n.**

ΔS is the area *ABC* .

h is the perpendicular distance from the origin *O* to ΔS.

ΔV = 1/3 h ΔS is the elemental volume of the tetrahedron *OABC*.

χ_i are the components of the body force vector per unit volume.

σ_{ij} are the components of the stress tensor.

Recall Newton's second law of motion, that is, the sum of forces acting on the tetrahedron is equal to the rate of change of linear momentum. In this context, letting v_i denote the components of the velocity vector, then, with reference to Figure 2.5, one can write that

$$\int_{ABC} T_i dS - \int_{OAB} \sigma_{3i} dS - \int_{OCA} \sigma_{2i} dS - \int_{OBC} \sigma_{1i} dS$$

$$+ \int_{\Delta V} \chi_i dV = \frac{d}{dt} \int_{\Delta V} \rho v_i dV \tag{2.15}$$

where ρ is the density. Equation (2.15) can be written with the inclusion of the components of the unit normal vector n, to the surface ABC, as

$$\int_{\Delta S} \left(T_i - \left[\sigma_{3i} n_3 + \sigma_{2i} n_2 + \sigma_{1i} n_1 \right] \right) dS$$

$$+ \int_{\Delta V} \chi_i dV = \frac{d}{dt} \int_{\Delta V} \rho v_i dV$$

or,

$$\int_{\Delta S} \left(T_i - \sigma_{ji} n_j \right) dS + \int_{\Delta V} \chi_i dV = \frac{d}{dt} \int_{\Delta V} \rho v_i dV \tag{2.16}$$

Both from the "*mean value theorem of calculus*" and the concept of "*continuity of momentum*", eqn. (2.11), one may write Eqn. (2.16) in the form

$$\left(<T_i> - <\sigma_{ji} n_j> \right) \Delta S = \left(<\rho \dot{v}_i> - <\chi_i> \right) \Delta V \tag{2.17}$$

where $< \cdot >$ indicates the mean value of the considered variable. Taking the limits as both ΔS and ΔV tend to zero, i.e., h also tends to zero, then, equation (2.17) may be approximated by

$$T_i = \sigma_{ji} n_j \tag{2.18}$$

where σ_{ji} is a second-order Cartesian tensor since both T_i and n_j are components of vectors.

Equation (2.18) establishes the *"stress boundary conditions"* on the free body (tetrahedron) considered and states that *"for every second order (symmetric) tensor σ_{ji}, defined at some point in the continuum, there is, associated with each direction (specified by the unit normal n at that point), a traction vector T given by the form of the equation (2.18)"*.

2.5.1. EQUATIONS OF MOTION. SYMMETRY OF THE STRESS TENSOR.

As dealt with in the foregoing, we consider an arbitrary region of a continuum. This region, Fig. 2.6, is of volume V which is enclosed by a surface S with an outward unit normal **n**. The surface forces acting on an elemental portion of S are T dS while the body forces on an elemental portion of V are χ dV.

Applying Newton's second law of motion to this arbitrary region, then,

$$\int_S T_i \, dS + \int_V \chi_i dV = \frac{d}{dt} \int_V \rho v_i \, dV \tag{2.19}$$

Thus, in view of the conservation of momentum relation (2.11) and the stress boundary conditions equation (2.18) above, one can write equation (2.19) as

$$\int_S \sigma_{ji} n_j dS + \int_V \chi_i dV = \int_V \rho \dot{v}_i dV \tag{2.20}$$

Applying Gauss's divergence theorem to the first integral of (2.20), it follows that

$$\int_V \left[\sigma_{ji,j} + \chi_i - \rho \dot{v}_i \right] dV = 0 \tag{2.21}$$

or

$$\sigma_{ji,j} + \chi_i = \rho \dot{v}_i \tag{2.22}$$

which is known as *"Cauchy's first equation of motion"*.

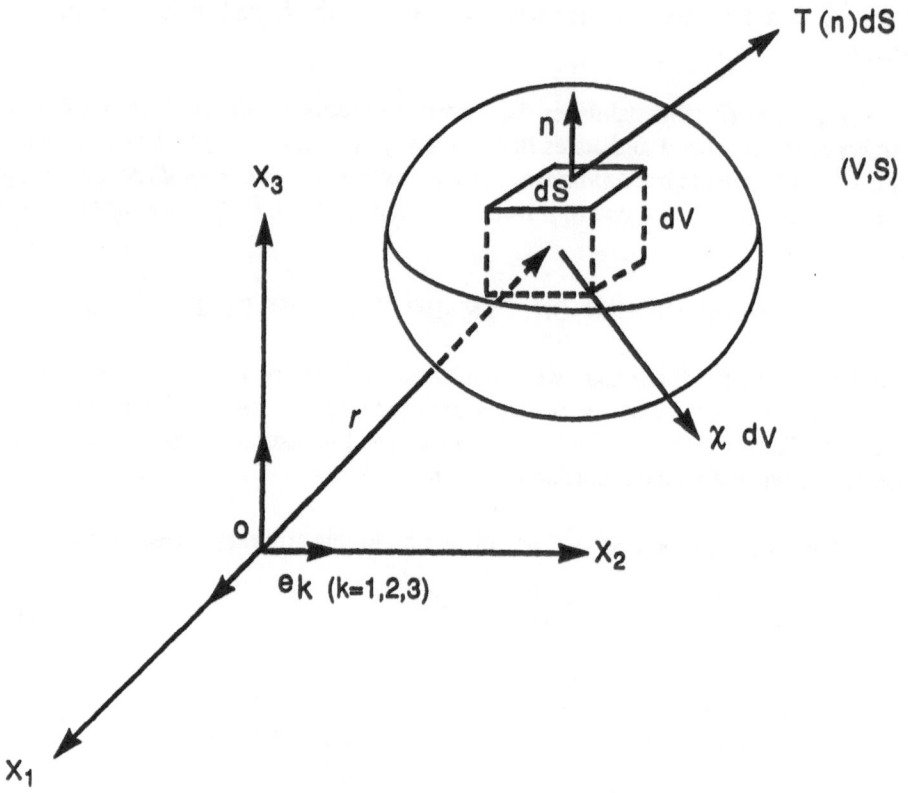

Figure 2.6. An elemental volume (*dV, dS*) of a continuous region (*V, S*).

With reference to Figure 2.6, one considers the sum of moments of forces about the origin O to be equal to the rate of change of angular momentum, then,

$$\int_S \mathbf{r} \times \mathbf{T} dS + \int_V \mathbf{r} \times \chi dV = \frac{d}{dt} \int_V \mathbf{r} \times (\rho \mathbf{v}) dV$$

or

$$\int_S \varepsilon_{ijk} x_i T_j \, \mathbf{e}_k \, dS + \int_V \varepsilon_{ijk} x_i \chi_j \, \mathbf{e}_k \, dV$$

$$= \frac{d}{dt} \int_V \rho \varepsilon_{ijk} x_i \dot{x}_j \, \mathbf{e}_k dV \qquad (2.23)$$

where ε_{ijk} is the alternating tensor (*see* Chapter 1) and $\mathbf{e}_k$ are the components of the unit vector associated with the coordinate system x_i as shown in Fig. 2.6.

Applying Gauss's divergence theorem and the previously established stress boundary conditions (2.18) to the left hand side of (2.23), the latter equation can be expressed as

$$\int_V e_k \left\{ \varepsilon_{ijk} \left[\left(x_i \sigma_{\ell j} \right)_{,\ell} + x_i \chi_j \right] \right\} dV = e_k \int_V \frac{\partial}{\partial t} \left(\rho \varepsilon_{ijk} x_i \dot{x}_j \right) dV$$
$$+ e_k \int_S \rho \varepsilon_{ijk} x_i v_j v_\ell n_\ell \, dS$$

(2.24)

Equation (2.24) above can be written with reference to equations (2.7) and (2.11) as

$$\int_V e_k \left\{ \varepsilon_{ijk} \left[(x_i \sigma_{\ell j})_{,\ell} + x_i \chi_j \right] \right\} dV = e_k \varepsilon_{ijk} \int_V \rho x_i \dot{v}_j \, dV$$

or,

$$e_k \, \varepsilon_{ijk} \int_V \left\{ \sigma_{ij} + x_i (\sigma_{\ell j,\ell} + \chi_j - \rho \dot{v}_j) \right\} dV = 0$$

(2.25)

In the above equation, since the expression in parentheses is equal to zero, Cauchy's first equation of motion (2.22), equation (2.25) becomes

$$e_k \int_V \varepsilon_{ijk} \, \sigma_{jk} \, dV = 0$$

(2.26)

which implies that

$$\varepsilon_{ijk} \, \sigma_{jk} = 0$$

(2.27)

or, alternatively, in view of the properties of the skew-symmetric tensor ε_{ijk} (*Chapter 1*),

$$\sigma_{ij} = \sigma_{ji}$$

(2.28)

It is apparent that equation (2.28) expresses the *"symmetry of the stress tensor"*. This equation is referred to as *"Cauchy's second law of motion"* and it implies the *"Conservation of moment of momentum"*.

2.6. Principal Axes of Stress, Principal Planes and Principal Stresses.

Regardless of the state of stress at a given point in a continuum, it is always possible to choose a special set of rectangular axes through this point so that the shear stress components vanish when the stress components are referred to this set of axes. This set of rectangular axes are referred to as *"principal axes"* or *"principal directions"*. Thus, on a plane perpendicular to a principal axis, the traction vector is entirely normal.

The principal planes, through the point, perpendicular to the three principal axes of stress are referred to as the *"principal planes of stress"*.

The normal stress components on the three principal planes are known as *"principal stresses"*. The principal stresses are physical quantities whose magnitudes do not depend on the particular coordinate system to which the stress components are referred. Accordingly, they are *"invariants"* pertaining to the stress state at the particular point under consideration.

Let the unit normal n defines the direction of a principal axis and σ designates the corresponding principal stress, then, the stress vector acting on the surface defined by n can be expressed, with reference to (2.18), by

$$T_i = \sigma\, n_i = \sigma_{ji}\, n_j \tag{2.29}$$

This is with the understanding that

$$n_i = \delta_{ji}\, n_j \tag{2.30}$$

where δ_{ji} is the Kronecker delta (*Chapter 1*);

$$\delta_{ij} = \begin{cases} 1 & \text{if } i=j \\ \\ 0 & \text{if } i \neq j \end{cases}$$

Combining equations (2.29) and (2.30), it follows that

$$(\sigma_{ji} - \sigma\,\delta_{ji})n_j = 0 \tag{2.31}$$

Expression (2.31) represents three equations to be solved for the components of the unit normal n, i.e., n_1, n_2, n_3. In this context, we search a set of nontrivial solutions for which

$$n_1^2 + n_2^2 + n_3^2 = 1 \tag{2.32}$$

Hence, equation (2.31), subject to (2.32), poses an eigenvalue problem. Since the matrix of σ_{ij} is real and symmetric, then, there exist three real valued principal stresses and a set of orthogonal principal axes. The three principal stresses are denoted by σ_1, σ_2, σ_3. The algebraically greatest of the three principal stresses is the algebraically greatest normal stress component acting on any plane through the point. At the same time, the algebraically smallest of the principal stresses is the algebraically smallest normal stress component acting on any plane through the point. Equation (2.31) has a set of non-vanishing solutions n_1, n_2, n_3 if, and only if, the determinant of its coefficients vanishes, i.e.,

$$\left| \sigma_{ij} - \sigma\,\delta_{ij} \right| = 0 \tag{2.33}$$

Equation (2.33) represents a cubic equation in σ. The roots of this equation are the principal stresses σ_1, σ_2 and σ_3. For each value of the principal stress, a unit normal n is involved.

Expansion of (2.33) leads to

$$\left| \sigma_{ij} - \sigma\,\delta_{ij} \right| =
\begin{vmatrix}
\sigma_{11} - \sigma & \sigma_{12} & \sigma_{13} \\
\sigma_{21} & \sigma_{22} - \sigma & \sigma_{23} \\
\sigma_{31} & \sigma_{32} & \sigma_{33} - \sigma
\end{vmatrix} \tag{2.34}$$

$$= -\sigma^3 + I_1\sigma^2 - I_2\sigma + I_3 = 0$$

where the coefficients I_1, I_2, I_3 denote, respectively, the following scalar expressions of the stress components

$$I_1 = \sigma_{11} + \sigma_{22} + \sigma_{33}$$
$$= \sigma_{kk} \tag{2.35a}$$

$$I_2 = \begin{vmatrix} \sigma_{22} & \sigma_{23} \\ \sigma_{32} & \sigma_{33} \end{vmatrix} + \begin{vmatrix} \sigma_{11} & \sigma_{13} \\ \sigma_{31} & \sigma_{33} \end{vmatrix} + \begin{vmatrix} \sigma_{11} & \sigma_{12} \\ \sigma_{21} & \sigma_{22} \end{vmatrix} \tag{2.35b}$$

$$I_3 = \begin{vmatrix} \sigma_{11} & \sigma_{12} & \sigma_{13} \\ \sigma_{21} & \sigma_{22} & \sigma_{23} \\ \sigma_{31} & \sigma_{31} & \sigma_{33} \end{vmatrix} \tag{2.35c}$$

It is recognized that both equation (2.34) and the expressions of the coefficients I_1, I_2, I_3, represented by (2.35), do not depend on the choice of the coordinate axes, hence, they are invariants of coordinate transformation. For this reason, the coefficients I_1, I_2, and I_3 of equation (2.34) are conventionally referred to as the "*invariants*" of the stress tensor, or, simply "*stress invariants*".

With reference to equations (2.34) and (2.35), the coefficient I_1 is the "*first invariant*" of the stress tensor. It is also referred to as the "*trace*" of the stress matrix. That is the sum of the elements on the main diagonal in the matrix of rectangular Cartesian components of σ_{ij}, i.e., the sum of the three normal stresses, Eqn. (2.35a). The coefficient I_2, "*second invariant*", is a homogeneous quadratic expression in the stress components. It is the sum of the three minor determinants of the three diagonal elements in the determinant of the stress matrix, Eqn. (2.35b). The "*third invariant*", I_3, is the determinant of the original stress matrix and it is a homogeneous cubic expression in the stress components, Eqn. (2.35c).

An important property of a symmetric second-order tensor:

"*For a symmetric second-order tensor, the three principal stresses are all real and the three principal planes are mutually orthogonal*".

It is left to the student to seek the proof of this property.

QUIZ 2.2

Show that the invariants I_1, I_2, and I_3 of σ_{ij} can be expressed in terms of the roots of the cubic Equation (2.34), i.e., the three principal stresses σ_1, σ_2 and σ_3. That is

$$I_1 = \sigma_1 + \sigma_2 + \sigma_3$$
$$I_2 = \sigma_1\sigma_2 + \sigma_2\sigma_3 + \sigma_3\sigma_1 \tag{2.36}$$
$$I_3 = \sigma_1\sigma_2\sigma_3$$

QUIZ 2.3

Show that if $\sigma_1 = \sigma_2 = \sigma_3$, then, any set of orthogonal axes may be considered as principal axes.

2.6.1. LINES OF PRINCIPAL STRESS "STRESS TRAJECTORIES"

If the stress field in the material specimen is non-uniform, the magnitudes and/or directions of the principal stresses would vary from a point to a point. If a curve be drawn in such a way that the direction of one of the principal stresses is tangential to it at every point along its length, then such a curve is referred to as "*line of principal stress*" or "*stress trajectory*" pertaining to the particular principle stress under consideration . Consequently, in a two-dimensional stress field, there will be two orthogonal sets of curves, corresponding to the maximum and minimum principal stresses. On the other hand, if the stress field is three-dimensional, then, there will be three orthogonal sets of stress trajectories pertaining to the existing three principal stresses. Figure 2. 7 illustrates schematically the two orthogonal sets of stress trajectories, corresponding to maximum and minimum principal stresses, in a two-

60

dimensional stress field.

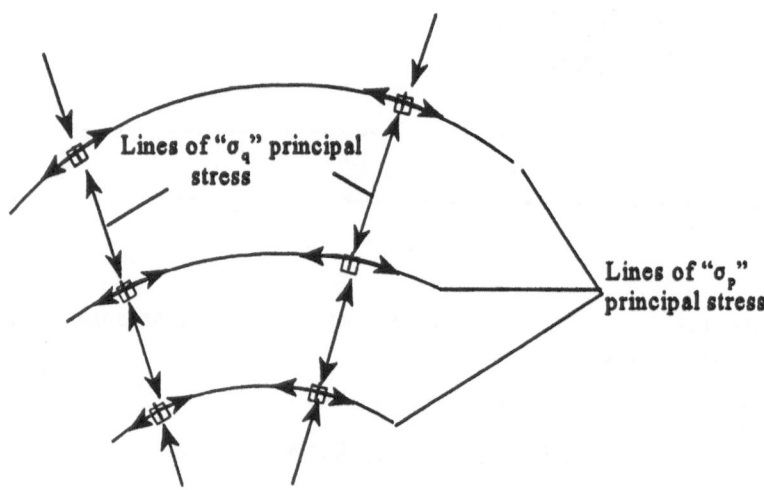

Figure 2.7. Two orthogonal sets of "stress trajectories", corresponding to maximum and minimum principal stresses, in a two-dimensional stress field (Adapted after Hendry, 1966).

The principal stress network for a frame model is shown in Figure 2.8 (Adapted after Hendry, 1966). The trajectories of maximum principal stress are shown as thin lines, whilst those of minimum principal stress are indicated by thick lines. The figure gives a valuable impression of the stress distribution in the model. For instance, the crowding together of the lines of minimum principal stress towards the inner edge at the right of the model points out to a situation of stress concentration there. Meanwhile, the cross-over of the two sets of stress trajectories in the horizontal part of the model indicates that the stress on the inner edge region changes from compression to tension.

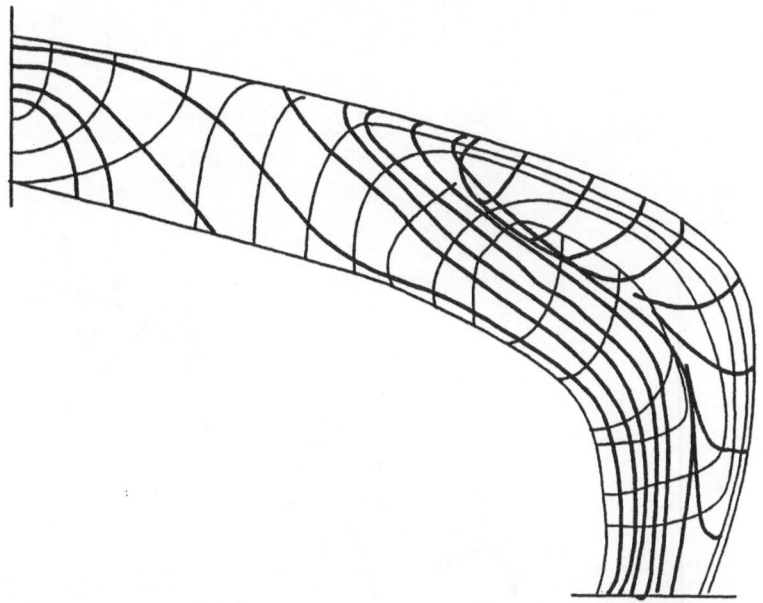

Figure 2.8. Hypothetical lines of principal stress in a frame model. Adapted after Hendry 1966).

Isoclinic Lines

The locus of points at which the principal stresses have the same inclination to an arbitrary reference axis is referred-to as "*isoclinic line*", Figure 2.9. The angle ϕ which the direction of the principal stress makes with the reference axis is termed the "*parameter*" of the isoclinic. Thus, in a given model, there will be an isoclinic line corresponding to every value of the parameter ϕ between 0 and 180°.

Figure 2.10 (adapted after Hendry, 1966), for instance, shows the isoclinic lines for parameters, of isoclinics, from 0 to 180° by 10° intervals, for the frame model presented earlier in Fig. 2. 7.

The isoclinic lines can be readily obtained for a given model by carrying out a photo-elastic experiment, and once they are known, one can easily construct the lines of principal stresses. This is illustrated schematically in Fig. 2. 11.

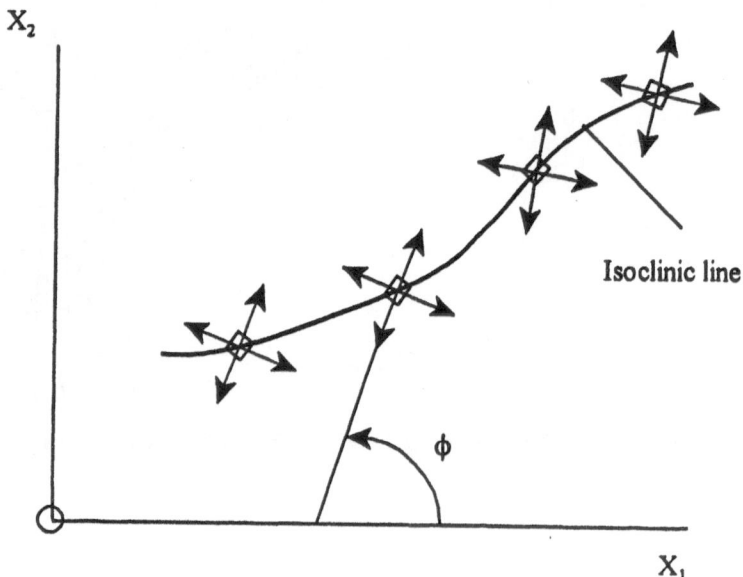

Figure 2.9. The isoclinic line (Adapted after Hendry, 1966).

Isotropic Points. In a system of plane stress, a point at which the principal stresses are equal is termed an "*isotropic point*". If the two principal stresses at a point are zero., i.e., $\sigma_1 = \sigma_2 = 0$, one has a special type of an isotropic point, usually referred to as a "*singular point*".

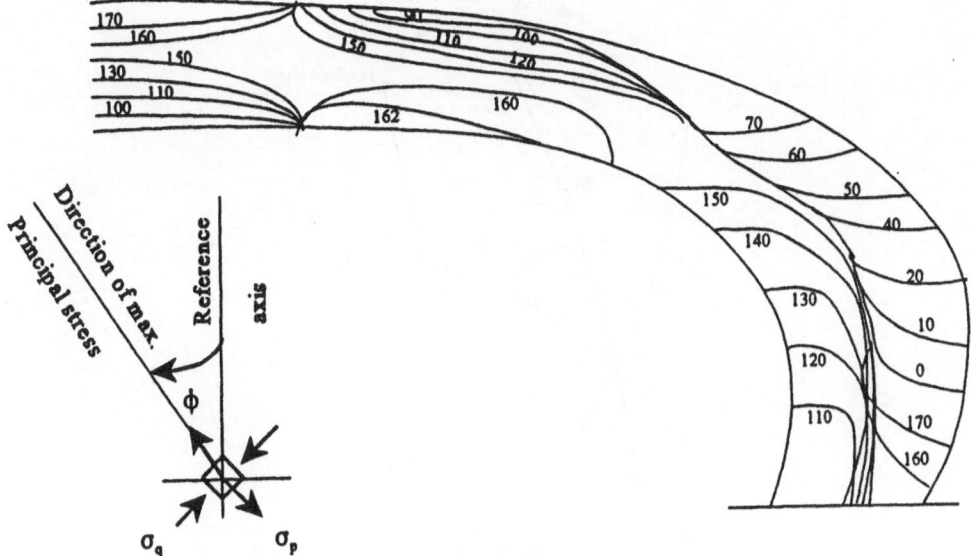

Figure 2.10. Schematics of the isoclinic lines in the frame model of Fig. 2. 8 (Adapted after Hendry, 1966).

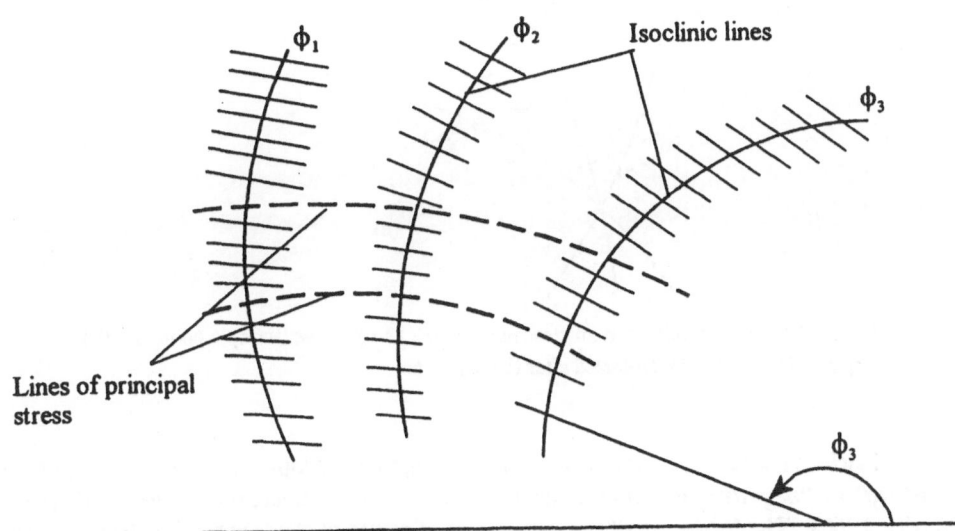

Figure 2.11. Construction of the principal stress trajectories from the isoclinic lines (Adapted after Hendry, 1966).

64

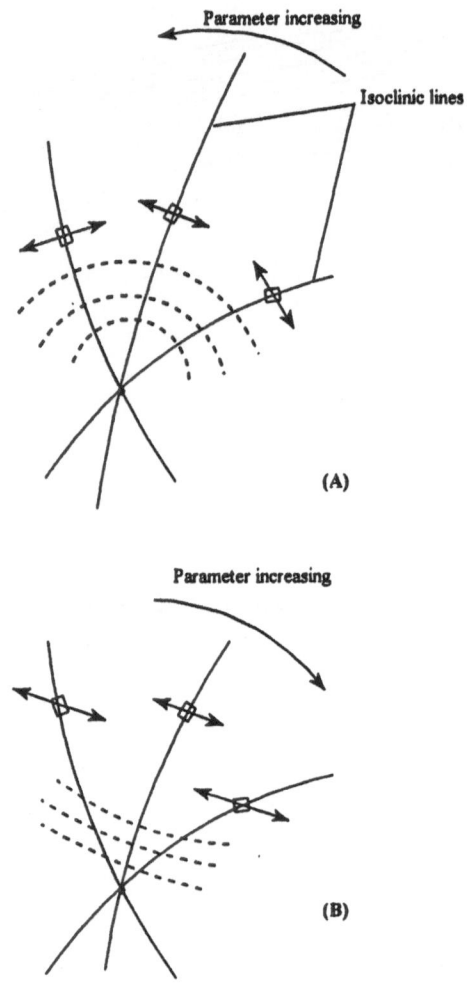

Figure 2.12. The definition of an "*isotropic point*"; (A) Positive isotropic point and (B) Negative isotropic point (Adapted after Hendry, 1966).

Isoclinic lines of many parameters pass through an isotropic point, which designated "*positive*" or "*negative*" isotropic point according to the convection of Fig. 2. 12 (e.g., Hendry, 1966) . The formation of the trajectories of principal stress around isotropic points is of particular interest. Figures 2. 13a and 2. 13b are typical of the systems formed around positive and negative isotropic points. While, there are many possible variations in detail from one model to another, they are similar in type, being classified either as "*interlocking*" or "*non-interlocking*" systems, e.g., Hendry, 1966.

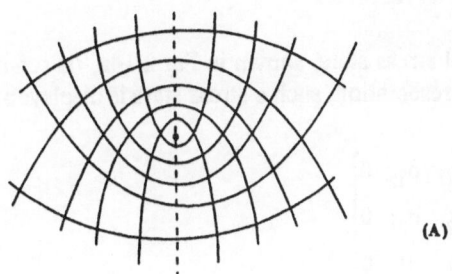

(A)

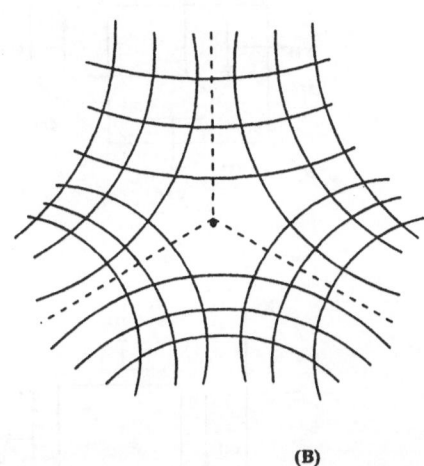

(B)

Figure 2.13. Arrangements of trajectories of principal stress at an isotropic point: (A) An *"interlocking"* at a positive isotropic point, and (B) A *"non-interlocking"* stress-trajectory system at a negative isotropic point (Adapted after Hendry, 1966).

2.6.2. STATE OF "NORMAL STRESS"

It is evident that if the reference axes x_1, x_2, x_3 were selected to coincide with the principal axes, then, the stress tensor would be expressed as

$$\sigma_{ij} = \begin{vmatrix} \sigma_1 & 0 & 0 \\ 0 & \sigma_2 & 0 \\ 0 & 0 & \sigma_3 \end{vmatrix} \qquad (2.37a)$$

2.6.3. STATE OF "PLANE STRESS"

The two-dimensional stress state, shown in Fig 2.14a, is referred to as a state of "plane stress". In matrix representation, such a stress state is displayed as

$$\sigma_{ij} = \begin{bmatrix} \sigma_{11} & \sigma_{12} & 0 \\ \sigma_{21} & \sigma_{22} & 0 \\ 0 & 0 & 0 \end{bmatrix} \tag{2.37b}$$

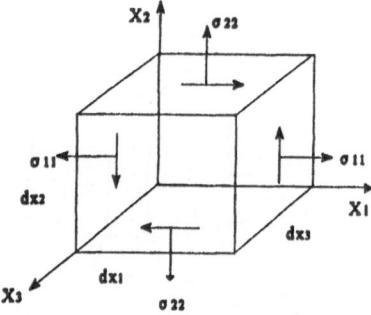

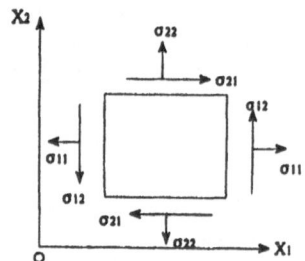

Figure 2.14a. State of "Plane Stress".

2.6.4. STATE OF "PURE SHEAR"

The state of pure shear is demonstrated by the elements shown in Fig. 2.14b. It is left to the student to display the corresponding matrix of stress components.

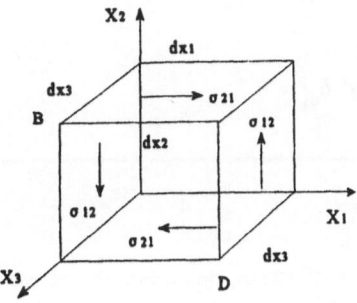

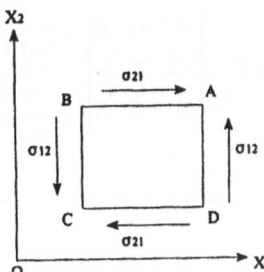

Figure. 2.14b. State of "Pure Shear".

$$p = \bar{\sigma} = \frac{1}{3}(\sigma_{11} + \sigma_{22} + \sigma_{33}) = \frac{1}{3}\sigma_{kk} = \frac{1}{3}I_1 \tag{2.38}$$

where $\bar{\sigma}$ is the mean stress and I_1 is the first invariant of σ_{ij} expressed earlier by (2.35a).

-    The second part, denoted below by σ_{ij}', is defining the deviator of the stress tensor as

$$\sigma'_{ij} = \sigma_{ij} - \bar{\sigma}\,\delta_{ij} \tag{2.39}$$

Thus,

$$\sigma_{ij} = \bar{\sigma}\,\delta_{ij} + \sigma'_{ij} \tag{2.40}$$

To determine the principal values of the stress deviator, the procedure illustrated earlier in Section 2.6 for the determination of the principal stresses of the stress tensor may be followed with replacing the determinant equation (2.33) by the corresponding one for the stress deviator, i.e.,

$$|\sigma'_{ij} - \sigma' \delta_{ij}| = 0 \qquad (2.41)$$

EXAMPLE PROBLEM 2.1

For the following state of stress, determine the stress invariants and the principal stress values. Display the matrix of the principal values of stress and determine the stress invariants for the latter case.

$$[\sigma_{ij}] = \begin{vmatrix} 6 & -3 & 0 \\ -3 & 6 & 0 \\ 0 & 0 & 6 \end{vmatrix}$$

Solution:

For the stress state

$$\sigma_{ij} = \begin{vmatrix} 6 & -3 & 0 \\ -3 & 6 & 0 \\ 0 & 0 & 6 \end{vmatrix}$$

the stress invariants, equations (2.35), are

$$I_1 = 6+6+6 = 18$$

$$I_2 = \det \begin{bmatrix} 6 & 0 \\ 0 & 6 \end{bmatrix} + \det \begin{bmatrix} 6 & -3 \\ -3 & 6 \end{bmatrix} + \det \begin{bmatrix} 6 & 0 \\ 0 & 6 \end{bmatrix} = 99$$

$$I_3 = \det \begin{bmatrix} 6 & -3 & 0 \\ -3 & 6 & 0 \\ 0 & 0 & 6 \end{bmatrix} = 162$$

Substituting the values of the invariants, as obtained above, into the characteristic equation (2.34), for the principal stresses, i.e.,

$$-\sigma^3 + I_1\sigma^2 - I_2\sigma + I_3 = 0$$
$$-\sigma^3 + 18\sigma^2 - 99\sigma + 162 = 0$$

The above cubic characteristic equation can be solved for its three roots, representing the principal values of stress. Thus, the principal values of stress are:

$$\sigma_1 = 9, \quad \sigma_2 = 6, \quad \sigma_3 = 3$$

Consequently, the matrix of principal stresses is

$$[\sigma_1, \sigma_2, \sigma_3] = \begin{bmatrix} 9 & 0 & 0 \\ 0 & 6 & 0 \\ 0 & 0 & 3 \end{bmatrix}$$

The student can prove that the values of the invariants for the matrix of principal stresses displayed above are the same as the values of the invariants for the original stress matrix.

EXAMPLE PROBLEM 2.2

For the following state of stress, determine the matrix of the spherical stress and that of the deviatoric stress. Calculate the value of the first invariant for the deviatoric stress state.

$$\sigma_{ij} = \begin{vmatrix} 12 & 4 & 0 \\ 4 & 9 & -2 \\ 0 & -2 & 3 \end{vmatrix}$$

Solution:

Recalling Eqn. (2.38), the mean stress $\sigma_m = -p = \dfrac{1}{3}[\,12 + 9 + 3\,] = 8$

Thus, the spherical stress state is expressed as

$$\sigma_m I = -p I = \begin{vmatrix} 8 & 0 & 0 \\ 0 & 8 & 0 \\ 0 & 0 & 8 \end{vmatrix}$$

where I is the unit matrix.

Recalling Eqn. (2.39), the deviatoric stress state is expressed by

$$\sigma'_{ij} = \sigma_{ij} - \sigma_m I$$

$$I' = 4 + 1 - 5 = \text{ trace } \sigma'_{ij}$$
$$= 0$$

The first invariant for the deviatoric stress is, thus, determined as

$$I'_i = \begin{vmatrix} 12 - 8 & 4 & 0 \\ 4 & 9 - 8 & -2 \\ 0 & -2 & 3 - 8 \end{vmatrix}$$

$$= \begin{vmatrix} 4 & 4 & 0 \\ 4 & 1 & -2 \\ 0 & -2 & -5 \end{vmatrix}$$

EXAMPLE PROBLEM 2.3

For the shown state of stress, determine the mean stress, the matrix of the deviatoric stress, and the pertaining invariant values.

$$\sigma_{ij} = \begin{vmatrix} 10 & -6 & 0 \\ -6 & 10 & 0 \\ 0 & 0 & 1 \end{vmatrix}$$

Solution:

Recalling Eqn. (2.38), the mean stress can be calculated for the stress state indicated above as

$$\sigma_m = -p = \frac{1}{3} [10 + 10 + 1] = 7$$

Meantime, the deviatoric stress is calculated as $\qquad \sigma'_{ij} = \sigma_{ij} - \sigma_m \, \delta_{ij}$

$$= \begin{vmatrix} 10 - 7 & -6 & 0 \\ -6 & 10 - 7 & 0 \\ 0 & 0 & 1 - 7 \end{vmatrix}$$

$$= \begin{vmatrix} 3 & -6 & 0 \\ -6 & 3 & 0 \\ 0 & 0 & -6 \end{vmatrix}$$

The principal deviatoric stress values are the roots of the characteristic equation:

$$\lambda^3 - I'_1 \lambda^2 + I'_2 \lambda - I'_3 = 0$$

where the invariants of the above equation are

$$I_1' = \text{trace } \sigma' = 0$$
$$I_2' = -63$$
$$I_3' = \det \sigma' = 162$$

Substituting the values of the invariants into the cubic (characteristic) equation above, i.e.,

$$\lambda^3 - 63\lambda - 162 = 0 \quad \rightarrow \quad (\lambda + 3)(\lambda - 9)(\lambda + 6) = 0$$

Thus, the principal deviatoric stress values are: +9, -3 and -6.

EXAMPLE PROBLEM 2.4

At a point p of a continuous medium, the stress tensor referred to the coordinate-axes $x_i (i=1,2,3)$ is given by

$$\sigma_{ij} = \begin{vmatrix} 15 & -10 & 0 \\ -10 & 5 & 0 \\ 0 & 0 & 20 \end{vmatrix}$$

If the new coordinate-axes x_i' are chosen by a rotation about the origin for which the transformation matrix is

$$A_{ij} = \begin{vmatrix} 3/5 & 0 & -4/5 \\ 0 & 1 & 0 \\ 4/5 & 0 & 3/5 \end{vmatrix}$$

Determine the traction vectors on each of the primed coordinate planes by projecting the traction vectors of the original axes onto the primed directions. Determine the stress deviator components σ_{ij}'. Verify your result using the transformation relation (*Consult Chapter 1*).

Solution:

The stress tensor at a point p, referred to the axes x_i $(i=1, 2, 3)$, is

$$\sigma_{ij} = \begin{vmatrix} 15 & -10 & 0 \\ -10 & 5 & 0 \\ 0 & 0 & 20 \end{vmatrix}$$

A new set of axes x_i' are obtained by a rotation about the original axes for which the transformation matrix is

$$A = [A_{ij}] = \begin{vmatrix} 3/5 & 0 & -4/5 \\ 0 & 1 & 0 \\ 4/5 & 0 & 3/5 \end{vmatrix}$$

The stress vector at a point of a surface with an outward unit normal n_i is a linear function of the stress vectors acting on the coordinate surfaces through the point under consideration p, the coefficients being the cosine directors of e_i, e.g.,

$$t_{(1)} = t_{k1} \ e_k = t_{11} \ e_1 + t_{21} \ e_2 + t_{31} \ e_3$$

where

$$t_{11} = \sigma_{11}, \quad t_{21} = \sigma_{21}, \quad t_{31} = \sigma_{31}$$
$$t(1) = 15 \, e_1 - 10 \, e_2 + 0 \, e_3$$
$$t(2) = -10 \, e_1 + 5 \, e_2 + 0 \, e_3$$
$$t(3) = 0 \, e_1 + 0 \, e_2 + 20 \, e_3$$

The transformation matrix $[A_{ij}]$ can transform the traction vectors into the new system of coordinate-axes x_i', i.e.,

$$t_1' = 3/5\ t_1 - 4/5\ t_3$$
$$= 9\,e_1 - 6\,e_2 - 16\,e_3$$
$$t_2' = t_2$$
$$= -10\,e_1 + 5\,e_2$$
$$t_3' = 4/5\ t_1 + 3/5\ t_3$$
$$= 12\,e_1 - 8\,e_2 + 12\,e_3$$

To obtain the stress tensor components σ_{ij}' (not deviatoric) with reference to the e_i' unit vectors,

$$\left[e_i'\right] = [A\,ij\,]\left[e_j\right] \rightarrow \left[e_j\right] = [A\,ij\,]^{-1}\left[e_i'\right]$$

$$\begin{bmatrix} e_1 \\ e_2 \\ e_3 \end{bmatrix} = \begin{bmatrix} 3/5 & 0 & 4/5 \\ 0 & 1 & 0 \\ -4/5 & 0 & 3/5 \end{bmatrix} \begin{bmatrix} e_1' \\ e_2' \\ e_3' \end{bmatrix}$$

since $[\,A_{ij}\,] \neq 0$,

Therefore,

$$\begin{bmatrix} t_1' \\ t_2' \\ t_3' \end{bmatrix} = \begin{bmatrix} 9 & -6 & -16 \\ -10 & 5 & 0 \\ 12 & -8 & 12 \end{bmatrix} \begin{bmatrix} 3/5 & 0 & 4/5 \\ 0 & 1 & 0 \\ -4/5 & 0 & 3/5 \end{bmatrix} \begin{bmatrix} e_1' \\ e_2' \\ e_3' \end{bmatrix}$$

$$= \begin{bmatrix} 91/5 & -6 & -12/5 \\ -6 & 5 & -8 \\ -12/5 & -8 & 84/5 \end{bmatrix} \begin{bmatrix} e_1' \\ e_2' \\ e_3' \end{bmatrix}$$

Thus, $\qquad \sigma_{ij}' = \begin{bmatrix} 91/5 & -6 & -12/5 \\ -6 & 5 & -8 \\ -12/5 & -8 & 84/5 \end{bmatrix}$

Alternatively, one may use the transformation relation $[\sigma_{ij}'] = A\ [\sigma_{ij}]\ A^T$, thus,

$$\sigma'_{ij} = \begin{bmatrix} 3/5 & 0 & -4/5 \\ 0 & 1 & 0 \\ 4/5 & 0 & 3/5 \end{bmatrix} \begin{bmatrix} 15 & -10 & 0 \\ -10 & 5 & 0 \\ 0 & 0 & 20 \end{bmatrix} \begin{bmatrix} 3/5 & 0 & 4/5 \\ 0 & 1 & 0 \\ -4/5 & 0 & 3/5 \end{bmatrix}$$

$$= \begin{bmatrix} 9 & -6 & -16 \\ -10 & 5 & 0 \\ 12 & -8 & 12 \end{bmatrix} \begin{bmatrix} 3/5 & 0 & 4/5 \\ 0 & 1 & 0 \\ -4/5 & 0 & 3/5 \end{bmatrix}$$

$$= \begin{bmatrix} 91/5 & -6 & -12/5 \\ -6 & 5 & -8 \\ -12/5 & -8 & 84/5 \end{bmatrix}$$

Note: The $[\sigma'_{ij}]$ is the same in both cases.

2.7. *Piola-Kirchhoff's* Stress Tensor

As presented earlier, the Cauchy's equations of motion (2.22) and (2.28) apply to the current deformed configuration. Here, the Cauchy's stress tensor field is a function of the spatial coordinate x_i and it was concluded by (2.28) to be a symmetric tensor for the case where there is no couple stress or assigned couples. In the applications of the theory of elasticity, however, it is often assumed that there exists a natural state to which the body would return when it is unloaded (e.g., Sokolnikoff, 1956). In this case, it is generally preferred that both the stress and the equations of motion be expressed as functions of the material point X_I (the reference state) and, hence, to derive the equations of motion in this state. The *first and second Piola-Kirchhoff's stress tensors* (due to Piola in 1833 and Kirchhoff in 1853) are two alternatives for the definition of the stress in the reference state; see Truesdell and Toupin (1960) and Malvern (1969). The two Piola-Kirchhoff's stress tensors are expressed in terms of the force per unit undeformed area.

The first Piola-Kirchhoff's stress tensor is the simpler one. However, it has the disadvantage of being antisymmetric. The second Piola-Kirchhoff's stress tensor, on the other hand, is symmetric, thus, it has been often used in the formulations of the theory of elasticity. In this context, we present below the definition of the "*second Piola-Kirchhoff's tensor*".

2.7.1. SECOND PIOLA-KIRCHHOFF'S STRESS TENSOR

The *second Piola-Kirchhoff's stress tensor* Σ gives a force dP_0 on a unit area dS in the undeformed configuration in relation to the actual force dP on the corresponding elemental area ds in the deformed configuration according to the following interpretations

$$dP_0 = g^{-1} \cdot dP \tag{2.42}$$

where $g^{-1} = X \nabla x$ is the inverse of the spatial deformation gradient (*see Chapter 3*).

The following relation can be proven (e.g., Malvern, 1969) between the *second Piola-Kirchhoff's stress* and the *Cauchy's stress*.

$$\Sigma = \frac{\rho_0}{\rho} g^{-1} \cdot \sigma \cdot (g^{-1})^T \tag{2.43a}$$

or, in indicial notations,

$$\Sigma_{Jl} = \frac{\rho_0}{\rho} X_{J,l} \sigma_{ji} \partial_i X_l \tag{2.43b}$$

which shows that the second Piola-Kirchhoff 's stress tensor Σ to be symmetric. In (2.43), ρ_0 and ρ are the material densities in the undeformed and deformed configurations, respectively.
The inverse relation to (2.43) is

$$\sigma = \frac{\rho}{\rho_0} g \cdot \Sigma \cdot g^T \tag{2.44a}$$

or,

$$\sigma_{ji} = \frac{\rho}{\rho_0} x_{j,J} \Sigma_{Jl} \partial_l x_i \tag{2.44b}$$

The symmetry of σ imposes the following condition on Σ so that the latter to be symmetric

$$\Sigma g^T = g \Sigma^T \tag{2.45}$$

where the superscript T indicates the transpose of the associated matrix. Meantime, the equation of motion corresponding to (2.22) in the reference state is expressed by

$$\nabla \cdot \left[\Sigma \cdot \mathbf{g}^{\mathsf{T}} \right] + \rho_0 \chi_0 = \rho_0 \frac{\partial^2 \mathbf{x}}{\partial t^2} \tag{2.46a}$$

or,

$$\partial_J \left[\Sigma_{JI} \, \partial_I x_i \right] + \rho_0 (\chi_0)_i = \rho_0 \frac{\partial^2 x_i}{\partial t^2} \tag{2.46b}$$

2.8. Problems

1. For the following stress state, determine the stress invariants, the principal stresses, and the deviatoric stress invariants. Display the matrix of principal stresses.

$$\sigma_{ij} = \begin{vmatrix} 6 & -3 & 0 \\ -3 & 6 & 0 \\ 0 & 0 & 8 \end{vmatrix}$$

2. For the following state of stress, determine the stress invariants and the principal stress values. Display the matrix of the principal values of stress and determine the stress invariants in the latter case.

$$[\sigma_{ij}] = \begin{vmatrix} 6 & -3 & 0 \\ -3 & 6 & 0 \\ 0 & 0 & 6 \end{vmatrix}$$

3. For the following stress matrix, determine the principal deviatoric stress values.

$$[\sigma_{ij}] = \begin{vmatrix} 12 & 4 & 0 \\ 4 & 9 & -2 \\ 0 & -2 & 3 \end{vmatrix}$$

4. For the following state of stress, determine the stress invariants and the principal stress values. Display the matrix of the principal values of stress and determine the stress invariants in the latter case.

$$[\sigma_{ij}] = \begin{vmatrix} 18 & 0 & 24 \\ 0 & -50 & 0 \\ 24 & 0 & 32 \end{vmatrix}$$

Display also the matrix of the corresponding stress deviators.

5. Evaluate the invariants for the stress tensor

$$\sigma_{ij} = \begin{vmatrix} 6 & -3 & 0 \\ -3 & 6 & 0 \\ 0 & 0 & 8 \end{vmatrix}$$

Determine the principal stress values for this state of stress and show that the diagonal form of the stress tensor yields the same values for the stress invariants.

6. Decompose the following stress tensor into its spherical and deviatoric parts and show that the first invariant of the deviator is zero.

$$\sigma_{ij} = \begin{vmatrix} 12 & 4 & 0 \\ 4 & 9 & -2 \\ 0 & -2 & 3 \end{vmatrix}$$

7. Determine the principal deviator stress values for the stress tensor

$$\sigma_{ij} = \begin{vmatrix} 10 & -6 & 0 \\ -6 & 10 & 0 \\ 0 & 0 & 1 \end{vmatrix}$$

2.9. Remarks on the Actual Three-Dimensional Stresses in Materials

Pindera (1998a) has dealt comprehensively with the problem of presentation of the actual three-dimensional stresses in engineering materials. In the mentioned article, Pindera advanced experimental evidence concerning the actual states of stress in the components of the homogeneous and adhesively bonded composite structures, as two separate illustrative issues. The presented evidence was obtained by using the advanced strain gages technique, and the theories and techniques of three-dimensional isodyne stress analysis. It is shown that a two-dimensional analytical treatment often yields errors not only in the magnitude but also in the sign of the evaluated stresses. In general, stress states in heterogenous engineering materials, such as composite and adhesively bonded structures, are noticeably three-dimensional. This is supported by the fact that the stress state in plates is noticeably three-dimensional in the presence of noticeable gradients of the in-plane stress components (e.g., Pindera and Krasnowski, 1982, Pindera, 1984, 1988, 1989, Pindera and Pindera, 1989, and Pindera and Liu, 1992).

It has been demonstrated that acceptance of the notion of a *generalized plane stress state*, which was introduced together with the notion of a *thin plate*, results in evaluation of stresses which are up to 30% lower than the actual stresses, particularly in the regions of local effects (Ladevese, 1985, Pindera and Krasnowski, 1982, Pindera and Liu, 1992, and Thum et al., 1960).

Thus, the stress states in the thin or thick plates are always three-dimensional in the presence of gradients of the in-plane stresses. The only question is how significant is the deviation from the plane stress state (Pindera, 1998a).

Pindera (1998a) also stressed the requirement that the notion of *"stress singularity"* to be treated differently. He argued that the concept of a finite load acting at a point is a mathematical abstraction, or a phenomenological assumption, which is theoretically (physically) inadmissible. This is always supported by the fact that the physical space could not support an infinite energy density, or an infinite power density. In practice, it means that the range of the inadmissible predicted stress values should be determined experimentally, unless a more general analytical solution would yield correct stress values (e.g., Ladevèse, 1985).

As stated by Pindera (1998a), all the notions related to the stress states in solids are of a major importance to a reliable stress analysis which is a major component of a modern designing process. The basic notion is the notion of *stress*, as a local quantity which could be defined as a *"stress at a point"*. In this context, Pindera (1998a&b) advanced that, in the experimental determination of the stress state, the range of validity of such a notion of stress is of major importance . This problem was considered not important when the procedures of theory of elasticity were being applied in mechanical design using metallic materials. At that time, because of the relatively low resolution of the measurement instruments, the grain size of a few hundredth of a millimeter was considered very small and of no consequence. Situations changed when the relations of theory of elasticity, or of engineering mechanics, were applied to such polymeric, natural composite materials, or to such a manmade composite material as, for instance, reinforced concrete. A complete alteration of the situation occurred when the contemporary instruments allowed, for the first time in history, to record positions of single atoms, or to move the selected atoms across the face of a specimen (Pindera, 1998a).

In the most cases, the thickness stresses in prismatic beams and in plates do not exceed 10 to 20 percent of the maximal surface stresses at the front faces of the objects, therefore they are often neglected. The situation is different in case of the laminated or composite structures, when the tensile strength in direction normal to the lamination plane is lower than the tensile strength in direction of lamination. Thus, it is worthwhile to test the three-dimensional stress states in terms of local effects, and to assess rationally the ranges of safe applicability of the pertinent simplified solutions. An example of such a test is given in (Pindera, 1989 and Pindera et al., 1989). The object of the referred-to test was a prismatic beam under a three-point load, Figure 2.15. The results show that at the distance from the loading force larger than 150% of the beam height the stress state in a beam is practically plane. Figure 2.16 illustrates the issue. It depicts the distribution of the in-plane shear stresses across the beam height at the distance from the loading force equal to 20% of the beam height, at various distances from the beam face (Pindera et al., 1989 and 1992). For further elaboration on the subject matter, the reader is referred to Pindera (1998 a&b).

2.10. References

Axelrad, D.R. and Haddad, Y.M. (1998) On the behaviour of materials with binary microstructures, in *Multilayered and Fibre-Reinforced Composites: Problems and Prospects*, NATO ARW, Kiev 1997, edited by Y. M. Haddad, *Kluwer*, Dordrecht, pp. 163-72.

Cauchy, A. L. (1827a) De la pression ou tension dans un corps solide, *Exercices de Mathématique* (see Love, 1944, pp. 8-9).

Cauchy, A. L. (1827b) Sur la condensation et la dilatation des corps solides, *Exercices de Mathématique* (see Love, 1944, pp. 8-9).

Cauchy, A. L. (1828) Sur les équations qui expriment les conditions d'équilibre ou les lois de mouvement intérieur d'un corps solide, *Exercices de mathématique* (see Love, 1944, pp. 8-9).

Flügge, W. (1972) *Tensor Analysis and Continuum Mechanics*, Springer-Verlag, Berlin.

Fung, Y.C. (1965). *Foundations of Solid Mechanics*, Prentice-Hall, Englewood Cliffs, N.J.

Haddad, Y.M. (1998) On the stochastic micromechanical approach to the response behaviour of engineering materials, in *Multilayered and Fibre-Reinforced Composites: Problems and Prospects*, NATO ARW, Kiev 1997, edited by Y. M. Haddad, *Kluwer*, Dordrecht, pp. 99-110.

Hendry, A. W. (1966) *Photo-Elastic Analysis*, Pergamon, New York.

Ladevèse, P. (ed.), (1985) *Local Effects in the Analysis of Structures*, Elsevier, New York.

Love, A.E.H. (1944). *A Treatise on the Mathematical Theory of Elasticity*, 4th edition, Dover, New York.

Malvern, L.E. (1969). *Introduction to the Mechanics of a Continuous Medium*, Prentice-Hall, Englewood Cliffs, N.J.

Pindera, J. T. (ed.) (1984) *Modelling problems in Crack Tip Mechanics*, Martinus Nijhoff Publishers, Dordrecht.

Pindera, J. T. (1988) Local effects - A major problem of contemporary stress/strength analysis of homogeneous and composite structures, in G. C. Sih, J. T. Pindera and S. V. Hoa (eds.), *Analytical and Testing Methodologies for Design with Advanced Materials*, North Holland, Amsterdam, pp. 9-55.

Pindera, J. T. (1989) Local effects and defect criticality in homogeneous and laminated structures, *Trans. ASME, J. Pressure Vessel Technology* 111, 136-50.

Pindera, J. T. (1993) On the limits of two-dimensional treatment of the actual three-dimensional stress states in adhesively bonded composite structures, *Second Canadian International Composites Conference and Exhibition*, W. Wallace, R. Gauvin, S. V. Hoa (eds.) *CANCOM '93*, Canadian Association for Composite Structures and Materials, Montreal, pp. 917-25.

Pindera, J. T. (1998a) Actual three-dimensional stresses in composite structures and in-local effects in homogeneous structures. Case studies, in *Multilayered and Fibre-Reinforced Composites: Problems and Prospects*, NATO ARW, Kiev 1998, edited by Y. M. Haddad, *Kluwer*, Dordrecht, pp. 57-84.

Pindera, J. T. (1998b) Principles and approaches of advanced experimental mechanics in service of modern technology, in *Multilayered and Fibre-Reinforced Composites: Problems and Prospects*, NATO ARW, Kiev 1997, edited by Y. M. Haddad, *Kluwer*, Dordrecht, pp. 13-56.

Pindera, J. T. and Krasnowski, B. R. (1982) Determination of stress intensity factors in thin and thick plate using isodyne photoelasticity, in L. A. Simpson (ed.), *Fracture Problems and Solutions in the Energy Industry*, Pergamon Press, pp. 147-56.

Pindera, J. T. and Liu, X. (1992) On the actual three-dimensional stresses in notches and cracks, *Composites Engineering* 1, 281-301.

Pindera, J. T. and Pindera, M.-J. (1989) *Isodyne Stress Analysis*, Kluwer Academic Publishers, Dordrecht.

Pindera, M.-J., Pindera, J. T. and Ji, X. (1989) Three-dimensional effects in beams-Isodyne assessment of a plane solution, *Experimental Mechanics* 29, 23-31.

Pindera, J. T. and Wang, G. (1992) Isodyne stress analysis of adhesively bonded symmetric joints, *Experimental Mechanics* 32, 348-56.

Sokolnikoff, I. S. (1956) *Mathematical Theory of Elasticity*, McGraw-Hill Book Co., New York.

Thum, A. et al. (1960) *Verformung, Spannung und Kerbwirkung* (Deformation, Stress and Notch Action), VDI-Verlag, Düsseldorf.

Truesdell, C. and Toupin, R.A. (1960) *The classical field theories*, Handbuch du Physik, Vol. III/1, S. Flügge (ed.), Springer-Verlag, Berlin.

2.11. Further Reading

Bowen, R. M. (1989) *Introduction to Continuum Mechanics for Engineers*, Plenum Press, New York.

Chung, T.J. (1988) *Continuum Mechanics*, Prentice-Hall, Englewood Cliffs, New Jersey.

Coleman, B.D. and Noll, W. (1960) An approximation theorem for functionals with applications in continuum mechanics, *Arch. Ration. Mech. Analysis* 6, 355-70.

Dantu, P. (1958) I - Étude des contraintes dans les milieux hétérogènes. Application au beton. II - Utilisation des réseaux pour l'étude des déformations. Laboratoire Central des Ponts et Chaussees, Publication 57-6. Annales de l'Institut Technique du Bâtiment et des Travaux Publics, Paris, pp.17-25.

Davis, J.L. (1987) *Introduction to Dynamics of Continuous Media*, Macmillan, New York.

Doeblin, E. O. (1983) *Measurement Systems. Application and Design*, McGraw-Hill Book Co., New York.

Durelli, A.J. and W.F. Riley (1965) *Introduction to Photomechanics*, Prentice-Hall Series in Solid and Structural Mechanics, edited by P. S. Symonds, Prentice-Hall, Englewood Cliffs.

Eringen, A.C. (1962) *Nonlinear Theory of Continuous Media*, McGraw-Hill, New York.

Eringen, A.C. (1967) *Mechanics of Continua*, John Wiley & Sons, New York.

Feynman, R. (1993) *The Character of Physical Law*, The MIT Press, Cambridge, Massachusetts.

Fredrick, D. and Chang, T.S. (1965) *Continuum Mechanics*, Allyn and Bacon, Boston.

Green, A.E. and Adkins, J.E. (1960) *Large Elastic Deformations and Nonlinear Continuum Mechanics*, Clavendon Press, Oxford.

Gurtin, M.E. (1981) *An Introduction to Continuum Mechanics*, Academic Press, New York.

Gurtin, M.E. and Williams, W.O. (1967) *An axiomatic foundation for continuum thermodynamics*, Arch. Rational Mech. Anal. 26, 83-117.

Haddad, Y. M. (1995) *Viscoelasticity of Engineering Materials*, Kluwer, Dordrecht.

82

Hunter, S.C. (1976) *Mechanics of Continuous Media*, Ellis Hordwood, Chister, England.

Jaunzemis, W. (1967) *Continuum Mechanics*, Macmillan, New York.

Jessop, H.T. (1950) The determination of the principal stress differences at a point in a three-dimensional photoelastic model, *Brit. J. Appl. Phys.* 1, 184-89.

Kac, M. (1969) Some mathematical models in science, *Science* 166, 695-99.

Kestin, J. (1979) *A Course in Thermodynamics, Volumes I and II*, Hemisphere Publishing Corp. and McGraw-Hill Book Company, New York.

Kuhn, T. S. (1985) *The Structure of Scientific Revolution*, University of Chicago Press, Chicago

Kuske, A. and Robertson, G. (1974) *Photoelastic Stress Analysis*, Wiley, New York.

Lai, W.M., Rubin, D. and Krempl, E. (1978) *Introduction to Continuum Mechanics (SI/Metric Units)*, Pergamon Press, New York.

Leipholz, H. H. E. (1983) On the role of analysis in mechanics, *Trans. of the CSME* 7, 3-7

Lewicki, B. and Pindera, J. T. (1956) Photoelastic model of reinforced structures (in Polish), *Archiwum Inżynierii Lądowej* 2, 381-418.

Mindlin, R.D. and Tiersten, H.F. (1962) Effect of couple-stresses in linear elasticity, *Arch. Rational Mech. Anal.* 11, 415-48.

Müller, R. (1964) Des einfluss der messlänge auf die ergebnisse bei dehnmessungen an beton, *Beton* 14, 205-8.

Noll, W. (1955) On the continuity of the solid and fluid states, *J. Rational Mech. Anal.* 4, 3-81.

Noll, W. (1958) A mathematical theory of the mechanical behaviour of continuous media, *Arch. Ration. Mech. Analysis* 2, 197-226.

Pindera, J. T. (1981) Foundations of experimental mechanics: Principles of modelling, observation and experimentation, in J. T. Pindera (ed.), *New Physical Trends in Experimental Mechanics*, International Centre for Mechanical Sciences, Udine, Springer-Verlag, Wien, pp. 188-236.

Pindera, J. T. (1987a) Advanced experimental mechanics in modern engineering science and technology, *Transactions of the CSME* 11, 125-38.

Pindera, J. T. (1987b) Advanced experimental mechanics and its components: Theoretical, physical, analytical and social aspects, in A. P. S. Selvadurai (ed.), *Developments in Engineering Mechanics*, Elsevier, New York, pp. 367-414.

Pindera, J. T. (1992) Actual three-dimensional stresses and related dynamic fractures in some adhesively bonded structures, in S. V. Hoa and R. Gauvin (eds.) *Composite Structures and Materials*, Elsevier Applied Science, London and New York, pp. 332-40

Pindera, J. T., Krasnowski, B. R. and Pindera, M.-J. (1985) Theory of elastic and photoelastic isodynes. samples of applications in composite structures, *Experimental Mechanics* 25, 272-81.

Popper, K. R. (1968) *The Logic of Scientific Discovery*, Harper and Row, New York.

Prager, W. (1961) *Introduction to Mechanics of Continua*, Ginn, Boston, Mass.

Sneddon, I.N. and Hill, R. (eds.), (1960-1963) *Progress in Solid Mechanics*, Vol 1 (1960), Vol. 2(1961), Vol. 3(1963), Vol. 4 (1963), North-Holland, Amsterdam.

Sommerfeld, A. (1950) *Mechanics of Deformable Bodies*, Academic Press, New York.

Stuart, H. A. (ed.) (1952-1956) *Die Physik der Hochpolymeren* I-IV, Springer-Verlag, Berlin.

Timoshenko, S. P. and Goodier, J. N. (1970) *Theory of Elasticity*, McGraw-Hill Book Company, New York.

Truesdell, C. (1965) The nonlinear field theories of mechanics, in *Encyclopedia of Physics*, Ed. S. Flügge, Vol. III/3, Springer-Verlag, Berlin.

Truesdell, C. (1966) *The Elements of Continuum Mechanics*, Springer-Verlag, New York.

Washizu, K. (1958) *A Note on the Condition of Compatability*, *J. Math. Phys.* 36, 306-12.

Williams, W.O. (1970) Thermodynamics of rigid continua, *Arch. Rational Mech. Anal.* 36, 270-84.

Życzkowski, M. (1981) *Combined Loading in the Theory of Plasticity*, PWN - Polish Scientific Publishers, Warszawa.

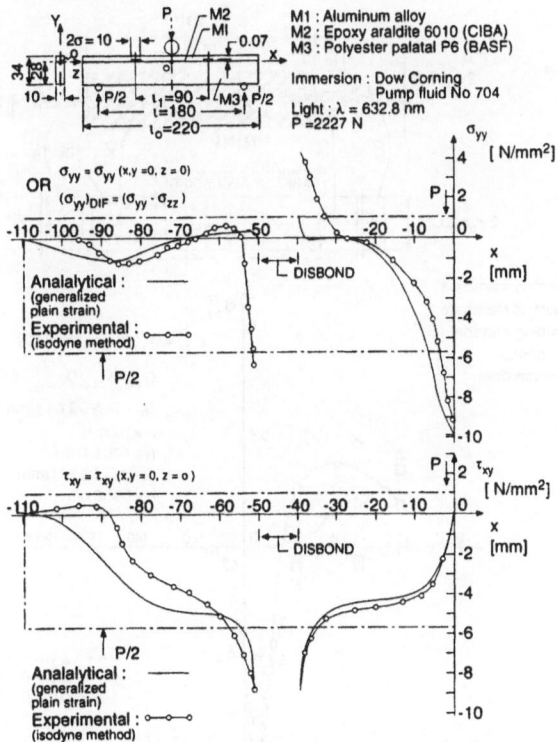

Figure 2.15. Stress analysis of a composite beam with two local disbonds, under three-point loading. Analytical and experimental results. From: Pindera, J. T. (1989) Local effects and defect criticality in homogeneous and laminated structures, *Trans. ASME, J. Pressure Vessel Technology* **111** (May 1989), 136-50. Reprinted with kind permission of ASME International.

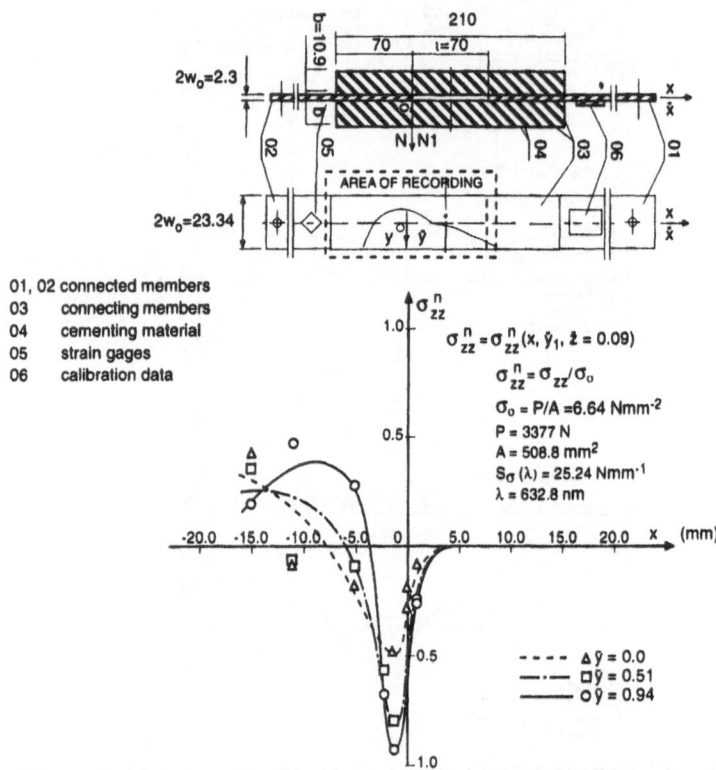

Figure 2.16. Stresses in the connecting member. Variation of the normal (zz)-stresses in the plane close to the lamination plane, along the connector, and in three planes: at the side face of the connector, in the symmetry plane of the connector, and between the symmetry plane and the side face of the connector. From: Pindera, J. T. and Wang, G. (1992) Isodyne stress analysis of adhesively bonded symmetric joints, *Experimental Mechanics* **32** (December 1992) *"published by the Society of Experimental Mechanics"*, 348-57. Reprinted with kind permission of the Society of Experimental Mechanics.

CHAPTER 3

DEFORMATION AND STRAIN. ANALYSIS OF MOTION.

3.1. Introduction

Engineering materials, when subjected to external loading, may undergo deformation and/or motion. In this chapter, we first consider, in Section 3.2, the kinematics of involved deformation in the continuous body and the determination of measures of the pertaining strain. In this context, we consider the deformation at two distinct times, namely $t = 0$ and $t = t$. Then, in Section 3.3, we analyze the motion, by developing relationships between the sequential configurations that the parts of a "*continuous*" material body may acquire with the passage of time.

3.2. Deformation Kinematics and Measures of Strain

The present section deals with the kinematics of involved deformation in the continuous body and the determination of pertaining strain. In the latter context, a number of nonlinear measures of strain are first introduced, then, subsequently, they are reduced to linear forms suitable for the treatment of infinitesimal deformation. Throughout the analysis, the relationships between the initial positions of the material points of the body and their subsequent positions are considered without taking into consideration the type of material that we are dealing with or the imposed boundary conditions.

3.2.1. "LAGRANGIAN" AND "EULERIAN" DESCRIPTIONS OF MOTION

In studying the motion of a continuous medium, we fix our attention on a single material point with which we associate the geometry of a mathematical Euclidean point and study its path (trajectory). Such trajectory can be established by determining the position vector p of the point at time t that was initially at a position characterized by the position vector P at time $t=0$, Figure 3.1.

This can be expressed with reference to the Cartesian coordinate system shown in Fig. 3.1 as

$$P = X_I e_I, \qquad p = x_i e_i \tag{3.1}$$

where X_I are the values of the rectangular Cartesian components of the position vector **P**, i.e., at time t=0 (the initial position) corresponding to a current position coordinates x_i at time t. In eqn. (3.1) above, e_I and e_i are the unit vectors associated with the rectangular Cartesian frames of reference in the undeformed and deformed states, respectively. As noticed, majuscules are used in this text to identify the undeformed configuration $X_I(I=1,2,3)$ and minuscules to designate the deformed configuration $x_i(i=1,2,3)$. Hence, the identity of the material point is X_I and its subsequent motion is described by

$$x_i = x_i(X_I,t) \; ; \; X_I = X_I(x_i,t) \tag{3.2}$$

The X_I-coordinates are referred-to in continuum mechanics as the *"material"* or *"Lagrangian"* description while the x_i-coordinates are called the *"spatial"* or *"Eulerian"* description. Equation (3.2) expresses the evolution of the deformation process in the body as a function of time. From a continuum mechanics point of view, the functions $x_i(X_I,t)$ are single-valued continuous functions whose *"Jacobian of transformation"* does not vanish, namely,

$$J = |x_{i,I}| = \frac{1}{6} \varepsilon_{ILM} \, \varepsilon_{ilm} \, x_{i,I} \, x_{i,L} \, x_{m,M} \neq 0 \tag{3.3}$$

where ε_{ILM}, ε_{ilm} are the alternating tensors (*see Chapter 1*). Equation (3.3) expresses the form of the *"implicit function theorem of calculus"*. It is a basic relation for securing the *"axiom of continuity"* in continuum mechanics. It points out the fact that the matter is *"indestructible"*, i.e., no region of positive, finite volume of matter may be deformed into a zero or an infinite volume. Another axiom secured by this equation is that the matter is *"impenetrable"*, that is, the motion carries every region into a region, every surface into a surface and every curve into a curve.

3.2.2. NONLINEAR MEASURES OF STRAIN

We derive below a number of non-linear measures of strain as based on the axioms of continuity introduced above.

With reference to Figure 3.1, one may consider a line element d**P** deformed into a line element d**p**. These two line elements are expressed, respectively, by

$$dP = e_I \, dX_I = e_I \, X_{I,i} \, dx_i$$
$$dp = e_i \, dx_i = e_i \, x_{i,I} \, dX_I \tag{3.4}$$

In the equations (3.4) above, $X_{I,i}$ and $x_{i,I}$ are the *"deformation gradients"* referred, respectively, to the spatial and material coordinates. Such deformation gradients are the most "*basic*" *or* "*primitive*" measures of strain.

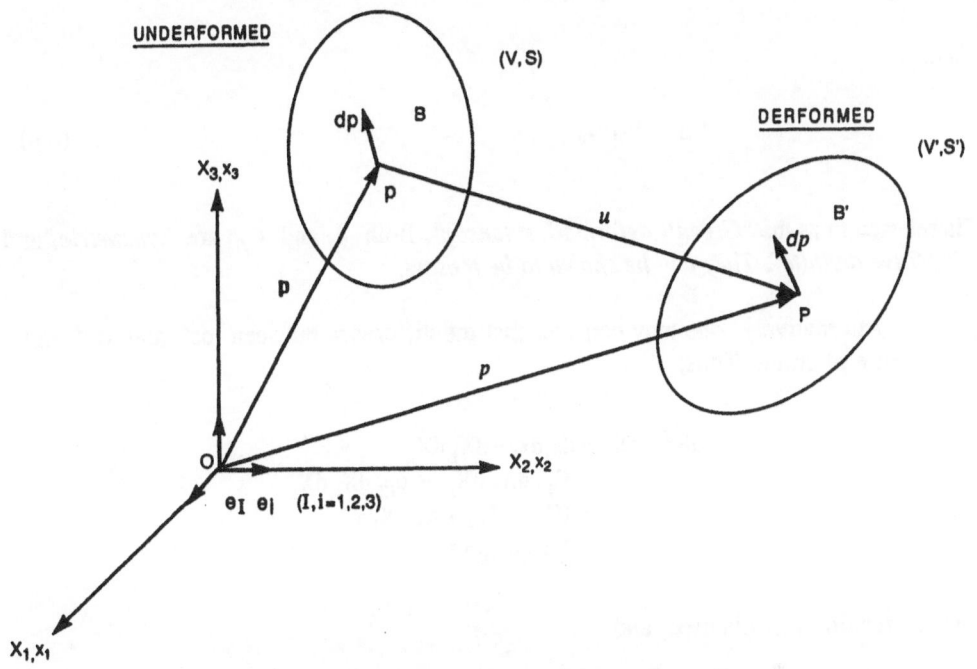

Figure 3.1. Deformation kinematics of a continuous body.

Alternative measures of strain can be expressed from the definition of the square of the line element. Thus, in the undeformed configuration, Figure 3.1, one has

$$dS^2 = dP \cdot dP = dX_K dX_K$$

The above relation can be written, using the chain rule of partial differentiation, as

$$
\begin{aligned}
dS^2 &= X_{K,j} X_{K,i} \, dx_j dx_i \\
&= c_{ji} \, dx_j \, dx_i
\end{aligned}
\tag{3.5}
$$

$$\text{where,} \quad c_{ji} = X_{K,j} X_{K,i}$$

is known as the "*Cauchy's deformation tensor*".

Similarly, one may express, with reference to Fig. 3.1, the square of the line element, in the deformed configuration as

$$ds^2 = dp \cdot dp = dx_k dx_k = x_{k,J} x_{k,I} dX_J dX_I$$
$$= C_{JI} dX_J dX_I$$

where,

$$C_{JI} = x_{k,J} x_{k,I} \qquad\qquad (3.6)$$

is referred to as the "*Green's deformation tensor*". Both c_{ji} and C_{JI} are *"symmetric"* and *"positive definite"*. They can be shown to be tensors.

Alternatively, one may consider that the difference between ds^2 and dS^2 to be a measure of strain. Thus,

$$ds^2 - dS^2 = dx_i dx_i - dX_L dX_L$$
$$= C_{KL} dX_K dX_L - \delta_{KL} dX_K dX_L$$
$$= (C_{KL} - \delta_{KL}) dX_K dX_L$$
$$= 2E_{KL} dX_K dX_L$$

where relation (3.6) is used, and

$$E_{KL} = \frac{1}{2} (C_{KL} - \delta_{KL}) \qquad\qquad (3.7)$$

is the "*material or Lagrangian strain tensor*".

Similarly,

$$ds^2 - dS^2 = dx_i dx_i - dX_L dX_L$$
$$= \delta_{ij} dx_i dx_j - c_{ij} dx_i dx_j$$
$$= (\delta_{ij} - c_{ij}) dx_i dx_j$$
$$= 2\epsilon_{ij} dx_i dx_j$$

where relation (3.5) is used, and

$$\epsilon_{ij} = \frac{1}{2}(\delta_{ij} - c_{ij})$$

(3.8)

is the *"spatial or Eulerian strain tensor"*.

QUIZ 3.1

Show that the strain tensors E_{KL} and ϵ_{kl} are related where,

$$E_{KL} = \epsilon_{kl} x_{k,K} x_{l,L}$$

and,

$$\epsilon_{kl} = E_{KL} X_{K,k} X_{L,l}$$

With reference to Figure 3.1, one may write the following relation:

$$p = P + u$$

or,

$$x_i e_i = X_K e_K + u_K e_K$$

and,

$$dx_i e_i = dX_K e_K + u_{K,L} dX_L e_K$$
$$= (\delta_{KL} + u_{K,L}) dX_L e_K$$

Hence,

$$ds^2 = dx_i dx_i = (\delta_{KL} + u_{K,L})(\delta_{KM} + u_{K,M}) dX_L dX_M$$

Now, if we choose $(ds^2 - dS^2)$ as a measure of strain, it can be shown that,

$$ds^2 - dS^2 = (u_{M,L} + u_{L,M} + u_{K,L} u_{K,M}) dX_L dX_M$$
$$= 2E_{LM} dX_L dX_M$$

where,

$$E_{LM} = \frac{1}{2}(u_{M,L} + u_{L,M} + u_{K,L}u_{K,M})$$
(3.9)

is an alternative expression for the *material* or *Lagrangian strain tensor* expressed previously by (3.7).

One may also choose, with reference to Fig. 3.1, the formulation

$$\mathbf{P} = \mathbf{p} - \mathbf{u}$$

i.e.,

$$X_l\, e_l = x_k\, e_k - u_k\, e_k$$

and,

$$dX_l\, e_l = dx_k\, e_k - u_{k,l}dx_l\, e_k$$
$$= (\delta_{kl} - u_{k,l})dx_l\, e_k$$

Thus,

$$dS^2 = dX_l dX_l = (\delta_{kl} - u_{k,l})(\delta_{km} - u_{k,m})dx_l dx_m$$
$$= (\delta_{lm} - u_{m,l} - u_{l,m} + u_{k,l}u_{k,m})dx_l dx_m$$

and one can show that

$$ds^2 - dS^2 = (u_{m,l} + u_{l,m} - u_{k,l}u_{k,m})dx_l dx_m$$
$$= 2\epsilon_{lm} dx_l\, dx_m$$

where,

$$\epsilon_{lm} = \frac{1}{2}(u_{m,l} + u_{l,m} - u_{k,l}u_{k,m})$$
(3.10)

is another expression for the spatial or *Eulerian strain tensor* given earlier by equation (3.8).

Similar to Cauchy's stress tensor (see Chapter 2), both the Lagrangian strain and the Eulerian strain tensors are symmetric, hence, each has only six independent components in a three-dimensional space.

3.2.3. INFINITESIMAL STRAIN AND ROTATION

We considered in the foregoing, the formulations of the three-dimensional nonlinear measures of strain as based only on the implications of assumed continuity. In the present sub-section, we introduce some of the simplifying assumptions in the theory of continuum mechanics with the aim of reducing the complexity of mathematics which otherwise would be involved. The definitions of infinitesimal strains and rotations depend upon the following two assumptions:

(a)   *The occurring deformations $u_i(x_{k,})$ or $u_l(X_{K,})$ are much smaller than the least dimension of the free body under consideration.*

(b)   *The deformation gradient $u_{i,j}$ or $u_{I,J} << 1$.*

Recalling the expression of nonlinear measure of strain given by equation (3.9), that is

$$E_{IJ} = \frac{1}{2}(u_{I,J} + u_{J,I} + u_{K,I}u_{K,J}) \tag{3.11}$$

This expression becomes the *infinitesimal* (or *linear*) *Lagrangian (material)* strain tensor $\hat{E}_{ij}$ by seeking the nonlinear term in the above expression (3.11) to be zero, i.e.,

$$u_{K,I}u_{K,J} = 0$$

Accordingly,

$$\hat{E}_{IJ} = \frac{1}{2}(u_{I,J} + u_{J,I}) = u_{(I,J)} \tag{3.12}$$

is the *"infinitesimal Lagrangian strain tensor"*.

It is apparent from the above expression that $\hat{E}_{ij}$ represents the symmetric portion of the deformation gradient tensor $u_{I,J}$. This is indicated in (3.12) above by $u_{(I,J)}$ following our notations in Chapter 1 (Section 1.6.2).

Introducing the *"Lagrangian infinitesimal rotation tensor"*:

$$\Omega_{IJ} = \frac{1}{2}(u_{I,J} - u_{J,I}) = u_{[I,J]} \tag{3.13}$$

where $u_{[I,J]}$ is the skew-symmetric portion of $u_{I,J}$. Thus,

$$u_{I,J} = E_{IJ} + \Omega_{IJ} = u_{(I,J)} + u_{[I,J]} \tag{3.14}$$

Repeating the same procedure for the *infinitesimal (or linear) Eulerian* strain tensor, then, with reference to (3.10), it follows that

$$\epsilon_{\ell m} = \frac{1}{2}(u_{\ell,m} + u_{m,\ell}) = u_{(\ell,m)} \tag{3.15}$$

where,

$$u_{\ell,m} = \epsilon_{\ell m} + \omega_{\ell m} \tag{3.16}$$

and

$$\omega_{\ell m} = \frac{1}{2}(u_{\ell,m} - u_{m,\ell}) = u_{[\ell,m]} \tag{3.17}$$

is the *"infinitesimal Eulerian rotation tensor"* which is skew-symmetric; see Chapter 1.

In three dimensions, it is possible to express a *"dual vector"* $\omega{:}\omega_k$ in terms of the skew-symmetric tensor ω_{ij}. That is

$$\omega_k = \epsilon_{kij}\,\omega_{ij} \tag{3.18}$$

where ϵ_{kij} is the alternating tensor.

At the same time, since ω_{ij} is skew-symmetric, it can be shown that (3.18) has a unique inverse, i.e.,

$$\omega_{ij} = \epsilon_{ijk}\,\omega_k \tag{3.19}$$

Hence, ω_{ij} may be called the *"dual tensor"* of a vector ω_k. The latter vector is referred to as the *"rotation vector of the displacement field "*.

QUIZ 3.2

Show that the infinitesimal Lagrangian and infinitesimal Eulerian strain tensors are equivalent. That is

$$E_{ij} = \epsilon_{ij} \tag{3.20}$$

At this point, it should be emphasized that both ϵ_{kl} and E_{KL} cannot be considered as strain measures and they are, in fact, only approximations of strain measures within the context of the infinitesimal strain theory.

Following (3.20), we shall adopt in the remainder of the book, the notation ϵ_{ij} to designate the infinitesimal strain, i.e.,

$$\epsilon_{ij} = \frac{1}{2}(u_{i,j} + u_{j,i})$$

(3.21)

3.2.4. PRINCIPAL STRAINS, PRINCIPAL DIRECTIONS AND STRAIN INVARIANTS

In Chapter 1 (Section 1.7), we dealt with the procedure of determining the principal values and principal directions of symmetric second-order tensors, whereby the invariants of the tensor are represented by the coefficients of the cubic equation (1.22). Following the same procedure, we have determined, in Chapter 2, Section 2.6, the invariants of the Cauchy's stress tensor and the pertaining principal values and principal directions. In an analogous manner, the reader can easily verify that the invariants of the strain tensor are expressed by

$$II_1 = \epsilon_{11} + \epsilon_{22} + \epsilon_{33} = \epsilon_{kk}$$

$$II_2 = \frac{1}{2}(\epsilon_{ij}\epsilon_{ij} - \epsilon_{ii}\epsilon_{jj})$$

(3.22)

$$= -(\epsilon_{11}\epsilon_{22} + \epsilon_{22}\epsilon_{33} + \epsilon_{33}\epsilon_{11}) + \epsilon_{23}^2 + \epsilon_{31}^2 + \epsilon_{12}^2$$

(3.23)

$$II_3 = \begin{vmatrix} \epsilon_{11} & \epsilon_{12} & \epsilon_{13} \\ \epsilon_{21} & \epsilon_{22} & \epsilon_{23} \\ \epsilon_{31} & \epsilon_{32} & \epsilon_{33} \end{vmatrix} = \det. \epsilon_{ij}$$

(3.24)

$$= \frac{1}{6}\varepsilon_{ijk}\varepsilon_{pqr}\epsilon_{ip}\epsilon_{jq}\epsilon_{kr}$$

The first strain tensor invariant, II_1, has a simple geometrical meaning in the case of infinitesimal strain. It represents the change in volume per unit volume, i.e.,

$$II_1 = \Delta V/V = \epsilon_{kk}$$

For this reason, ϵ_{kk} is called the *"cubical dilatation"*. If two-dimensional state (plane strain) is considered, the first invariant represents the change of area per unit area of the surface under strain. In the definition of finite (nonlinear) strain, the sum of principal strains does not have such a simple interpretation.

we aim below to approach the problem of determining the strain invariants from a different point of view. Again, the analysis would apply to any symmetric second order tensor, e.g., the stress tensor.

Consider the differential element dX_1 which sweeps out a sphere of radius dS at X_1 and holds dS as a fixed number. The element dX_1 is deformed into the differential element dx_i which sweeps out a quadric surface or *"strain ellipsoid"*, in the interpretation presented earlier by (3.5):

$$dS^2 = dX_1 dX_1 = c_{k\ell} dx_k dx_\ell$$

where $c_{k\ell}$ is the Cauchy's deformation tensor.

Choose one vector dX_1' and let dX_1'' be any vector in a plane perpendicular to dX_1', then,

$$dX_1' dX_1'' = 0$$

i.e., $\quad 0 = dX_1' dX_1'' = X_{1,k} X_{1,\ell} dx_k' dx_\ell'' = c_{k\ell} dx_k' dx_\ell''$

Accordingly, one can write the following relationship:

$$\frac{d(dS^2)}{dx_\ell'} = 2 c_{k\ell} dx_k'$$

which expresses the gradient vector of dS^2 at the end of dx_i'. The relation $c_{k\ell} dx_k' dx_\ell'' = 0$ asserts that dx_ℓ'' is perpendicular to this gradient. In this context, the following theorem may be introduced:

> **Theorem**: "Perpendicular diameters of an infinitesimal sphere at the point X_I are deformed into conjugate diameters of the strain ellipsoid at x_i and conversely".

This theorem is referred to as *"Cauchy's First Fundamental Theorem"*.

QUIZ 3.3

Show that the following hypothesis is correct: *"Any quadric has three diameters that are perpendicular to their conjugate planes, i.e. its axes."*

Following the hypothesis stated in Quiz 3.3, above, one may conclude that the axes of the *"strain ellipsoid"* at X_I and x_i are the principal axes of strain at X_I and x_i, respectively. Thus, one advance the following postulate:

Postulate: "There exists an orthogonal triad at X_I which deforms into an orthogonal triad at x_i and conversely. Furthermore, any deformation rotates the principal axes of strain at X_I into the principal axes of strain at x_i".

The above postulate is illustrated schematically, for the two dimensional case, in Figure 3.2.

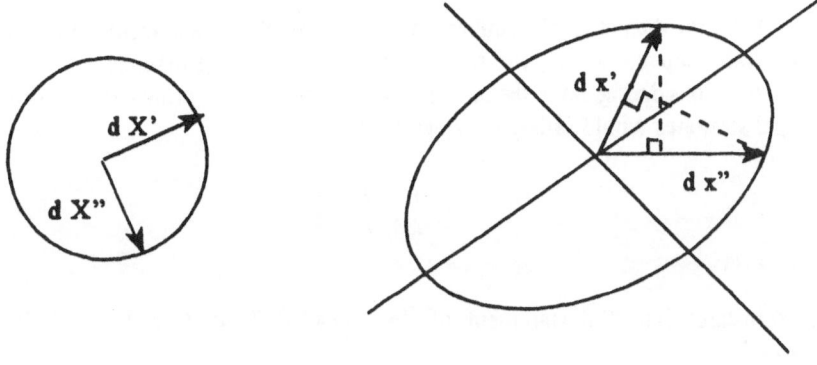

Figure 3.2. Correspondence between orthogonal triads in the undeformed and deformed configurations.

In the mechanics of deformable media, the ratio ds/dS is often considered as a primitive measure of strain, and is referred-to as the *"stretch"*. It may be expressed either in terms of the deformation ratio dx_k/dS or dX_K/dS. The latters are given the designations g_k and G_K, respectively. That is:

$$g_k = dx_k / dS \qquad \text{and} \qquad G_K = dX_K / dS$$

Meantime, recalling the relations (3.5) and (3.6), i.e.,

$$dS^2 = c_{k\ell}\, dx_k\, dx_\ell \quad ; \quad ds^2 = C_{KL}\, dX_K\, dx_L\,,$$

Thus, the resulting material stretch may be expressed either in terms of either G_K or g_k , respectively as

$$\Lambda_{(G)} = C_{KL}\, G_K\, G_L \quad ; \quad \lambda_{(g)} = \frac{1}{\sqrt{c_{k\ell}\, g_k\, g_\ell}}$$

It is noted that the above two expressions of the "stretch" have the same physical significance. They are related to the classical Lagrangian and Eulerian strain measures, previously introduced in Section (3.2.3), via the following expressions:

$$E_{(G)} = \Lambda_{(G)} - 1 \quad ; \quad \epsilon_{(g)} = \frac{\lambda_{(g)} - 1}{\lambda_{(g)}}$$

Since the diameters of an infinitesimal sphere at X_I are deformed into the diameters of the strain ellipsoid at x_I , the ratios of the corresponding diameters are the pertaining "*stretches*". Considering, now, the three mutually perpendicular principal directions, then, the principal stretches would satisfy the relationship

$$\overset{1}{\lambda} \geq \overset{2}{\lambda} \geq \overset{3}{\lambda}$$

This translates into the statement of the so-called "*Cauchy's Second Fundamental Theorem*".

Returning back to our discussion concerning the strain invariants, it is evident that

$$\frac{1}{\lambda_{(g)}^2} = C_{k\ell}\, g_k\, g_\ell \quad ; \text{ with } \quad g_k\, g_k = 1\,,$$

from which one is able to demonstrate expressions (3.22), (3.23) and (3.24) for the strain invariants. It is left to the student to demonstrate that the latter referred-to expressions are valid for any of the five different measures of strain we already introduced in this chapter.

3.2.5. SOME EXAMPLES OF STRAIN

Rigid Deformation
The deformation of a body is said to be "*rigid*" if the distance between every pair of material points remains unchanged. In this case, one can show that

$$II_1 = II_2 = 3 \quad ; \quad II_3 = 1.$$

where II_1, II_2 and II_3 are the strain invariants. It is left to the student to prove this argument.

Isochoric Deformation
When the volume element remains unaltered during a deformation process, the ongoing deformation is referred-to as "*isochoric*". The necessary and sufficient conditions for this type of deformation may be stated as follows:

$$J = \left| x_{i,I} \right| = 1 \quad ; \quad II_3 = 1 ;$$

or:

$$II_1 = 0.$$

Homogeneous Deformation
A large number of important types of deformation of this class is included in the so called "*homogeneous affine transformations*". The latter is defined to be in accordance with:

$$x_k = \Theta_{kK} X_K$$

where Θ_{kK} is a constant matrix which admits an inverse. By its nature of definition, an affine transformation carries straight lines into straight lines, ellipses into ellipses and ellipsoids into ellipsoids. The class of such deformations is referred to as "*homogeneous strain*". As introduced earlier in this chapter, this class of deformation satisfies the axiom of continuity, i.e.,

$$x_k = x_{k,K} X_K$$

Uniform Dilation or Dilatation
A homogeneous deformation in which the principal stretches have the same value, *say* λ, is referred to as "*uniform dilation*" or "*dilatation*", previously introduced in Section (3.2.4). For this type of deformation:

$$\Theta_{kK} = \lambda\, \delta_{kK} \sim \begin{vmatrix} \lambda & 0 & 0 \\ 0 & \lambda & 0 \\ 0 & 0 & \lambda \end{vmatrix} ; \ 0 < \lambda < \infty$$

It is evident that, for $\lambda > 1$ one has uniform extension, whilst for $0 < \lambda < 1$ one has uniform compression. Clearly, for the case of *"dilatation"*, the material sphere, $dS^2 = dX_l \, dX_l$ would deform into a spatial sphere and not an ellipsoid.

Simple Extension

It is considered as homogeneous deformation, whereby two of the three principal stretches are equal, but the third principal extension is of a different value. Consequently, the pertaining transformation matrix may be expressed as

$$\Theta_{kK} = \sim \begin{vmatrix} a\lambda & 0 & 0 \\ 0 & a\lambda & 0 \\ 0 & 0 & \lambda \end{vmatrix} \; ; \; 0 < a < \infty \quad ; \quad 0 < \lambda < \infty$$

whereby, the x_3 or X_3 has been set equal to the axis of extension. Furthermore, by considering extensions or classical strain measures:

$$E_{(G)} = \frac{ds - dS}{dS}$$

thus, in the x_1 or x_2 directions, one can illustrate that:

$$E_{(1)} = E_{(2)} = a\lambda - 1$$

whilst, in the x_3 direction, it follows that

$$E_{(3)} = \lambda - 1$$

Accordingly, the well-known *"Poisson's ratio"*, v, is determined by the absolute numerical value of the following *"transverse contraction ratio"*:

$$v = \frac{E_{(1)} \text{ or } E_{(2)}}{E_{(3)}} = \frac{1 - a\lambda}{\lambda - 1}$$

from which, one can write that

$$a = \frac{1}{\lambda}(1 + v) - v$$

It is noted that *Poisson's ratio* has the range

$$-1 \le v \le \frac{1}{2}$$

where $\nu = -1$ represents a situation of *uniform dilatation* or *dilation*, whilst $\nu = 1/2$ corresponds to *isochoric* deformation. The case where $\nu < 0$ is not observed for homogenous materials under tensile loading, although, in some recent studies concerning composite materials, negative values of *Poisson's ratio* have been reported.

Simple Shear

It is a homogeneous strain, whereby:

      - two orthogonal families of parallel planes are preserved,
      - the lines normal to one family remain normal to that family,
      - the lines common to both families are not stretched.

Thus, simple sheer is described, as demonstrated in Figure 3.3, by the relations:

$$x_1 = X_1 + k X_2; \quad x_2 = X_2; \quad x_3 = X_3,$$

and, hence, the matrix of deformation gradients $\Theta_{kK} = x_{k,K}$ is given by

$$x_{i,J} = \begin{vmatrix} 1 & k & 0 \\ 0 & 1 & 0 \\ 0 & 0 & 1 \end{vmatrix}$$

In this case, the Eulerian strain tensor is given by:

$$\epsilon_{ij} \sim \begin{vmatrix} 0 & k/2 & 0 \\ k/2 & -k^2/2 & 0 \\ 0 & 0 & 0 \end{vmatrix}$$

In the case of simple shear, one can demonstrate that the strain invariants have the values:

$$II_1 = II_2 = 3 + k^2, \qquad II_3 = 1,$$

Accordingly, since $II_3 = 1$, it follows that simple shear is *"isochoric"*.

QUIZ 3.4

Using the matrix for the deformation gradient in simple shear:

$$x_{i,J} = \begin{vmatrix} 1 & k & 0 \\ 0 & 1 & 0 \\ 0 & 0 & 1 \end{vmatrix}$$

where k is a constant, determine the Green, Cauchy, Lagrangian, and Eulerian strain tensors.

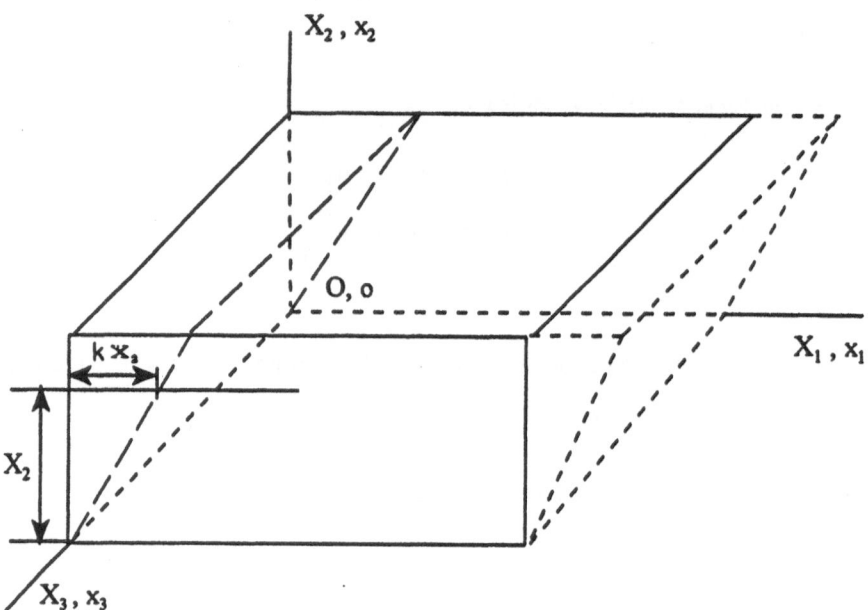

Figure 3.3

3.2.6. COMPATIBILITY CONDITIONS

We deal now with the problem of determining the displacement components u_i when the strain components are known. Considering, for the simplification of presentation, the expression of linear strain as presented by (3.21), i.e.

$$\epsilon_{ij} = \frac{1}{2}(u_{i,j} + u_{j,i}) \tag{3.25}$$

The problem is how one would integrate the differential equation (3.25) above to determine the displacement components u_i.

Since we have six equations corresponding to the six independent components of ϵ_{ij} for three unknown functions u_i, the system of Eqn. (3.25) above will not have a single-valued solution in general if the functions ϵ_{ij} were arbitrarily assigned. However, we would expect that a solution for this equation may exist only if the functions ϵ_{ij} satisfy certain conditions. The conditions of integration of (3.25) are referred to as the "*compatibility conditions*". The latter are to be satisfied by the strain components and may be determined by eliminating the components u_i from (3.25).

Differentiating (3.25) twice with respect to the coordinates, it follows that

$$\epsilon_{ij,kl} = \frac{1}{2}(u_{i,jkl} + u_{j,ikl}) \tag{3.26}$$

Interchanging the subscripts in the above equation, i. e.,

$$\epsilon_{kl,ij} = \frac{1}{2}(u_{k,lij} + u_{l,kij})$$

$$\epsilon_{jl,ik} = \frac{1}{2}(u_{j,lik} + u_{l,jik}) \tag{3.27}$$

$$\epsilon_{ik,jl} = \frac{1}{2}(u_{i,kjl} + u_{k,ijl})$$

leads to the following expression specifying the compatibility conditions:

$$\epsilon_{ij,kl} + \epsilon_{kl,ij} - \epsilon_{ik,jl} - \epsilon_{jl,ik} = 0 \tag{3.28}$$

Equation (3.28) was first obtained by St. Venant (1860) and is named after him. This equation represents 81 equations of which six only are essential. The remaining equations are repetitions due to the symmetry of the strain tensor.

In the solution of a boundary value problem, within the realm of continuum mechanics, if the displacement field is unknown and the displacement components u_i are required to be continuous and single-valued functions of the coordinates, the compatibility requirement would then be fulfilled. On the other hand, if the displacements are not explicitly retained as unknowns, the compatibility conditions must then be imposed on the strain field to ensure that there exists a continuous single-valued displacement distribution corresponding to the strain distribution.

EXAMPLE PROBLEM 3.1

Determine the principal strains and the corresponding principal directions for the following strain tensor:

$$\epsilon_{ij} = 10^{-2} \begin{vmatrix} 2 & 3 & 2 \\ 3 & 2 & 1 \\ 2 & 1 & 2.5 \end{vmatrix}$$

Solution:

Principal strains:

From the given strain tensor

$$\epsilon_{ij} = 10^{-2} \begin{pmatrix} 2 & 3 & 2 \\ 3 & 2 & 1 \\ 2 & 1 & 2.5 \end{pmatrix}$$

We determine the strain invariants

$$II_1 = 10^{-2} \, (\, 2 + 2 + 2.5) = 6.5 \times 10^{-2}$$

$$II_2 = 10^{-2} \, [(4 + 5 + 5) - 1 - 4 - 9] = 0$$

$$II_3 = 10^{-2} \begin{vmatrix} 2 & 3 & 2 \\ 3 & 2 & 1 \\ 2 & 1 & 2.5 \end{vmatrix} = 10^{-2} \, [2\,(5-1) - 3\,(7.5-2) + 2\,(3-4)] = -10.5 \times 10^{-2}$$

The characteristic equation can be, then, written as

$$-\epsilon^3 + 6.5 \times 10^{-2} \epsilon^2 - 10.5 \times 10^{-2} = 0$$

Solving the above characteristic equation, thus, the values of the principal strains are

$$\epsilon_1 = -1.17 \times 10^{-2}$$
$$\epsilon_2 = 1.4406 \times 10^{-2}$$
$$\epsilon_3 = 6.2994 \times 10^{-2}$$

The corresponding principal directions:

$$10^{-2} \begin{pmatrix} 2-\epsilon & 3 & 2 \\ 3 & 2-\epsilon & 1 \\ 2 & 1 & 2.5-\epsilon \end{pmatrix} \begin{pmatrix} n_1 \\ n_2 \\ n_3 \end{pmatrix} = \begin{pmatrix} 0 \\ 0 \\ 0 \end{pmatrix}$$

$$(2-\epsilon)\, n_1 + 3\, n_2 + 2 n_3 = 0$$
$$3 n_1 + (2-\epsilon)\, n_2 + n_3 = 0$$
$$2 n_1 + n_2 + (2.5-\epsilon)n_3 = 0$$

For $\epsilon_1 = -1.17 \times 10^{-2}$:

$$3.17 n_1 + 3 n_2 + 2 n_3 = 0$$
$$3 n_1 + 3.17 + n_3 = 0$$
$$2 n_1 + n_2 + 3.67 n_3 = 0$$

Solving the above three equations, together with

$$n_1^2 + n_2^2 + n_3^2 = 1 \qquad \text{(I)}$$

One obtains the following principal directions for $\epsilon_1 = -1.17 \times 10^{-2}$:

$$n_1 = 0.742$$
$$n_2 = -0.6286$$
$$n_3 = -0.233$$

For $\epsilon_2 = 1.4406 \times 10^{-2}$:

$$0.5594\,n_1 + 3\,n_2 + 2\,n_3 = 0$$
$$3\,n_1 + 0.5594\,n_2 + n_3 = 0$$
$$2\,n_1 + n_2 + 1.0594\,n_3 = 0$$

Solving the above three equations, together with Eqn. (I), one obtains the following principal directions for $\epsilon_2 = 1.4406 \times 10^{-2}$:

$$n_1 = -0.1805$$
$$n_2 = -0.5221$$
$$n_3 = 0.8336$$

For $\epsilon_3 = 6.2994 \times 10^{-2}$:

$$-4.994\,n_1 + 3\,n_2 + 2\,n_3 = 0$$
$$3\,n_1 - 4.994\,n_2 + n_3 = 0$$
$$2\,n_1 + n_2 - 4.494\,n_3 = 0$$

Solving the above three equations, together with (I), we obtain the following principal directions for the principal strain $\epsilon_3 = 6.299 \times 10^{-2}$:

$$n_1 = 0.645$$
$$n_2 = 0.576$$
$$n_3 = 0.500$$

3.3. Problems

1. For the following deformation gradient tensor, determine the strain matrix, the rotation matrix, the volume strain, the deviatoric strain matrix, and the three invariants of the deviatoric strain.

$$\begin{vmatrix} 9 & -10 & -14 \\ 10 & 18 & -18 \\ -14 & -18 & 27 \end{vmatrix}$$

2. Show that the strain tensor using the Eulerian approach, i.e., $x_i = x_i(X_j)$, is given by:

$$E_{jk} = 1/2\,[u_{j,k} + u_{k,j} - u_{i,j}\,u_{i,k}]$$

Compare with Lagrangian strain tensor and show why no distinction between the two tensors is necessary in the case of small displacement theory.

3. The vector $t_i = \varepsilon_{ijk}\,T_{jk}$ is called a "*dual vector*" of the tensor T_{jk}. Show that if the dual vector vanishes, the tensor is symmetric.

4. Show that the strain tensor using the Eulerian approach can be expressed as

$$\epsilon_{jk} = \frac{1}{2}\Big[u_{j,k} + u_{k,j} - u_{i,j}\,u_{i,k}\Big]$$

Compare with the Lagrangian strain tensor and show why no distinction between the two is necessary in the case of small displacement theory.

5. Show that the large deformation strain is given by:

$$\epsilon_L = \epsilon_{ij}\,\ell_i\,\ell_j$$

where,

$$\epsilon_L = \frac{ds - dS}{dS} \qquad \text{and}$$

$$\ell_i = \frac{dx_i}{dS}$$

6. The most general form of a linear relationship between the stress and strain components of an elastic isotropic solid is represented by the generalized Hooke's Law as follows:

$$\sigma_{ij} = A_{ijkl}\,\epsilon_{kl}$$

Prove that the elastic constants are the components of a Cartesian tensor of the fourth order (*Consult Chapter 1*).

7. Let a_i be the original direction cosines of an "undeformed" line element dS in a solid and a_j^* be the final direction cosines of the "deformed" element ds. Show that:

$$a_j^* = \frac{(\delta_{jk} + u_{j,k})a_k}{(1 + \epsilon_L)^{1/2}}$$

where $\quad \epsilon_L = \frac{ds - dS}{dS}$

Show also that the rotation of a line element can be written as:

$$\omega_i = \frac{1}{2} \text{ curl } u_i = \frac{1}{2} e_{ijk}(\partial_j u_k - \partial_k u_j)$$

8. For the following two displacement fields, find:

  a. The Green strain tensor C_{KL}
  b. The Cauchy strain tensor c_{kl}
  c. The Lagrangian strain tensor E_{KL}
  d. The Eulerian strain tensor ϵ_{kl}
  e. The infinitesimal strain tensor.

A. Simple Extension:

$$z_1 = (1+\epsilon)Z_1 \; ; \; z_2 = Z_2 \; ; \; z_3 = Z_3$$

B. Simple Shear:

$$z_1 = Z_1 + kZ_2 \; ; \; z_2 = Z_2 \; ; \; z_3 = Z_3.$$

9. The displacement field in a continuous body is described by:

$$u_1 = A\frac{z_1 z_3}{r^3} \; ; u_2 = A\frac{z_2 z_3}{r^3} \; ; u_3 = A\left(\frac{z_3^2}{r^3} + \frac{\lambda + 3\mu}{\lambda + \mu}\frac{1}{r}\right)$$

where $r = (z_i z_i)^{1/2}$ and A, λ and μ are constants.

a. Determine the infinitesimal strain ϵ_{kl} and rotations ω_{kl}

b. Sketch the deformed shape of a spherical cavity $r = r_0$.

c. Determine the principal strains.

d. Determine the principal axes.

3.4. Analysis of Motion

In the previous section, we devoted our analysis to the kinematics of deformation and analysis of strain. In this context, we considered the picture of deformation at two distinct times, namely $t = 0$ and $t = t$. In the present section, we analyze the relationships between the sequential configurations that the parts of a "*continuous*" material body may acquire with the passage of time.

3.4.1. THE MATERIAL DERIVATIVE

We have already seen that a material point identified by X at $t = 0$ is deformed into a spatial point x at time $t = t$; its motion being described either by $x = x(X ; t)$ or by $X = X(x ; t)$.

The "*material time rate*" or "*material derivative*" of any arbitrary tensor field is defined either by:

$$\frac{D \sigma_{IJ...}}{dt} (X ; t) = \frac{\partial \sigma_{IJ...}}{\partial t} \bigg|_{X = constant} , \tag{3.29}$$

if $\sigma_{IJ...} (X ; t)$ is a material function, or as:

$$\frac{D \sigma_{ij...}}{dt} (x ; t) = \frac{\partial \sigma_{ij...}}{\partial t} \bigg|_{x = constant} + \frac{\partial \sigma_{ij...}}{\partial x_k} \frac{\partial x_k}{t} , \tag{3.30}$$

if $\sigma_{IJ...} (x ; t)$ is a spatial function.

The above expression can also be written as:

$$\frac{D \sigma_{ij...}}{Dt} \equiv \dot{\sigma}_{ij...} \equiv \frac{d \sigma_{ij...}}{dt} \equiv \frac{d \sigma_{ij...}}{\partial t} + \sigma_{ij,k} v_k \tag{3.31}$$

where the minuscules denote the spatial reference, or by:

$$\frac{D \sigma_{IJ...}}{Dt} \equiv \dot{\sigma}_{IJ...} \equiv \frac{d \sigma_{IJ...}}{\partial t} \equiv \frac{d \sigma_{IJ...}}{dt} \tag{3.32}$$

where the majuscules indicate the material reference. In (3.31) the *velocity v*, or material derivative of the position vector $p = x_i (X_I ; t) e_I$ is utilized; since

$$v = \frac{dp}{dt} = \frac{Dx_i}{Dt} e_I = \dot{x}_i e_I = \frac{\partial x_i}{\partial t} e_I \tag{3.33}$$

Recalling, from Figure 3.1, that $u = p - P$, then the velocity may be alternatively defined by

$$v = \dot{u} = \frac{\partial u_i}{\partial t} e_I = \frac{\partial U_i}{\partial t} e_I = V \tag{3.34}$$

The *acceleration* a is the material derivative of the velocity vector for a given particle, i.e.,

$$a = \frac{dv}{dt} \equiv \frac{\partial v_i}{\partial t} e_I + v_{i,k} v_k e_I \tag{3.35}$$

where $v_{i,k} v_k$ is the *"convective"* term, or it is

$$a = \frac{dv}{dt} = \frac{\partial v_I}{\partial t} e_I \tag{3.36}$$

In the spatial description, the velocity and acceleration at time t at a specific spatial point are known but the particle occupying this point is not known. As each particle moves through a spatial point, it acquires the velocity and acceleration associated with that point at that time. In the material description, the material particle is identifiable and has a given velocity and acceleration.

3.4.2. DEFORMATION-RATE, SPIN, VORTICITY AND STRAIN RATE

First, we consider the material derivative of a spatial line element dx. Hence:

$$\frac{D}{Dt}(dx_k) = \frac{D}{Dt}(x_{k,K} dX_K) = v_{k,K} dX_K$$
$$= v_{k,\ell} x_{\ell,K} dX_K = v_{k,\ell} dx_\ell \tag{3.37}$$

from which one can further writes that:

$$\frac{D}{Dt}(x_{k,K}) = v_{k,\ell} x_{\ell,K} \tag{3.38}$$

Furthermore, the material derivative of the square of the spatial line element is expressed as

$$\frac{D}{Dt}(ds^2) = \frac{D}{Dt}(dx_k dx_k) = 2\frac{D}{Dt}(dx_k) dx_k = 2 v_{k,\ell} dx_\ell dx_k \tag{3.39}$$

which is symmetric in the indices k & ℓ. Thus, one may write that:

$$\frac{D}{Dt}(ds^2) = (v_{k,\ell} + v_{\ell,k})\,dx_k\,dx_\ell = 2\,d_{k\ell}\,dx_k\,dx_\ell \tag{3.40}$$

in which the *"deformation-rate"* tensor is

$$d_{k\ell} = \frac{1}{2}(v_{k,\ell} + v_{\ell,k}) = v_{(k,\ell)} \tag{3.41}$$

Meantime, the *"spin tensor"* is defined as:

$$\omega_{k\ell} = \frac{1}{2}(v_{k,\ell} - v_{\ell,k}) = v_{[k,\ell]} \tag{3.42}$$

which, upon adding to the deformation-rate tensor (3.41), results in:

$$v_{k,\ell} = d_{k\ell} + \omega_{k\ell} \tag{3.43}$$

As mentioned earlier (Section 3.2.3), one can construct an *"axial vector"* ω_m from the spin tensor by setting:

$$\omega_{km} = \epsilon_{k\ell m}\,\omega_{k\ell} = \epsilon_{k\ell m}\,v_{k,\ell} \quad ; \quad \omega = \text{curl } v \tag{3.44}$$

where ω is the *"vorticity vector"*.

QUIZ 3.5

Show that the relation between the *Lagrangian strain-rate*, the *Green strain-rate* and the *deformation rate* can be expressed as

$$\dot{E}_{KL} = \frac{1}{2}\,\dot{C}_{KL} = d_{k\ell}\,x_{k,K}\,x_{\ell,L} \tag{3.45}$$

But, since

$$x_{\ell,L} X_{L,k} = \delta_{\ell k}$$

$$\frac{D}{Dt}(x_{\ell,L}) X_{L,k} + x_{\ell,L} \dot{\overline{X_{L,k}}} = 0$$

$$\text{Then,} \quad x_{\ell,L} \dot{\overline{X_{L,k}}} = - v_{\ell,m} x_{m,L} X_{L,k} = - v_{\ell,k}$$

(3.46)

$$\text{Thus,} \quad \dot{\overline{X_{K,k}}} = - v_{\ell,k} X_{K,\ell}$$

Substituting this into the previous expression then by utilizing (3.45), it follows that

$$\dot{\epsilon}_{k\ell} = d_{mn} x_{m,K} x_{n,L} X_{K,k} X_{L,\ell}$$
$$- E_{KL} v_{\ell,k} X_{K,\ell} X_{L,\ell} - E_{KL} X_{K,k} v_{m,\ell} X_{L,m}$$

$$\therefore \dot{\epsilon}_{k\ell} = d_{k\ell} - \epsilon_{m\ell} v_{m,k} - \epsilon_{k,m} v_{m,\ell}$$

(3.47)

Substituting $\dot{\epsilon}_{k\ell} = -\frac{1}{2} \dot{c}_{k\ell}$ and $\epsilon_{k\ell} = \frac{1}{2}(\delta_{k\ell} - c_{k\ell})$ into (3.38) results in

$$\dot{c}_{k\ell} = - c_{m\ell} v_{m,k} - c_{mk} v_{m,\ell}$$

(3.48)

QUIZ 3.6

Show that the expressions for the Eulerian and Cauchy strain-rates, corresponding to (3.45) can be obtained as

$$\dot{\epsilon}_{k\ell} = -\frac{1}{2} \dot{C}_{k\ell} = \frac{\partial \epsilon_{k\ell}}{\partial t} + \frac{\partial \epsilon_{k\ell}}{\partial x_m} v_m$$

3.5. Objective Tensors

The response of any physical material must be independent of how it is observed and hence must be the same in all frames of reference. If this view is accepted, then the measurements made in one frame of reference are sufficient to determine the material properties in all other frames that are in rigid motion with respect to one another. Hence, in the formulation of physical laws, it is desirable to employ tensorial quantities that are independent of the motion of the observer. Such quantities are termed *"objective"*

Consider a rectangular coordinate-frame $\mathscr{F}'$ which is in relative rigid motion with respect to another rectangular coordinate-frame $\mathscr{F}$. A point with coordinates x_k at time t in $\mathscr{F}$ has the coordinates x_k at time t' in $\mathscr{F}'$. Hence, since the two coordinate-frames are in rigid motion with respect to each other, then,

$$x_k'(t') = \alpha_{\ell k}(t)\, x_\ell(t) + b_k(t) \quad ; t' = t + \tau \tag{3.49}$$

where τ is an arbitrary constant allowing us to select the time difference between x' and x, b_k (t) and $\alpha_{\ell k}$ (t) are functions of time alone and

$$\alpha_{\ell k}\, \alpha_{\ell m} = \alpha_{k\ell}\, \alpha_{m\ell} = \delta_{km}$$

Thus, any tensorial quantity is said to be "objective" if, in any two objectively equivalent motions, this tensor obeys its appropriate tensor transformation law".

Example Problem 3.2

Consider the time dependent velocity field $v(t) = \dot{x}(t)$. Upon taking the material derivative of the basic transformation (3.49) it follows that:

$$v_k'(t') = \frac{D}{Dt'}\, x_k'(t') = \alpha_{\ell k}(t)\, v_\ell(t) + \dot{\alpha}_{\ell k}(t)\, x_\ell(t) + \dot{b}_k(t) \tag{3.50}$$

which indicates in general that the velocity is not objective. Similarly, the acceleration is not objective.

Example Problem 3.3

Consider the deformation rate tensor $d_{k\ell} = \frac{1}{2}(v_{k,\ell} + v_{\ell,k})$.

First, consider from the previous relation (3.50) that

$$\begin{aligned} v_{k,m}'(t') &= \alpha_{\ell k}(t)\, v_{\ell,m}'(t,t') + \dot{\alpha}_{\ell k}(t)\, x_{\ell,m}(t,t') \\ &= \alpha_{\ell k}(t)\, v_{\ell,n}(t)\, x_{n,m}(t,t') + \dot{\alpha}_{\ell k}\, x_{\ell,m}(t,t') \end{aligned} \tag{3.51}$$

Now, from the basic transformation (3.49), one obtains

$$\alpha_{\ell k}(t) x_{\ell}(t) = x_k'(t') - b_k(t)$$
$$\therefore \alpha_{nk}(t) \alpha_{\ell k}(t) x_{\ell}(t) = \alpha_{nk}(t) [x_k'(t') - b_k(t)]$$
$$\therefore \frac{\partial x_n(t)}{\partial x_k'(t)} = x_{n,k}(t, t') = \alpha_{nk}(t) \qquad (3.52)$$

Substituting back into (3.51), then,

$$v_{k,m}'(t') = \alpha_{\ell k}(t) \alpha_{nm}(t) v_{\ell,n}(t) + \dot{\alpha}_{\ell k}(t) \alpha_{\ell m}(t)$$

It is, however, usual to set:

$$\Omega_{km}(t) = \dot{\alpha}_{\ell k}(t) \alpha_{\ell m}(t) \quad ; \quad \Omega_{km}(t) = - \Omega_{mk}(t)$$

and, thus, $v_{km}(t)$ is **not objective** since:

$$v_{k,m}'(t') = \alpha_{\ell k}(t) \alpha_{nm}(t) v_{\ell,n}(t) + \Omega_{km}$$

Further, we see that:

$$d_{k\ell}'(t') = \frac{1}{2} [v_{k,\ell}'(t') + v_{\ell,k}'(t')]$$
$$= \frac{1}{2} [\alpha_{mk} \alpha_{n\ell} v_{m,n} + \Omega_{k\ell} + v_{n,m} \alpha_{mk} \alpha_{n\ell} + \Omega_{\ell k}] \qquad (3.53)$$
then, $d_{k\ell}'(t') = \frac{1}{2} \alpha_{mk} \alpha_{n\ell} (v_{m,n} + v_{n,m})$
Thus, $d_{k\ell}'(t') = \alpha_{mk}(t) \alpha_{n\ell}(t) d_{mn}(t)$

Hence, the deformation rate is objective.

QUIZ 3.7

Show that while the Green and Lagrangian strain rates, namely $\dot{C}_{KL}$ and $\dot{E}_{KL}$, respectively, are objective, the Cauchy and Eulerian strain rates are not objective,

Example Problem 3.4

Consider the spin tensor $\omega_{k\ell} = \frac{1}{2}(v_{k,\ell} - v_{\ell,k})$.

From (3.42) and (3.50), it follows that

$$\begin{aligned}
\omega'_{k\ell}(t') &= \frac{1}{2}(v'_{k,\ell}(t') - v'_{\ell,k}(t')) \\
&= \frac{1}{2}[\,\alpha_{ik}\,\alpha_{j\ell}\,v_{i,j} + \Omega_{ij} - \alpha_{j\ell}\,\alpha_{ik}\,v_{j,i} - \Omega_{ji}\,] \qquad (3.54) \\
&= \alpha_{ik}(t)\,\alpha_{j\ell}(t)\,\omega_{ij}(t) + \Omega_{ij}(t)
\end{aligned}$$

Hence, the spin is not objective. Furthermore, from this relation it is apparent that $\Omega_{ij}(t)$ represents the relative angular velocities of the two coordinate-frames.

3.6. Problems

1. For each of the following displacement gradient matrices, determine the strain matrix, the rotation matrix, the volume strain, the deviatoric strain matrix, and the three invariants of the strain and the deviatoric strain.

    (i) $10^{-4}\begin{vmatrix} 9 & 10 & -4 \\ -10 & 18 & -18 \\ -14 & -18 & 27 \end{vmatrix}$ (ii) $10^{-4}\begin{vmatrix} 4 & 1 & 4 \\ -1 & -4 & 0 \\ 0 & 2 & 6 \end{vmatrix}$

2. (a) If the displacement field is given by

$$u_1 = k X_1 X_2 \; ; \quad u_2 = k X_1 X_2 \; ; \quad u_3 = 2 k (X_1 + X_2) X_3$$

where k is a constant small enough to ensure applicability of small deformation theory.

(i) Express the small strain tensor components and rotation components as functions of X_1, X_2, X_3 and display them in matrices.

(ii) At the point (1, 1, 0), determine the principal strains and principal axes of strain for the strain matrix of (i).

(b) If the motion of a continuum form the initial position X to the current position x is defined by $x = (1 + B) . X$, where 1 is the unit tensor and B is a second order tensor whose components are all constants, small in comparison to unity, display in matrices the components of the displacement u, small strain ϵ, rotation Ω and the strain deviator ϵ'. What is the volume strain?

3.   The motion of a continuum is defined by the equations

$$x_1 = \frac{1}{2} (X_1 + X_2) e^t + \frac{1}{2} (X_1 - X_2) e^{-t}$$

$$x_2 = \frac{1}{2} (X_1 + X_2) e^t - \frac{1}{2} (X_1 - X_2) e^{-t}$$

$$x_3 = X_3$$

(a)   Express the velocity components in terms of the material coordinates and time.

(b)   Express the velocity components in terms of the spatial coordinates and time.

(c)   Calculate the components of the rate-of-deformation matrix.

(d)   Express the displacement components u_1, u_2, u_3 in terms of the material coordinates and time.

(e)   Express the small strain components in terms of material coordinates and time. Evaluate these components at $t = 0$ and at $t = 0.05$.

(f)   Calculate the rate of change $d\epsilon_{ij} /dt$ of the small strain components and compare the obtained results with the rate of deformation tensor d_{ij} at t=0 and at $t = 0.05$.

4.   (a)   Sketch the initial and final positions of an element which was initially a square ABCD in the X -Y plane of side length dL with AB parallel to the X-axis and

AD parallel to the Y-axis, which undergoes the following simple shear:

$$x = X + kY, \quad y = Y, \quad z = Z.$$

(b)　　Calculate and display in three matrices the components of the deformation gradients $\mathbf{F}$ and $\mathbf{F}^T$ and the Green deformation tensor $\mathbf{C}$.

Partial Answer
$$| C_{IJ} | = \begin{vmatrix} 1 & k & 0 \\ k & 1+k^2 & 0 \\ 0 & 0 & 1 \end{vmatrix}$$

(c)　　Compare the matrices of rectangular Cartesian components of the finite strain E_{IJ} and the small strain ϵ_{IJ} for (a). Use the finite strain tensor to calculate the change in squared length of edges AB and AD and diagonals AC and DB.

Partial answer for AC is:

$$(ds)^2 - (dS)^2 = (2k + k^2)(dL)^2$$

(d)　　Calculate the stretch $\Lambda_{(G)}$ and the unit elongation $E_{(G)}$ for AB, AD, AC and DB. Use the following equation to calculate the angle change at A and check your result by trigonometry on the sketch of (a).

$$\cos \theta_{12} = \frac{C_{12}}{\Lambda_{(1)} \Lambda_{(2)}} = \frac{C_{12}}{\sqrt{C_{11} C_{22}}} = \frac{2E_{12}}{\sqrt{(1+2E_{11})(1+2E_{22})}}$$

Partial Answer:
$$\Lambda_{AC} = \sqrt{1 + k + \frac{1}{2}k^2}$$

5.　　In a certain region of a continuous medium the flow velocity components are as follows:

$$v_x = -A(x^3 + xy^2)e^{-kt}$$

$$v_y = A(x^2 y + y^3)e^{-kt}$$

$$v_z = 0.$$

where A, k are known constants and x, y, z are spatial coordinates and t designates

116

the time.

(a) Determine the acceleration components at the point (1, 1, 0) at t= 0.

(b) Show the matrices of the rectangular Cartesian components of the spatial gradient **F**, the rate of deformation **d** and the spin ω at the point (1, 1, 0) at t = 0.

3.7. Further Reading

Bowen, R. M. (1989) *Introduction to Continuum Mechanics for Engineers*, Plenum Press, New York.
Chung, T.J. (1988) *Continuum Mechanics*, Prentice-Hall, Englewood Cliffs, New Jersey.
Coleman, B.D. and Noll, W. (1960) An approximation theorem for functionals with applications in continuum mechanics, *Arch. Ration. Mech. Analysis* **6**, 355-70.
Davis, J.L. (1987) *Introduction to Dynamics of Continuous Media*, Macmillan, New York.
Doeblin, E. O., (1983) *Measurement Systems. Application and Design*, McGraw-Hill Book Co., New York.
Eringen, A.C. (1962) *Nonlinear Theory of Continuous Media*, McGraw-Hill, New York.
Eringen, A.C. (1967) *Mechanics of Continua*, John Wiley & Sons, New York.
Feynman, R., (1993) *The Character of Physical Law*, The MIT Press, Cambridge, Massachusetts.
Flügge, W. (1972) *Tensor Analysis and Continuum Mechanics*, Springer-Verlag, Berlin.
Fredrick, D. and Chang, T.S. (1965) *Continuum Mechanics*, Allyn and Bacon, Boston.
Fung, Y.C. (1965) *Foundations of Solid Mechanics*, Prentice-Hall, Englewood Cliffs, N.J.
Green, A.E. and Adkins, J.E. (1960) *Large Elastic Deformations and Nonlinear Continuum Mechanics*, Clavendon Press, Oxford.
Gurtin, M.E. (1981) *An Introduction to Continuum Mechanics*, Academic Press, New York.
Gurtin, M.E. and Williams, W.O. (1967) An axiomatic foundation for continuum thermodynamics, *Arch. Rational Mech. Anal.* **26**, 83-117.
Haddad, Y. M. (1995) *Viscoelasticity of Engineering Materials*, Kluwer, Dordrecht.
Hunter, S.C. (1976) *Mechanics of Continuous Media*, Ellis Hordwood, Chister, England.
Jaunzemis, W. (1967) *Continuum Mechanics*, Macmillan, New York.
Kac, M. (1969) Some mathematical models in science, *Science* **166**, 695-9.
Kuhn, T. S. (1985) *The Structure of Scientific Revolution*, University of Chicago Press, Chicago.
Ladevèse, P. (ed.) (1985) *Local Effects in the Analysis of Structures*, Elsevier, New York.
Lai, W.M., Rubin, D. and Krempl, E. (1978) *Introduction to Continuum Mechanics (SI/Metric Units)*, Pergamon Press, New York.
Leipholz, H. H. E. (1983) On the role of analysis in mechanics, *Trans. of the CSME* **7**, 3-7
Love, A.E.H. (1944) *A Treatise on the Mathematical Theory of Elasticity*, 4ᵗʰ edition, Dover, New York.
Malvern, L.E. (1969) *Introduction to the Mechanics of a Continuous Medium*, Prentice-Hall, Englewood Cliffs, N.J.
Mindlen, R.D. and Tiersten, H.F. (1962) Effect of couple-stresses in linear elasticity, *Arch. Rational Mech. Anal.* **11**, 415-48.
Noll, W. (1955) On the continuity of the solid and fluid states, *J. Rational Mech. Anal.* **4**, 3-81.
Noll, W. (1958) A mathematical theory of the mechanical behaviour of continuous media, *Arch. Ration. Mech. Analysis* **2**, 197-226.
Popper, K. R. (1968) *The Logic of Scientific Discovery*, Harper and Row, New York.
Prager, W. (1961) *Introduction to Mechanics of Continua*, Ginn, Boston, Mass.
Sneddon, I.N. and Hill, R. (Eds.), (1960-1963) *Prog s in Solid Mechanics*, Vol 1 (1960), Vol. 2(1961), Vol. 3(1963), Vol. 4 (1963), North-Holland Pub. Co., Amsterdam.
Sokolnikoff, I. S. (1956) *Mathematical Theory of Elasticity*, McGraw-Hill Book Co., New York.

Sommerfeld, A. (1950) *Mechanics of Deformable Bodies*, Academic Press, New York.

Timoshenko, S. P. and Goodier, J. N. (1970) *Theory of Elasticity*, McGraw-Hill Book Company, New York.

Truesdell, C., (1965a) The nonlinear field theories of mechanics, in *Encyclopedia of Physics*, Ed. S. Flügge, Vol. III/3, Springer-Verlag, Berlin.

Truesdell, C. (1965b) *The Elements of Continuum Mechanics*, Springer-Verlag, New York.

Truesdell, C. and Toupin, R.A. (1960) The classical field theories, *Handbuch du Physik*, Vol. III/1 (Ed. S. Flügge), Springer-Verlag, Berlin.

Washizu, K. (1958) A note on the condition of compatability, *J. Math. Phys.* 36, 306-12.

Williams, W.O. (1970) Thermodynamics of rigid continua, *Arch. Rational Mech. Anal.* 36, 270-84.

Życzkowski, M. (1981) *Combined Loading in the Theory of Plasticity*, PWN - Polish Scientific Publishers, Warszawa.

CHAPTER 4

THERMOMECHANICAL CONTINUA

4.1. Introduction

In this Chapter, we present a brief summary of the basic structure of the classical theory of thermodynamics. Within this frame, we attempt to study the restrictions that classical thermodynamics impose on the theory of solids, and to seek information concerning the thermodynamics of continuous media (see, e.g., Fung, 1965).

4.2. The Laws of Thermodynamics

4.2.1. DEFINITIONS AND CONCEPTS

A system? Closed system? Isolated system?
A "*system*" is often defined as a particular collection of matter under observation. Systems which do not exchange matter with their surroundings, are referred to as "*closed systems*". An additional restriction may be made, that is no interaction between the system and its surroundings would occur; the system is then said to be "*isolated*".

State? State variables?   Equation of state?   State function?
For a given system, when all the quantities required for a complete characterization of the system, for a particular consideration, are available, it is said that the "*state*" of the system is defined. Such quantities, under certain restrictions, may be referred to as "*state variables*". If a certain state variable can be expressed as a simple-valued function of a set of other state variables, then, the functional relationship is referred-to as an "*equation of state*", and the variable is said to be described by a "*state function*".

Homogeneous system?
A system is said to be "*homogeneous*" if the state variables do not depend on space coordinates.

Thermodynamic equilibrium?   Process?
If for a given system, the values of the state variables are time-independent, the system is said to be in "*thermodynamic equilibrium*". Meanwhile, if the state variables are evolving with the time, then, the system is said to undergo a "*process*". The number of state variables required to describe a process may be different from that required to describe the system at

thermodynamic equilibrium.

A thermally insulated system?   Adiabatic process?

The *"boundary"* or *"wall"* separating two systems is said to be *"insulating"* if it has the following property: If any system in complete internal equilibrium is completely surrounded by an insulating wall, then no change can be produced in the system by an external agency except by movement of the wall or long range forces such as gravitation. A system-surrounded by an insulating wall is said to be *"thermally insulated"*, and any process taking place in the system is called *"adiabatic"*.

The concept of temperature?

In thermodynamics, it is postulated that if each of two systems is in *thermal equilibrium* with a third system, then, they are in thermal equilibrium with each other. From this, it can be demonstrated that the condition of thermal equilibrium between several systems is the equality to a certain simple-valued function of the thermodynamic states of the systems which may be called the *temperature* θ, called the *"empirical temperature"*. The latter is measured on a scale which is determined by the arbitrary choice of the system being used as a *"thermometer"*.

First law of thermodynamics?

The *"first law of thermodynamics"* can be formulated as follows:

> *"If a thermally insulated system can be taken from a state I to a*
> *state II by alternative (adiabatic) paths, the work done on the system*
> *has the same value for every such path".*

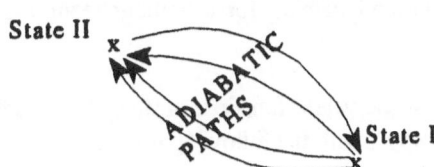

One can, thus, deduce that there exits a single-valued function of the state of a system, called its *"energy"*, such that for any adiabatic process, the increase of the energy is equal to the work done on the system. Thus,

$$E_{II} - E_I = \Delta \text{ energy } = \text{ work done (: adiabatic process)} \tag{4.1}$$

It is to be noticed that for this definition of energy to be valid, it is necessary and sufficient to be able to change the system, by an adiabatic process, either from State I to State II or from State II to State I.

We now define the heat Q absorbed by a system as the increase in energy of the system minus the work done on this system. Thus,

$$Q = \Delta \text{ energy } - \text{ work done (all processes)} \tag{4.2}$$

or,

$$\Delta \text{ energy } = Q + \text{ work done (all processes)} \tag{4.3}$$

If (4.3) is regarded as a statement of the conservation of energy and is compared with (4.1), we observe that the energy of the system can be increased either by the work done on it or by absorption of heat.

It is customary to identify several types of energy which makes up the total: the kinetic energy K, the potential (gravitational) energy G, and the internal energy ζ. Thus

$$\text{Energy} = K + G + \zeta$$

Second law of thermodynamics (for a homogenous system)?
The second law of thermodynamics, for a homogeneous system, may be formulated as follows:

There exist two single-valued functions of state; T, called the "*absolute temperature*", and S, called the "*entropy*", with the following properties:

a) The absolute temperature T is a positive number, which is a function of the empirical temperature θ only.

b) The entropy of a system is equal to the sum of entropies of its parts.

c) The entropies of a system can change in two distinct ways; namely, by interactions with the surroundings and by changes taking place inside the system, i.e.,

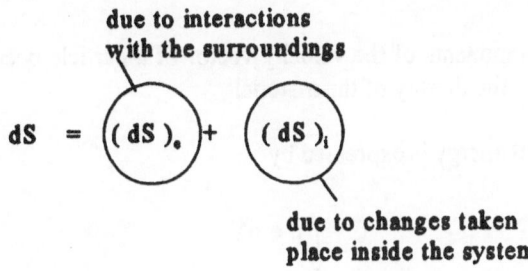

where dS denotes the increase of entropy of the system, $(dS)_e$ denotes the part of this increase due to interaction with the surroundings of the system, and $(dS)_i$ denotes the part of this increase due to changes taking place inside the system.

Then, if dQ denotes the heat absorbed by the system from its surroundings, we have

$$(dS)_e = \frac{dQ}{T} \tag{4.4}$$

The change $(dS)_i$ is never negative. If $(dS)_i$ is zero the process is said to be *"reversible"*. If $(dS)_i$ is positive the process is said to be *"irreversible"*. Thus,

$(dS)_i > 0$ *(irreversible process),*
$(dS)_i = 0$ *(reversible process), and*
$(dS)_i < 0$ *(never occurs in nature)*

4.2.2. THE ENERGY EQUATION

In this section, we derive the equation of *"balance of energy"* in a form convenient for continuum mechanics.

For a body of particles in a configuration occupying a region V bounded by a surface S, the kinetic energy is defined by

$$K = \int_V \frac{1}{2} \rho \, v_i \, v_i \, dV$$

(4.5)

where v_i are the components of the velocity vector of a particle occupying an elemental volume dV, and ρ is the density of the material.

Meantime, the internal energy is expressed by

$$\zeta = \int_V \rho \, e \, dV$$

(4.6)

where e is the internal energy per unit mass.

The heat input into the body must be imported through the boundary. Let ds be a surface element of the body, with an outward unit vector normal n_i. Then, the rate at which heat is transmitted across the surface element ds in the direction of n_i is assumed to be given by $q_i \, n_i \, ds$ where q_i are the components of the heat flux vector q. To be specific, we postulate (e. g., Fung, 1965), when defining the heat flux in the case of a moving medium, that the surface element ds be composed of the same particles. The rate of heat input may be, thus, expressed as

$$\dot{Q} = - \int_S q_i \, n_i \, ds = - \int_V q_{j,j} \, dV$$

(4.7)

The rate at which work is done on the body by the volumetric (body) force χ_i in V and surface traction T_i on S is the "power". The latter is given the symbol P in the following expression:

Rate of work done — Body force per unit volume

$$
\begin{aligned}
\boxed{P} &= \int \boxed{\chi_i} v_i \, dV + \int T_i \; v_i \, ds \\
&= \int \chi_i \, v_i \, dV + \int \sigma_{ij} n_i \; v_i \; ds \\
&= \int \chi_i \, v_i \, dV + \int (\sigma_{ij} v_i)_{,j} \, dV
\end{aligned}
\tag{4.8}
$$

The first law states that

$$
\dot{K} + \zeta = \dot{Q} + P \tag{4.9}
$$

where the dot denotes the material derivative "D(.)/Dt" (*introduced earlier in Chapter 2*).

It can be shown that

$$
\begin{aligned}
\frac{1}{2} \rho \frac{D v^2}{D t} &+ \frac{v^2}{2} \frac{D \rho}{D t} + \frac{v^2}{2} \rho \; \text{div } v + \rho \frac{D e}{D t} + e \frac{D \rho}{D t} \\
&+ e \rho \; \text{div } v \\
&= - q_{j,j} + \chi_i v_i + \sigma_{ij,j} v_i + \sigma_{ij} v_{i,j}
\end{aligned}
\tag{4.10}
$$

On substituting the continuity of mass equation (2.8) and the equation of motion (2.22), i.e.,

$$
\frac{D \rho}{D t} + \rho \; \text{div } v = 0, \qquad \rho \frac{D v_i}{D t} = \chi_i + \sigma_{ij,j}
$$

respectively, into (4.10), one obtains

$$
\rho \frac{D e}{D t} = - \frac{\partial q_j}{\partial x_j} + \sigma_{ij} d_{ij} \tag{4.11}
$$

where

$$d_{ij} = \frac{1}{2} \left(\frac{\partial v_i}{\partial x_j} + \frac{\partial v_j}{\partial x_i} \right) \qquad (4.12)$$

is the symmetric part of the tensor $v_{i,j}$, the *"rate of deformation tensor"*; see *Chapter 3*.

In classical thermodynamics, we are concerned with the small neighborhood of the thermodynamic equation. It suffices here to consider the case of an infinitesimal strain, which is imposed slowly. In this case, equation (4.11) can be written as

$$\rho \, de = dQ + \sigma_{ij} \, d\epsilon_{ij} \qquad (4.13)$$

If there is no internal entropy production in the process, then, the second law gives

$$dQ = T \, \rho \, d\varsigma$$

where ς denotes the *"specific entropy"*, or the entropy per unit mass. Hence, by combining the above equation with eqn. (4.13), it follows that

$$d\zeta = T \, d\varsigma + \frac{1}{\rho} \sigma_{ij} \, d\epsilon_{ij} \qquad (4.14)$$

4.2.3. THE CONDITIONS OF THERMODYNAMIC EQUILIBRIUM

In the foregoing, we have defined thermodynamic equilibrium as a situation in which no state variable changes with time. For a system to be in thermodynamic equilibrium no change in boundary conditions is permissible and no spontaneous process will occur in the system.

Gibbs gave the following statement concerning thermodynamic equilibrium:

> *"For the equilibrium of any isolated system, it is necessary and sufficient that in all possible variations in the state of the system which do not alter its entropy, the variation of its energy shall either vanish or be positive".*

That is

$$(\Delta\zeta)_\varsigma \geq 0 \quad , \quad (\delta\zeta)_\varsigma \geq 0 \qquad (4.15)$$

where $\delta\zeta$ denotes the first-order infinitesimal term of the internal energy and $\Delta\zeta$ includes the variation in the same including the first, second, and higher-order infinitesimal terms. In other words, the internal energy shall be a minimum with respect to all neighboring states which have the same entropy.

The derivation of Gibbs conditions for thermodynamic equilibrium would require a little more than the first and second laws of thermodynamics. The conditions also guarantee more than equilibrium in the ordinary sense. They guarantee "*stable*" equilibrium, stable in the sense that a neighboring disturbed state will actually tend to return to the equilibrium state (e.g., Fung, 1965).

4.3. Thermodynamics of Continuous Media

4.3.1. PRINCIPLE OF CONSERVATION OF ENERGY

The "*principle of conservation of energy*" may be stated as follows

> "*The time rate of change of the kinetic plus internal energy is equal to the sum of the rate of work W of the external forces plus the thermomechanical energy Q (heat) that enters or leaves the body per unit time*".

That is

$$\frac{d}{dt}(K + \zeta) = W + Q \qquad (4.16)$$

The "*mechanical rate of work done*" on the volume V with surface S is:

$$W = \int_V \rho\, \chi_i\, v_i\, dV + \int_S T_i\, v_i\, ds = \int_V \rho\, \chi_i\, v_i\, dV + \int_S \sigma_{ij}\, n_j\, v_i\, ds \qquad (4.17)$$

In a continuous medium, heat may enter the body through its surface S or it may be generated internally at a rate of h per unit mass. If q_i denotes the heat vector per unit area acting at a point x_i of the surface S directed outward, then, the total heat input is given by

$$Q = \int_V \rho \, h \, dV - \int_S q_i \, n_i \, ds \qquad (4.18)$$

Hence, the principle of conservation of energy (4.16) may be stated as:

$$\frac{D}{Dt} \int_V \rho \left(\frac{1}{2} v_i v_i + e \right) dV = \int_V \rho \left(\chi_i v_i + h \right) dV + \int_S \left(\sigma_{ij} n_i v_i - q_i n_i \right) ds \qquad (4.19)$$

Now, we examine each integral of (4.19) in turn. For the first integral, one has

$$\frac{D}{Dt} \int_V \rho \left(\frac{1}{2} v_i v_i + e \right) dV = \int_V \frac{\partial}{\partial t} \left[\rho \left(\frac{1}{2} v_i v_i + e \right) \right] dV + \int_S \left[\rho \left(\frac{1}{2} v_i v_i + e \right) \right] v_i n_j \, ds$$

$$= \int_V \left\{ \frac{\partial}{\partial t} \left[\rho \left(\frac{1}{2} v_i v_i + e \right) \right] + \left[\rho \left(\frac{1}{2} v_i v_i + e \right) \right]_{,j} v_j + \left[\rho \left(\frac{1}{2} v_i v_i + e \right) \right] v_{j,j} \right\} dV$$

$$= \int_V \left\{ \frac{D}{Dt} \left[\rho \left(\frac{1}{2} v_i v_i + e \right) \right] + \left[\rho \left(\frac{1}{2} v_i v_i + e \right) \right] v_{j,j} \right\} dV \qquad (4.20)$$

$$= \int_V \left\{ \left[\overbrace{\frac{D\rho}{Dt} + \rho \, v_{j,j}}^{=0 \;\; \text{continuity}} \right] \left(\frac{1}{2} v_i v_i + e \right) + \rho \left(v_i a_i + \dot{e} \right) \right\} dV$$

$$= \int_V \rho \left(v_i a_i + \dot{e} \right) dV$$

The last integral of (4.19) can be converted into a volume integral using the Gauss's divergence theorem. Thus,

$$\int_S \left(\sigma_{ij} v_i - q_j \right) n_j \, ds = \int_V \left[\left(\sigma_{ij} v_i \right)_{,j} - q_{j,j} \right] dV \qquad (4.21)$$

Hence, upon substituting (4.20) and (4.21) back into (4.19), the latter reduces to

$$\int_V \left\{ \rho \left[v_i \, a_i + \dot{e} - \chi_i \, v_i - h \right] - \sigma_{ij,j} \, v_i - \sigma_{ij} \, v_{i,j} + q_{i,i} \right\} dV = 0$$

$$\therefore \int_V \left\{ v_i \left[\overset{=0 \ \text{Eqn. of Motion}}{\underset{\rho \, a_i - \chi_i - \sigma_{ij,j}}{\underline{\hspace{3.5cm}}}} \right] + \rho \, (\dot{e} - h) - \sigma_{ij} \, v_{i,j} + q_{i,i} \right\} dV = 0 \qquad \textbf{(4.22)}$$

$$\therefore \rho \, \dot{e} = \rho \, h + \sigma_{ij} \, d_{ij} - q_{i,i}$$

which is the *"local conservation of energy equation"*.

4.3.2. ENTROPY: THE CLAUSIUS-DUHEM INEQUALITY

The thermodynamics of continuous media is formulated under the quasi-static approximation of real irreversible processes. If the thermostatic changes taking place in a system are considered as time-independent, then, we may write:

$$\dot{K} = 0 \ ; \quad \dot{\zeta} \, dt = d\zeta \ ; \quad \dot{W} \, dt = \overline{d} \, W \quad \text{and} \quad \dot{Q} \, dt = \overline{d} \, Q \qquad \textbf{(4.23)}$$

where $\overline{d}$ is being used to denote an "inexact differential". Hence, the principle of conservation of energy (4.16) reduces to the *"first law of thermodynamics"*, namely:

$$d\zeta = \overline{d} \, W + \overline{d} \, Q \qquad \textbf{(4.24)}$$

In order to convert the inexact differential of heat into an exact one, it has been shown that an integrating factor, $1/T$, being the same for all materials, is required to be introduced so that we may define the total differential of entropy, dS, by:

$$dS = \overline{d} \, Q / T \qquad \textbf{(4.25)}$$

This gives the *entropy S* the role of a *"state function"*.

The situation is, however, quite different when non-equilibrium thermodynamics is being considered since no definition of entropy has ever been given in this case. Clausius only postulated that during a process an entropy exists such that the inequality

$$S(B) - S(A) \geq \int_A^B \frac{\overline{d}Q}{T} \qquad (4.26)$$

exists, which refers to an *"irreversible process"* starting from an equilibrium state A and ending in an equilibrium state B, and may be referred to as the *"Second law of thermodynamics"*.

The thermomechanical energy, heat (Q), has previously been defined for the body by (4.18) and hence, since $\overline{d}Q$ can be replaced by Q dt, as in (4.23), it follows, with reference to (4.18) and (4.26), that

$$S(B) - S(A) \geq \int_A^B \left[\int_V \frac{\rho h}{T} dV - \int_S \frac{q_i}{T} n_i \, ds \right] dt \qquad (4.27)$$

Taking the limit as the equilibrium states A & B become closer together as $(t_B - t_A) \to 0$, then, the inequality (4.27) becomes

$$\frac{DS}{Dt} - \int_V \frac{\rho h}{T} dV + \int_S \frac{q_i}{T} n_i \, ds \geq 0 \qquad (4.28)$$

which is the *"integral form"* of the *"Clausius-Duhem inequality"*. The differential form is obtained if a *specific entropy* ς can be defined such that:

$$S = \int_V \rho \varsigma dV \qquad (4.29)$$

Upon applying Gauss's divergence theorem to the Clausius inequality (4.27), it follows that

$$\rho [\varsigma(B) - \varsigma(A)] \geq \int_A^B \left[\frac{\rho h}{T} - \left(\frac{q_i}{T} \right)_{,i} \right] dt \qquad (4.30)$$

Again, by taking the limit as the two equilibrium states A and B get closer together as

$t_B - t_A \to 0$, the differential form of the *Clausius - Duhem inequality* is obtained.

4. 4. Thermodynamics of the Deformation Process

4.4.1. A THERMODYNAMIC PROCESS

We consider a solid body B occupying a regular domain of volume V with surface boundary S. A material particle of B is defined in the reference configuration by the position vector X. A thermodynamic process in B is described by the following eight functions of X and the time parameter t (Coleman, 1963, 1964 a&b):

- The spatial particle position $x=x(X,t)$. This describes the evolution of the deformation process in the medium.
- The symmetric stress tensor: Cauchy stress $\sigma=\sigma(X,t)$ or the second Piola-Kirchhoff stress $\Sigma = \Sigma(X,t)$.
- The body force (per unit mass) $\chi=\chi(X,t)$, exerted on the body at X by outside bodies not intersecting the body B.
- The specific internal energy (per unit mass) $e=e(X,t)$.
- The specific entropy (per unit mass) $\varsigma=\varsigma(X,t)$.
- The local absolute temperature $T=T(X,t)$; assumed to be positive.
- The heat flux vector $q=q(X,t)$.
- The heat supply $r=r(X,t)$. It is the radiation energy (per unit mass and per unit time) that is absorbed by the body B at X and supplied by the environment or any other bodies not intersecting with B.

Couple stresses, body couples and other mechanical interactions not included in the stress tensor σ or the body force χ are assumed to be absent or at least negligible. In order for the above set of eight functions to be called a *"thermodynamic process"*, it must be compatible with both the law of balance of linear momentum and the law of balance of energy. This is with the understanding that the balance of moment of momentum is satisfied by the assumed symmetry of the stress tensor σ (*Chapter 2*).

In a differential form, the laws of balance of linear momentum and balance of energy can be expressed, respectively, as

$$\text{div } \sigma + \rho\chi = \rho \ddot{x} \tag{4.31}$$

$$\text{tr}\{\sigma L\} - \text{div } q - \rho\dot{e} = -\rho r \tag{4.32}$$

In the above two equations, ρ is the mass density, L is the velocity gradient ($L= grad\ \dot{x}$), "*tr*" is the trace operator, a super-imposed dot designates the material derivative, i.e., the derivative with respect to the time parameter t keeping the position vector X fixed, and the operators "*grad*" and "*div*" refer to spatial derivatives, i.e., the gradient and divergence with respect to the current position vector x with the time parameter t kept fixed. Thus, to specify a thermodynamic process, it would suffice to prescribe only the six functions x, σ, e, q, S and T. The remaining functions χ and r are, then, determined by the two laws (4.31) and (4.32).

The law of balance of energy (4.24) is a form of *"First law of thermodynamics"*. The latter, as mentioned earlier, states that the rate at which the total energy of the body increases is balanced by the power of the external forces and the rate at which heat is supplied to the body. In other words, the quantity of heat supplied to the body is measured as the difference between change of the total energy and work done by the external forces (Rivlin, 1975).

Consider that heat enters the body throughout its volume V at a rate H per unit mass and through its surface S at a rate h per unit area. Both H and h are measured in the reference configuration, whereby the superimposed dot indicates differentiation with respect to time. Thus, according to the *First law of thermodynamics*, the rate of change of the total energy is given by

$$\Xi = \int_V \rho \chi \cdot \dot{x} \, dV + \int_S F \cdot \dot{x} \, ds + \int_V \rho \, \dot{H} \, dV + \int_S \dot{h} \, ds \tag{4.33}$$

where F is the force per unit area acting on the body and measured in the reference configuration. Let ψ denote the total energy of the body per unit mass, then, the specific internal energy e is defined by

$$e = \psi - \frac{1}{2} \dot{x} \, \dot{x} \tag{4.34}$$

i.e., e, defined by the equation above, is the total energy per unit mass less the kinetic energy per unit mass.

4.4.2. RESTRICTIONS IMPOSED BY SECOND LAW OF THERMODYNAMICS

Regarding q/T to be a vectorial flux of entropy due to heat flow and r/T to be a scalar supply of entropy from radiation, Coleman (1964a&b) defined the time-rate of entropy production in a part ΔB of the body B to be

$$S = \frac{d}{dt} \int_{\Delta B} \varsigma \, dm - \int_{\Delta B} \frac{r}{T} \, dm + \int_{\partial s} \frac{1}{T} \, q \cdot n \, ds \tag{4.35}$$

where,

dm   is an element of mass of the body B,

n   is the outward unit normal to the surface ds of ΔB,

ds   is the element of surface area in the current configuration of the body, at time t.

Under an appropriate smoothness (Coleman, 1964a&b), equation (4.35) may be expressed as

$$\Gamma = \int_P \gamma \, dm \tag{4.36}$$

in which,

$$\begin{aligned}
\gamma &= \zeta - \frac{r}{T} + \frac{1}{\rho} \, \text{div}(q/T) \\
&= \zeta - \frac{r}{T} + \frac{1}{\rho T} \, \text{div } q - \frac{1}{\rho T^2} \, q \cdot g(T)
\end{aligned} \tag{4.37}$$

is the specific rate of entropy production and where $g(T)$ is the gradient of the current temperature. In this context, Coleman and Noll (1963) and Coleman (1964a&b) gave the following postulate as a mathematical expression of the '*Second law of thermodynamics*'.

Postulate: For every admissible thermodynamic process in a body B, the following "*Clausius-Duhem inequality*" must hold for all t and all parts P of B.

$$\Gamma \geq 0 \tag{4.38}$$

where Γ is the rate of production of entropy in a part P of the body B, eqn. (4.36).

It is evident that the postulate above places restrictions on the format of the constitutive equation of the material (*see Chapter 6*).

Thermomechanical systems must satisfy the same general conservation laws, concerning mass and momentum, that were introduced earlier in Chapter 2. The law of conservation of energy, however, contains both mechanical and thermal energies. Since the change of thermal energy is related to the change of entropy, a description of the evolution of a thermomechanical system requires a knowledge of the entropy production.

Recalling the conservation of mass principle as expressed by the equation of continuity (*Chapter 2, Eqn. 2.6*). That is

$$\frac{\partial \rho}{\partial t} + \frac{\partial(\rho v_i)}{\partial x_j} = 0 \tag{4.39}$$

Meantime, recalling Cauchy's expressions (2.22), (2.18) and (2.28). They are, respectively,

$$\rho \dot{v}_i = \sigma_{ij,j} + \rho \chi_i \qquad \text{within } V \tag{4.40}$$

$$T_i = \sigma_{ij} n_j$$
$$\sigma_{ij} = \sigma_{ji} \tag{4.41}$$

The conservation of energy is given by

$$\rho \dot{e} = \sigma_{ij} v_{i,j} - q_{i,i} \tag{4.42}$$

where the superimposed dot indicates the material derivative defined earlier by eqn. (2.4), *Chapter 2*, i.e.,

$$(\cdot) = \frac{\partial}{\partial t} + v_j \frac{\partial}{\partial x_j} \tag{4.43}$$

and v_j are the velocity components.

In order to establish the entropy balance equation, one may assume (e.g., Fung, 1965) that the specific entropy ς is a function of both the internal energy per unit mass e and the strain ϵ_{ij} irrespective of the equilibrium of the system. That is

$$\varsigma = \varsigma(e, \epsilon_{ij}) \tag{4.44}$$

This is in agreement with the expression for the total differential of the specific entropy ς as given by Gibb's relation

$$\rho T d\varsigma = \rho de - \sigma_{ij} d\epsilon_{ij} \tag{4.45}$$

Thus, along the path of the motion, one may write that

$$\rho T \dot{\varsigma} = \rho \dot{e} - \sigma_{ij} v_{ij} \tag{4.46}$$

where the superimposed dot indicates the material derivative, Eqn. (4.43), and v_{ij} is the rate of deformation tensor, i.e.,

$$v_{ij} = \frac{1}{2}(v_{i,j} + v_{j,i}) \tag{4.47}$$

Thus, it follows that

$$\rho \, T \, \zeta = - \, q_{i,i} \tag{4.48}$$

which can be equivalently written as

$$\rho \, \zeta = - \, \frac{q_{i,i}}{T} = - \left(\frac{q_i}{T}\right)_{,i} - q_i \, \frac{T,i}{T^2} \tag{4.49}$$

The first term on the right hand side of (4.49) is the divergence of the entropy flow and the second term is the entropy production which must be positive as previously discussed, equation (4.38).

A constitutive law, defining the relationship between the stress tensor σ_{ij} and the strain tensor ϵ_{ij} must be further included (*Chapter 5*) so that the strain field is uniquely defined. Thus, a sufficient number of differential equations are obtained for which a boundary value problem may be formulated.

4.5. Problems

1. Comment on the definition of a thermodynamic process and the required functions to specify such a process from a continuum mechanics point of view.

2. Using the energy equation,

$$\rho \, \frac{du}{dt} = T : D + \rho \, r - \nabla \cdot q$$

or

$$\rho \, \frac{du}{dt} = T_{ij} \, D_{ij} + \rho \, h - \frac{\partial q_j}{\partial x_j} \quad \text{(Cartesian components)}$$

134

show that the Clausius-Duhem inequality

$$\dot{e} \equiv \frac{d\varsigma}{dt} - \frac{r}{\theta} + \frac{1}{\rho\,\theta} \operatorname{div} q - \frac{q}{\rho\,\theta^2} \cdot \operatorname{grad} \theta \geq 0$$

takes the form $\rho\,(\,\theta\,\varsigma - \dot{u}\,) + \mathbf{T} : \mathbf{D} - \left(\dfrac{1}{\theta}\right) q \cdot \operatorname{grad} \theta \geq 0$

where $\dot{e}$ = internal entropy production rate per unit mass
ς = specific entropy (per unit mass).

4.6. References

Coleman, B.D. (1963) The thermodynamics of elastic materials with heat conduction and viscosity, *Arch. Ration. Mech. Anal.* 13, 167-78.
Coleman, B.D. (1964a) Thermodynamics of materials with memory, *Arch. Ration. Mech. Anal.* 17, 1-46.
Coleman, B.D. (1964b) On thermodynamics, strain impulses, and viscoelasticity, *Arch. Ration. Mech. Anal.* 17, 230-54.
Fung, Y.C. (1965) *Foundations of Solid Mechanics*, Prentice-Hall, Englewood Cliffs, N.J., pp. 377-446.
Rivlin, R.S. (1975) The thermodynamics of materials with fading memory, in *Theoretical Rheology*, edited by J.F. Hutton, J.R.A. Pearson and K. Walters, Applied Science Publishers, London, 83-103.

4.7. Further Reading

Bataille, J. and Kestin, J. (1979) Irreversible processes and physical interpretation of rational thermodynamics, *J. Non Equil. Thermodynamics* 4, 229-58
Biot, M.A. (1954) Theory of stress-strain relationship in anisotropic viscoelasticity and relaxation phenomena, *J. Applied Physics* (25)11, 1385-91.
Biot, M.A. (1958) Linear thermodynamics and the mechanics of solids, Proc. 3rd U.S. Nat. Congr. Appl. Mech., pp. 1-18.
Biot, M.A. (1973) Nonlinear thermoelasticity, irreversible thermodynamics and elastic instability, *Indi. Math. J.* 23, 309-35.
Blatz, P.J. (1956) Rheology of Composite Solid Propellants, *Industrial and Engineering Chemistry* 48 (4), 727-29.
Bodner, S. R. (1968) Constitutive equations for dynamic material behaviour, in: *Mechanical Behaviour of Materials under Dynamic Loads*, edited by U. S. Lindholm, Symposium held in San Antonio, Texas, Sept. 6-8, 1967, Springer-Verlag New York Inc., New York, pp. 176-90.
Breuer, S. (1969) Lower bounds on work in linear viscoelasticity, *Quart. Appl. Math.* 27(2), 139-46.
Breuer, S. and Onat, E.T. (1964) On the determination of free energy in linear viscoelastic solids, *Z. Angew Math. Phys.* 15, 184-91.
Callen, H. B. (1963) *Thermodynamics*, John Wiley and Sons, New York.
Christensen, R.M. (1971) *Theory of Viscoelasticity*, Academic Press, New York.
Christensen, R.M. and Naghdi, P.M. (1967) Linear non-isothermal viscoelastic solids, *Acta Mechanca* 3, 1-12.
Coleman, B.D. and Gurtin, M.E. (1967) Thermodynamics with internal state variables, *J. Chem. Phys.* 47,

597-613.

Coleman, B.D. and Gurtin, M.E. (1967) Equipresence and constitutive equations for rigid heat conductors, *Z. Angew Math. Phys.* **18**, 199-208.

Coleman, B.D. and Mizel, V. J. (1963) Thermodynamics and departures from Fourier's law of heat conduction, *Arch. Ration. Mech. Anal.* **13**, 245-61.

Coleman, B.D. and Mizel, V. J. (1964) Existence of coloric equations of state in thermodynamics, *J. Chem. Phys.* **40**, 1116-25.

Coleman, B.D. and Mizel, V.J. (1968) On the general theory of fading memory, *Arch. Ration. Mech. Anal.* **29**, 18-31.

Coleman, B.D. and Noll, W. (1960) An approximation theorem for functionals with applications in Continuum Mechanics, *Arch. Ration. Mech. Anal.* **6**, 355-70.

Coleman, B.D. and Noll, W. (1963) The thermodynamics of elastic materials with heat conduction and viscosity, *Arch. Ration. Mech. Anal.* **13**, 167-78.

Coleman, B.D. and Noll, W. (1964) Simple fluids with fading Memory, Proc. Int. Sympos., *Second Order Effects*, Heifa, 1962, McMillan, New York, pp. 530-52.

Coleman, B.D. and Owen, D.R. (1970) On the thermodynamics of materials with memory, *Arch. Ration. Mech. Anal.* **36**, 245-69.

Crochet, M.J. (1975) A non-isothermal theory of viscoelastic materials, in: *Theoretical Rheology*, Eds. J.F. Hutton, J.R.A. Pearson and K. Walters, Applied Science Publishers, London, pp. 111-22.

Crochet, M.J. and Naghdi, P.M. (1969) A Class of simple solids with fading memory, *Int. J. Eng. Sci.* **7**, 1173-98.

Crochet, M.J. and Naghdi, P.M. (1974) On a restricted non-isothermal theory of simple materials, *J. de Mécanique* **13**, 97-114.

Crochet, M.J. and Naghdi, P.M. (1979) On 'thermo-rheologically simple' solids, Proc. IUTAM Symp., *Thermoelasticity*, Springer, New York, 59-86.

Drucker, D. C. (1968) Closing Comments by Session Chairmen, in: *Mechanical Behavior of Materials under Dynamic Loads*, edited by U. S. Lindholm, Symposium held in San Antonio, Texas, Sept. 6-8, 1967, Springer-Verlag New York Inc., New York, pp. 403-19.

Dugdale, J.S. and McDonald, D. (1953) The thermal expansion of solids, *Phys. Rev.* **89**, 832.

Eringen, A.C. (1960) Irreversible thermodynamics and continuum mechanics, *Phys. Rev.* **117**, 1174-83.

Feynman, R. (1993) *The Character of Physical Law*, The MIT Press, Cambridge, Massachusetts.

Green, A.E. and Naghdi, P.M. (1965) A general theory of an elastic-plastic continuum, *Arch. Ration. Mech. Anal.* **18**, 251-81.

Green, A.E., Rivlin, R.S. and Spencer, A.J.M. (1959) The mechanics of nonlinear materials with memory, Part II, *Arch. Ration. Mech. Anal.* **3**, 82-90.

Gurtin, M.E. (1965) Thermodynamics and the possibility of spatial interaction in elastic materials, *Arch. Ration. Mech. Anal.* **19**, 339-52.

Gurtin, M.E. and Williams, W.O. (1966) On the inclusion of the complete symmetry group in unimodular group, *Arch. Ration. Mech. Anal.* **23**, 163-72.

Gurtin, M.E. and Williams, W.O. (1967) An axiomatic foundation for continuum thermodynamics, *Arch. Rational Mech. Anal.* **26**, 83-117.

Kac, M. (1969) Some mathematical models in science, *Science* **166**, 695-9.

Kestin, J. (1966) *A Course in Thermodynamics*, Vol. 1, Blaisdell, Waltham, Mass.

Kestin, J. (1979) *A Course in Thermodynamics*, Volumes I and II, Hemisphere Publishing Corp. and McGraw-Hill Book Company, New York

Kuhn, T. S. (1985) *The Structure of Scientific Revolution*, University of Chicago Press, Chicago.

Haddad, Y. M. (1995) *Viscoelasticity of Engineering Materials*, Kluwer, Dordrecht.

Ladevèse, P. (ed.), (1985) *Local Effects in the Analysis of Structures*, Elsevier, New York.

Landel, R.F. and Peng, S.T.T. (1986) Equations of state and constitutive equations, *J. Rheology* **30**(4), 741-65.

136

Laws, N. (1967) On the thermodynamics of certain materials with memory, *Int. J. Eng. Science* **5**, 427-34.

Leipholz, H. H. E. (1983) On the role of analysis in mechanics, *Trans. of the CSME* **7**, 3-7.

Meixner, J. (1969) Processes in simple thermodynamic materials, *Arch. Ration. Mech. Anal.* **33**, 33-53.

Müller, I. (1967) On the entropy inequality, *Arch. Ration. Mech. Anal.* **26**, 118-41.

Noll, W. (1958) A mathematical theory of the mechanical behaviour of continuous media, *Arch. Ration. Mech. Anal.* **2**, 197-226.

Owen, D.R. (1968) Thermodynamics of materials with elastic range, *Arch. Ration. Mech. Anal.* **31**, 91-112.

Owen, D.R. (1970) A mechanical theory of materials with elastic range, *Arch. Ration. Mech. Anal.* **37**, 85-110.

Popper, K. R. (1968) *The Logic of Scientific Discovery*, Harper and Row, New York.

Prager, W. (1968) Closing Comments by Session Chairmen, in: *Mechanical Behavior of Materials under Dynamic Loads*, edited by U. S. Lindholm, Symposium held in San Antonio, Texas, Sept. 6-8, 1967, Springer-Verlag New York Inc., New York, pp. 403-19.

Rivlin, R.S. (1972) On the Principles of Equipresence and Unification, *Quart. App. Math.* **30**, 227-28.

Schapery, R.A. (1968) On a thermodynamic constitutive theory and its application to various nonlinear materials, *Proc. IUTAM Symp.*, East Kilbride, pp. 259-85.

Schapery, R.A. (1969) On a thermodynamic constitutive theory and its application to various nonlinear materials, *Proc. IUTAM Symp. on Thermoinelasticity*, Springer, Berlin.

Sneddon, I.N. and Hill, R. (Eds.), (1960-1963) *Progress in Solid Mechanics*, Vol 1 (1960), Vol. 2(1961), Vol. 3(1963), Vol. 4 (1963), North-Holland Pub. Co., Amsterdam.

Truesdell, C. and Toupin, R.A. (1960) Classical field theories, in: *Handbuch der Physik*, ed. S. Flügge, Vol. III/1. Springer, Berlin, 226-90.

Wang, C.C. and Bowen, R.M. (1966) On the thermodynamics of non-linear materials with quasi-elastic response, *Arch. Ration. Mech. Anal.* **22**, 79-99.

Williams, W.O. (1970) Thermodynamics of rigid continua, *Arch. Rational Mech. Anal.* **36**, 270-84.

Wilson, A.H. (1957) *Thermodynamics and Statistical Mechanics*, Cambridge Univ. Press, Cambridge.

CHAPTER 5

TRANSITION TO THE RESPONSE BEHAVIOUR OF ENGINEERING MATERIALS

5.1. Introduction

Different materials of the same geometry may respond differently under identical external effects. Such difference in response is often attributed to the inherent constitution of the material. Consequently, the response behaviour of a particular material, or of a class of such material, is described mathematically by so-called "*constitutive relations*". These constitutive relations define the response behaviour of *"idealized"* media within a *"specific range"* of external effects. Accordingly, they only approximate the response characteristics of real materials, within a specified domain of actual service conditions. Constitutive relations establish, under certain physical and thermodynamical restrictions, the connection between the stimuli acting on the material specimen and the evolution of the occurring response.

5.2. The Constitutive Equation

5.2.1. RESTRICTIONS DUE TO PHYSICAL PRINCIPLES

In the majority of situations, the stimuli are the external forces, or the stresses caused by them, and the evolution of the response is expressed by the histories of both the deformation, or the calculated strain, and the temperature. In a continuum mechanics sense, a general form of a constitutive relation may be expressed as (e.g., Hunter, 1976),

$$\sigma_{ij} = f_{ij} \text{ (history of deformation, history of temperature)} \tag{5.1}$$

where σ_{ij} is the stress tensor and f_{ij} are the components of a second-order tensorial response function. In continuum mechanics formulations, constitutive relations are "*deterministic*", i.e., they are single-valued functions of the input (stimulus). The constitutive relations of different classes of engineering materials, within various domains of application, are particular forms of the constitutive relation (5.1) shown above. For elastic materials, for instance, the stress tensor σ_{ij} is a function of the current strain and temperature (e.g., Sokolnikoff, 1956). In case of viscoelastic materials, constitutive equations must account for the time-history of the deformation process and that of the temperature (e.g., Haddad, 1995). In the general form of the constitutive relation (5.1) above, the choice of the independent variables pertaining to f_{ij} is usually guided by the experimental results, but, in most situations, such choice is also restricted by a number of physical principles. In this context, a properly formulated constitutive equation must satisfy certain invariance principles (e.g., Eringen, 1962,1967 and Hunter, 1976). They are:

(i)   Invariance of the constitutive relation with respect to different stationary coordinate systems. This requirement is readily satisfied by expressing the constitutive law in tensorial form.

(ii)   Invariance of the constitutive relation with respect to coordinate systems in an arbitrary relative motion. This condition is usually dealt with within the context of *"material frame indifference"*; e. g., Hunter, 1976. This is translated into the requirement that the transformation law relating the components of the tensorial function f_{ij} in different co-ordinate frames in relative various motions to be exactly the same as the ordinary tensor transformation law.

Conformity of the form of the constitutive relation to the invariance requirements mentioned above, together with the (frequently employed) assumption of isotropy of the continuous body impose restrictions on the form of (5.1) and lead to explicit forms of constitutive equations for particular materials, or classes of such materials, under specific conditions. The material functions or parameters, characterizing the explicit form of the constitutive relation would be then characteristic of the response of the particular material, or its class, under consideration, within a specified range of applicability. In the latter context, Prager (1968), for instance, emphasized that the explicit statement of a domain of validity of a constitutive equation is an integral part of its validity. That is, an assumption that may be eminently reasonable in one set of circumstances could be obviously wrong in others. According to Prager (1968), two facts appear to be important in this context:

First, if a constitutive relation is to be useful in the solution of practical problems, it can, at best, apply only to limited ranges of strain, strain rate, temperature, etc.

Second, even though a constitutive equation is an essential ingredient of a continuum theory, it may well contain functions that are not continuous or continuously differentiable, e.g., for the purpose of obtaining solutions to boundary value problems that may not be solved otherwise.

5.2.2. RESTRICTIONS IMPOSED BY THERMODYNAMICS

Significant research efforts have been undertaken in the last four decades, or so, towards the development of a rigorous thermomechanical theory concerning the response behaviour of materials with memory as based on phenomenological considerations. In this context, theories have been presented from different points of view by Biot (1958,1973), Coleman and Noll (1963), Coleman (1964a&b), Schapery (1964), Crochet and Naghdi (1974) and Rivlin (1975), among others. Other work of interest includes that of Coleman and Mizel (1963,1964) and Day (1972) concerning the free energy concept, recoverable work and related work bounds. The reader is also referred to Müller (1967), Meixner (1969) and others for developments in the subject of continuum thermodynamics.

Coleman and Noll (1963) adopted the Clausius-Duhem inequality (*see Chapter 4*), as an expression for the second law of thermodynamics, to determine the validity of the constitutive relations of a body of material. In this, their paper demonstrates that the second law of thermodynamics requires the Clausius-Duhem inequality to be satisfied in a process that is compatible with the balance laws of mass, momentum, moment of momentum and energy. This translates into the requirement that the constitutive relations must be compatible with the Clausius-Duhem inequality in order for such constitutive relations be able to describe the response behaviour of a material under the restrictions imposed by thermodynamics.

Following the above approach, Coleman and Mizel (1963) studied heat conduction in rigid bodies and, in (1964), these authors established the existence of caloric equations of state for materials of the rate type. For applications of the Coleman and Noll's approach (1963) to different classes of materials, the reader is referred to the research works by Green and Naghdi (1965), Gurtin (1965), Gurtin and Williams (1966, 1967), Wang and Bowen (1966), Coleman and Gurtin (1967), Green and Laws (1967), Laws (1967), Coleman and Mizel (1967,1968), Owen (1968, 1970) and Coleman and Owen (1970), among others.

As a continuation of the work of Coleman and Noll (1963), Coleman (1964a&b) dealt with the foundations of a thermodynamic theory of materials with memory; from a macroscopic point of view and based on the principles of continuum physics. Again, Coleman takes the Clausius-Duhem inequality to be the expression of the second law of thermodynamics, and establishes the restrictions for reducing the constitutive equations to forms compatible with thermodynamics. As the statement of the Clausius-Duhem inequality involves the entropy of the body, one must acquire, in Coleman's approach, the entropy from the beginning to deal with (Day, 1972).

Simple Materials
Neglecting any thermodynamic effect, a substance for which the stress $\sigma(t)$ is determined by the history of a measure of strain is referred to, from continuum mechanics point of view, as "*simple material*". The response equation of such material may be written (Coleman, 1964a&b) as

$$\sigma(t) = \prod_{\tau=0}^{\infty} (F(t-\tau)) \tag{5.2}$$

In this equation, $F(t-\tau)$ is the deformation gradient at time $(t-\tau)$, $0 \le (t,\tau) < \infty$ and Π is a functional mapping the function $F(t-\tau)$ into tensor $\sigma(t)$. The functional Π may be considered as a general functional subject to the requirements of *"material symmetry"* (Noll, 1958; Coleman and Noll, 1964 and Coleman 1964a&b), the principle of *"material objectivity"* (Noll, 1958 and Coleman, 1964a&b) and the *"principle of fading memory"* (Coleman and Noll, 1960 and Coleman 1964a&b).

In the more general case, i.e. when one includes thermodynamic effects, the stress $\sigma(t)$ would depend on both the deformation gradient history $F(t-\tau)$ as well as the temperature history $T(t-\tau)$. Thus, a more generalized form of (5.2) may be written as

$$\sigma(t) = \prod_{\tau=0}^{\infty} (F(t-\tau), T(t-\tau)) \tag{5.3}$$

One may also assume, Coleman (1964a&b), that the specific internal energy per unit mass e is determinable, similar to $\sigma(t)$, as illustrated above, by the histories $F(t-\tau)$ and $T(t-\tau)$, i.e.

$$e(t) = \mathop{e}_{\tau=0}^{\infty} (F(t-\tau), T(t-\tau)) \tag{5.4}$$

On the other hand, the heat flux vector q is dependent on the temperature gradient $g(T)$ during the thermodynamic process. Since, according to (5.3), the stress is assumed to depend on $F(t-\tau)$ and $T(t-\tau)$, it is likely that these histories would influence the dependence of the heat flux q on the temperature gradient $g(T)$. Accordingly, the constitutive equation for the heat flux may be expressed, (Coleman, 1964a&b), as

$$q(t) = \mathop{q}_{\tau=0}^{\infty} (F(t-\tau), T(t-\tau), g(T)) \tag{5.5}$$

where $q(t)$, in (5.%), is interpreted as a functional whose arguments are the histories $F(t-\tau)$, $T(t-\tau)$ and the gradient of the current temperature, i.e., $g(T)$. Recalling at this point the *"principle of equipresence"* (Truesdell, 1951, Truesdell and Toupin, 1960, Coleman and Mizel, 1964 and Coleman and Gurtin, 1967) which may be stated, (Coleman, 1964a&b), as:

> *"An independent variable present in one constitutive equation of a material should be assumed to be so present in all, until its presence is shown to be in*

*direct contradiction to the assumed symmetry of the material, the principle
of material objectivity, or the laws of thermodynamics."*

Thus, the constitutive equations (5.3) and (5.4) should also include in their argument the
dependence on the temperature gradient $g(T)$ present in (5.5). Accordingly, one replaces
(5.3) and (5.4), respectively, by

$$\sigma(t) = \overset{\infty}{\underset{\tau=0}{\Pi}} \; (F(t-\tau), \; T(t-\tau), \; g(T)) \tag{5.6}$$

$$e(t) = \overset{\infty}{\underset{\tau=0}{e}} \; (F(t-\tau), \; T(t-\tau), \; g(T)) \tag{5.7}$$

In order to include the restrictions imposed by the Clausius-Duhem inequality on the above-
mentioned constitutive relations (5.5), (5.6) and (5.7), an expression for the specific entropy
(per unit mass) is introduced in the form

$$\varsigma(t) = \overset{\infty}{\underset{\tau=0}{\varsigma}} \; (F(t-\tau), \; T(t-\tau), \; g(T)) \tag{5.8}$$

which also satisfies the *"principle of equipresence"*.

In Coleman's theory (1964a&b), it is assumed that the four functionals q, σ, e and ς
corresponding respectively to equations (5.5) to (5.8) are given at each point X of the
material. These functions, in general, depend on the choice of the reference configuration.
However, if there exists a reference configuration that would render these functionals to be
independent of X for all material points in the body B, then one may consider B to be
materially homogeneous.

Admissibility. A thermodynamic process is said to be admissible in B if it is compatible with
the constitutive relations (5.5) to (5.8) at each material point X of B and all times t. In this
context, Coleman (1964a&b), showed that the following remark to be valid.

Remark: To every choice of the deformation function $x(X,t)$ and the
temperature distribution $T(x,t)$, $(x,X$ in B; $-\infty < t < \infty)$, there
corresponds a unique admissible thermodynamic process in B.

Coleman (1964a&b), also, showed that the Clausius-Duhem inequality requires that the
temperature gradient $g(T)$ to be dropped out from relations (5.6) to (5.8). Accordingly, the
new set of constitutive equations are expressed as

$$\sigma(t) = \prod_{\tau=0}^{\infty} (F(t-\tau), T(t-\tau)) \tag{5.9a}$$

$$e(t) = \overset{\infty}{\underset{\tau=0}{e}} (F(t-\tau), T(t-\tau)) \tag{5.9b}$$

$$q(t) = \overset{\infty}{\underset{\tau=0}{q}} (F(t-\tau), T(t-\tau), g(T)) \tag{5.9c}$$

$$\varsigma(t) = \overset{\infty}{\underset{\tau=0}{\varsigma}} (F(t-\tau), T(t-\tau)) \tag{5.9d}$$

Equations (5.9b) and (5.9d) may be also used to express a constitutive equation based on the specific "*Helmholtz free energy*". Denoting the latter by A, it is defined by

$$A = e - T\varsigma$$

Since both e and ς are given in (5.9b) and (5.9d), respectively, by functionals of $F(t-\tau)$ and $T(t-\tau)$, it follows that

$$A = \overset{\infty}{\underset{\tau=0}{A}} (F(t-\tau), T(t-\tau)) \tag{5.9e}$$

The reader is referred to Coleman (1964a&b) for theorems and remarks concerning the set of constitutive relations (5.9a-e).

Entropy as an Independent Variable. Recalling the specific entropy constitutive equation (5.9d). That is

$$\varsigma(t) = \overset{\infty}{\underset{\tau=0}{\varsigma}} (F(t-\tau), T(t-\tau))$$

Assume that the above functional transformation is invertible in the sense that there exists a functional T(t) such that

$$T(t) = \underset{\tau=0}{T} \; (\mathbf{F}(t-\tau), \; \varsigma(t-\tau)) \tag{5.10}$$

Accordingly, one may rewrite the other constitutive equations (5.9a-c) and (5.9e), respectively, as

$$\sigma(t) = \overset{\wedge}{\underset{\tau=0}{\prod}} \; (\mathbf{F}(t-\tau), \; \varsigma(t-\tau)) \tag{5.11a}$$

$$e(t) = \overset{\wedge}{\underset{\tau=0}{e}} \; (\mathbf{F}(t-\tau), \; \varsigma(t-\tau)) \tag{5.11b}$$

$$q(t) = \overset{\wedge}{\underset{\tau=0}{q}} \; (\mathbf{F}(t-\tau), \; \varsigma(t-\tau), \; g(T)) \tag{5.11c}$$

$$A(t) = \overset{\wedge}{\underset{\tau=0}{A}} (\mathbf{F}(t-\tau), \; \varsigma(t-\tau)) \tag{5.11d}$$

Internal Energy as an Independent Variable. Consider the constitutive equation for internal energy (5.9b), i.e.

$$e(t) = \overset{\infty}{\underset{\tau=0}{e}} \; (\mathbf{F}(t-\tau), \; T(t-\tau))$$

Assume now that the above functional transformation is invertible, i.e., there exists a functional $T(t)$ such that

$$T(t) = \overset{\vee}{\underset{\tau=0}{T}} (\mathbf{F}(t-\tau), \; e(t-\tau)) \tag{5.12}$$

Accordingly, one may rewrite the rest of the constitutive equations (5.9), in sequence, as

$$\sigma(t) = \overset{\overset{\displaystyle\text{v}}{}}{\underset{\tau=0}{\Pi}}(F(t-\tau), \ e(t-\tau)) \tag{5.13a}$$

$$q(t) = \overset{\overset{\displaystyle\text{v}}{}}{\underset{\tau=0}{q}}(F(t-\tau), \ e(t-\tau), \ g(T)) \tag{5.13b}$$

$$\varsigma(t) = \overset{\overset{\displaystyle\text{v}}{}}{\underset{\tau=0}{\varsigma}}(F(t-\tau), \ e(t-\tau)) \tag{5.13c}$$

$$A(t) = \overset{\overset{\displaystyle\text{v}}{}}{\underset{\tau=0}{A}}(F(t-\tau), \ e(t-\tau)) \tag{5.13d}$$

Coleman (1964a&b) showed that the following remarks to be valid:

Remark: In every admissible process

$$\varsigma \geq \frac{1}{T}(\dot{e} - \frac{1}{\rho}\text{tr } \sigma \ L) \tag{5.14a}$$

$$\varsigma \geq \frac{1}{T}(h - \frac{1}{\rho}\text{div } q) \tag{5.14b}$$

where **L** is the velocity gradient, i.e., **L** = grad $\dot{x}$.

The inequality in (5.14a&b) above is referred to as the *"principle of positive internal production of entropy"*.

Remark: Whenever the strain and internal energy are held constant, the entropy cannot decrease, regardless of the past history.

Remark: Whenever the strain and entropy are held constant, the internal energy cannot increase regardless of the past history.

Remark: In an admissible thermodynamic process, the material time derivative of the free energy obeys the inequality

$$\dot{A} \le \frac{1}{\rho} \text{ tr } \sigma \text{ L} - \varsigma \dot{T} \tag{5.15}$$

Thus,

$$\text{if } \mathbf{L=0}, \quad \dot{T} = 0 \quad \text{then, } \dot{A} \le 0 \tag{5.16}$$

i.e., if at a given instant of time, a material point X be held at constant strain and temperature (e.g. isothermal stress relaxation), the free energy at X at that instant cannot increase regardless of the past history.

The significance of Coleman's work (1964a&b) in establishing the restrictions imposed by thermodynamics on the constitutive equations for materials with memory is apparent. However, as pointed out by Rivlin (1975), no prescription was given in Coleman's work for determining the actual form of the constitutive functional either analytically, or by deduction from experiment. Rivlin (1975), criticized further Coleman's approach for considering the entropy as a "primitive quantity": In this context, Rivlin argued that the entropy "*is not in the same category as the primitive mass, length and time because it cannot be measured and nothing can be done with it*". Rivlin (1975), on the other hand, defined materials with memory as materials for which the Piola-Kirchhoff's stress Σ (see *Chapter 2*) and the empirical temperature θ are functions of the histories of the deformation gradient tensor $F(\tau)$ and the specific internal energy $e(\tau)$, with support within the limits $(-\infty, t^+)$, i.e.,

$$\sum(t) = \sum(F(\tau), e(t)); \qquad \theta(t) = \theta(F(\tau), e(\tau)) \tag{5.17}$$

Alternatively, one may consider $\Sigma(t)$ and $e(t)$ as functionals of the histories $F(\tau)$ and $\theta(\tau)$ with support $(-\infty, t^+)$. That is

$$\sum(t) = \sum(F(\tau), \theta(\tau)); \qquad e(t) = e(F(\tau), \theta(\tau)) \tag{5.18}$$

The support in (5.17) and (5.18) is taken by Rivlin (1975) to be $(-\infty, t^+)$, rather than $(-\infty, t)$, in order to include the possibility that Σ and θ may depend on the instantaneous values of the time derivatives of $F(\tau)$ and $e(\tau)$ at the instant t, even though these may change discontinuously at time t.

Rivlin (1975) considered the material to have "*fading memory*" if the functional in (5.17) and (5.18) is such that, for two histories which differ only up to time $(t-\tau)$, the difference in the corresponding functionals decreases to zero as τ increases to infinity. Coleman (1964a&b) considered also a similar assumption for the definition of the "*fading memory*" of simple materials, that is the memory of such materials fades in time. Coleman's assumption implies the assertions that deformations and temperatures experienced in the distant past should have less effect on the present values of the entropy, energy, stress and heat flux than deformations and temperatures which occurred in the recent past. In this context, Coleman (1964a&b) introduced an *"influence function"* $C(\tau)$, $0 \leq \tau < \infty$, which would characterize the rate at which the memory fades. The influence function $C(\tau)$ is assumed to be a positive monotonic decreasing and continuous function for the time parameter τ (*see* Coleman and Noll, 1960&1964).

With reference to (5.18), a material is said (Rivlin, 1975) to be perfectly elastic if Σ and θ depend only on the instantaneous values of F and e. At this case, Σ and θ are ordinary functions of F and e, i.e.

$$\Sigma = \Sigma(F,e); \quad \theta = \theta(F,e) \tag{5.19}$$

Alternatively, Σ and e will be ordinary functions of F and θ and (5.19) can be replaced by

$$\Sigma = \Sigma(F,\theta); \quad e(F,\theta) \tag{5.20}$$

Accordingly, in the case of materials with fading memory, if we restrict ourselves to processes carried out quasi-statically, the constitutive equations (5.17) and (5.18) will assure the forms of the constitutive equations (5.19) and (5.20), respectively. In other words, materials with fading memory behave as perfectly elastic materials with respect to quasi-static processes.

Rivlin (1975) adopted the *"Carathéodory's principle"* as a form of the second law of thermodynamics: *"There are states of a system, differing infinitesimally from a given state, which are unattainable from that state by any adiabatic process whatever"*. Here, *"state"* is used in the sense of *"equilibrium state"* and it is postulated that the materials considered can always be taken from any such state to another state by a quasi-static process. As a consequence of the above, Rivlin (1975) asserted the existence of the *"specific entropy"* which is a function of the variables used to describe the state, and of the absolute temperature which is a function of the empirical temperature. The function through which the specific entropy relates to the state variables depends on the material considered, while the function through which the absolute temperature associates with the empirical temperature is independent of this material.

As an illustration of the above arguments, Rivlin (1975) considered a body of material with fading memory to be in equilibrium with uniform empirical temperature θ, specific internal energy e and deformation gradient F. The constitutive equations describing the response of such material are assumed to be given by (5.17). Rivlin, then, assumed that the body is taken from a homogeneous equilibrium state by a "homothermal" quasi-static process, to a neighbouring equilibrium state, in which θ, e and F are changed, respectively, to $\theta + d\theta$, e + de and F + dF. Thus, letting dH be the amount of heat (per unit mass) which is absorbed by the body in this process, it can be shown, following the first law of thermodynamics, that

$$\rho \; dH = \rho \; de - tr \left(\sum , \; dF \right) \tag{5.21}$$

In (5.21), ρ is the material density in the fixed reference state with respect to which F is measured, and *de* is the increase in specific internal energy, of the material, in the process.

With equations (5.17) to (5.20), equation (5.21) yields

$$\rho dH = \rho \left(\frac{\partial e}{\partial \theta} \right)_F d\theta + tr \left\{ \rho \left(\frac{\partial e}{\partial F} \right)_\theta^{\cdot} - \sum \right\} F \tag{5.22}$$

Based on the Carathéodory's principle and on the assumption that the process to be quasi-static, then there must exist values of $d\theta$ and dF for which $dH \neq 0$ (Rivlin, 1975). In other words, the process must not be adiabatic for the transition between the two neighbouring states to take place. From this fact, Rivlin (1975) asserts, with the support of the work of Wilson (1957) and Kestin (1966), that there exists an integrating factor $1/T(F,\theta)$ such that ρ dH/T, (5.22), is a perfect differential.

Following Rivlin (1975), one may consider unit mass of a material with fading memory to be taken by a quasi-static homothermal process from an equilibrium state A to an equilibrium state B. Let the states A and B be identified, respectively, by the two sets of values (θ_A, F_A, e_A, ς_A) and (θ_B, F_B, e_B, ς_B). Let, also, dH denote the heat fed into the body in an infinitesimal step of the process. Since

$$d\varsigma = dH/T(\theta) \tag{5.23}$$

and ς is a function of θ and F only, then,

$$\varsigma_B - \varsigma_A = \int_A^B dH/T(\theta) \tag{5.24}$$

where the integration is carried out along the path in the ten-dimensional space (θ, F) followed by the process.

Consider, now, a body of a material with fading memory to be taken from an equilibrium state A to an equilibrium state B by a process which is not necessarily quasi-static. It can be shown (Rivlin, 1975 and Kestin, 1966), by application of Carathéodory's principal, that if dH is the amount of heat (per unit mass) entering the system in an infinitesimal step of the process, at an instant at which the empirical temperature of the system is θ, then

$$\int_A^B \frac{dH}{T(\theta)} \leq \varsigma_B - \varsigma_A \qquad (5.25)$$

In the above relation, the equality sign applies if the process is quasi-static. Formula (5.25) is the "*Clausius inequality*" or "*Clausius-Planck inequality*" whereby the integral is referred to as the "*Clausius integral*".

Recalling (5.23). That is

$$d\varsigma = \frac{dH}{T(\theta)}$$

where ς, the specific entropy, is a function of the instantaneous values of F and $T(\theta)$. This equation is valid, for example, for an infinitesimal step of a homothermal process, whether quasi-static or not, in a perfectly elastic material.

For a material with fading memory, the path in (F, T) space which may be followed by the non-quasi-static process could also be followed by a quasi-static process. Accordingly, at each point of an arbitrary homothermal process in a material with fading memory, the Clausius inequality (5.25), i.e.,

$$\int_A^B \frac{dH}{T(\theta)} \leq \varsigma_B - \varsigma_A$$

can be replaced by the "*Clausius-Duhem inequality*",

$$\dot{\varsigma} \leq \dot{H}/T \qquad (5.26)$$

where the dot designates material differentiation with respect to time, thus, $\dot{H}$ denotes the rate at which heat enters the body at the instant considered. There is an essential physical difference between the Clausius and Clausius-Duhem inequalities. This may be illustrated by the following comparison given by Rivlin (1975):

Consider a body of material with fading memory to be taken from an equilibrium state A to an equilibrium state *B* by quasi-static and non-quasi-static, isothermal, homothermal processes which follow the same paths in *(F,T)* space. *The Clausius inequality* states that less heat is fed into the system in the non-quasi-static process than in the quasi-static process. The Clausius-Duhem inequality asserts, however, that the amount of heat fed into the system, in each infinitesimal step of the non-quasi-static process, is no greater than that for the corresponding step of the quasi-static process.

Rivlin (1975), however, showed by an example, that the Clausius-Duhem inequality may not be valid for all materials and all processes.

Instantaneous response behaviour. In case of materials with fading memory, instantaneous changes in the deformation gradient tensor *F* and the empirical temperature θ result in instantaneous changes in the Piola-Kirchhoff stress Σ and in the specific internal energy *e* which could be followed by further changes in these quantities.

In order to describe the type of behaviour above, Rivlin (1975), following Green et al (1959), made explicit the dependence of Σ and *e*, at time *t*, on the instantaneous values of the deformation gradient tensor *F* and of the empirical temperature at time t. Accordingly, the following constitutive equations may be written

$$\sum = \sum (F(\tau), \quad \theta(\tau); \quad F, \theta) \tag{5.27}$$

$$e = e(F(\tau); \quad \theta(\tau); \quad F, \theta) \tag{5.28}$$

indicating that Σ and *e* are functionals of the histories $F(\tau)$ and $\theta(\tau)$ with support $(-\infty,t)$ and ordinary functions of *F* and θ.

For materials possessing instantaneous elasticity, for which the constitutive equations (5.27) and (5.28) are valid, Σ and *e* are functions of F and θ only. In this case, we would restrict ourselves to processes for which the histories $F(\tau)$ and $\theta(\tau)$ are fixed functions of τ, $-\infty < \tau < t$, while only *F* and θ may change.

Drucker (1968) commented that, for the purpose of solving nonlinear boundary value problems, one may need to idealize the constitutive relations or the involved thermodynamics quite drastically. As commented by Drucker (1968), one must worry far less about general mathematical consistency and to concentrate on the physical aspects. The latter must be put down in clear and reasonably acceptable form before subjecting them to formal mathematical treatment. On the other hand, there is the continuous effort by which researchers aim at a rather general thermodynamic approach to the inelastic behavior of materials; *see* Drucker (1968) and Bodner (1968).

5.3. Pertinent Notions of Analytical (Phenomenological) Mechanics

Pindera (1992, 1998a&b) advanced the following arguments as pertaining to the formulation and the validity of the constitutive equation following the axioms of phenomenological mechanics:

- It is known that all our basic concepts, notions and solutions are formulated and developed within a chosen theoretical framework. In other words, it is theoretically impossible to describe a body or a process as it is. Reality is perceived and recognized within the framework of accepted models. Thus, for instance, *Young's modulus E* is not an inherent property of a certain class of materials, but it is a parameter in Hooke's mathematical model. It is valid only as long as the Hooke's model is acceptable, and as long as the theory of experimental determination of the E modulus is acceptable. The same pertains to the notions of stress and strain. A more detailed presentation of this issue is given by Feynman (1993), Kac (1969), Pindera (1981, 1987a&b), Pindera (1989) and Popper (1968).

- Due to various reasons, *see Chapter 2*, the methods and relations of the mathematical theory of elasticity have been developed on the basis of a notion of a *"material continuum"*. This material continuum is, in its basic form, homogeneous, isotropic, and isothermal (Sokolnikoff, 1956, and Timoshenko and Goodier, 1970). A similar model is accepted in viscoelasticity and in plasticity (Haddad, 1995 and Zyczkowski, 1981). The existence of the *thermoelastic effect*, which is responsible for the fact that the strain energy is not equal to the external work, is still treated as a second order phenomenon represented by the *coupled equation of thermoelasticity*. It is common to ignore the fact that the reversible strain-induced temperature alteration may result in an additional thermal strain of an order comparable with the load-induced strain (Kestin, 1966). As a routine, the local thermal strain/stress states caused by local inelastic deformation are ignored, as are ignored the thermal strain/stress states caused by phase transformation. This is, however, described by Pindera as "strange", as the related alteration of the thermal expansion coefficients of the commercial polymeric materials by the factor 2 or 3 often causes unexpected internal fractures or decohesions when the temperature of an object increases by as little as 5-10 degrees Kelvin (Pindera, 1989, and Pindera and Pindera, 1989).

- It is also known that the stress state in plates is noticeably three-dimensional in the presence of noticeable gradients of the in-plane stress components (Pindera and Krasnowski, 1982, Pindera, 1984, 1988, 1989, Pindera and Pindera, 1989, and Pindera and Liu, 1992). Also, in general, stress states in the composite and adhesively bonded structures are noticeably three-dimensional; as we discussed earlier in Chapter 2.

- The above discussed, load-induced stress states are conveniently analyzed within the framework of a material continuum, elastic or linear viscoelastic. The linear theory of elasticity (e.g., Sokolnikoff, 1956), and of viscoelasticity (e.g., Haddad, 1995) supply convenient general frameworks, provided that the notion of a *generalized plane stress state*, which was introduced together with the notion of a *thin plate*, are not accepted. It has been demonstrated that acceptance of those notions results in evaluation of stresses which are up to 30% lower than the actual stresses, particularly in the regions of local effects (Ladevèse, 1985, Pindera and Krasnowski, 1982, and Pindera and Liu, 1992). It appears that a contributing factor is the popular belief that the three-dimensional compatibility conditions (3.28) can be reduced to a two-dimensional compatibility condition in the case of the so-called thin plates. That is

$$\nabla^2 \left(\sigma_{xx} + \sigma_{yy} \right) = 0 \tag{5.29}$$

This is, as argued by Pindera (1998a), impossible, even when the faces of a plate are free of tractions (Pindera and Pindera, 1989). The stress states in the thin or thick plates are always three-dimensional in the presence of gradients of the in-plane stresses. The only question is how significant is the deviation from the plane stress state.

- Meantime, the notion of *stress singularity* should be treated differently. The concept of a finite load acting at a point is a mathematical abstraction, or a phenomenological assumption, which is theoretically (physically) inadmissible. The physical space could not support an infinite energy density, or an infinite power density. Different limiting processes might yield different singular solutions, so the validity of any given singular solution should be experimentally tested. In practice, it means that the range of the inadmissible predicted stress values should be determined experimentally, unless a more general analytical solution would yield correct stress values (Ladevèse, 1985).

- Within the framework of Continuum Mechanics, analysis of three-dimensional (elastic) stress and strain states requires data on, at least, two basic materials coefficients, e. g., Young's modulus E, and Poisson's ratio v, which must be understood as parameters of the Hooke's model, but not as inherent properties of solid bodies. Determination of those, so-called, *materials elastic constants* according to the various testing standards, is simple and straightforward. The obtained data are unequivocal. The situation is entirely different regarding the time-dependent materials (e.g., Haddad, 1995). Mechanical testing standards for various classes of engineering materials,e. g., plane and fiber-reinforced polymers are designed most for industrial purposes. Thus, it is normal that the pertaining specifications list different values of the E-modulus, for instance, depending on the mode of loading, e.g., tension, compression, bending, or torsion. The common practice of introducing an averaged value of E-modulus, for example, results in errors of unknown magnitude. In this context, Pindera (1998a) reasonably argued that such a presentation could be very misleading when some major coupled materials responses are to be presented in a coherent manner, because it gives no insight into the actual mechanisms of the involved processes. Thus, a phenomenological approach in mechanics is able to provide a general methodological analytical framework, but is not able to provide an insight into, and understanding of the mechanisms of involved processes. However, such an insight and understanding are prerequisites for a rational optimization of performances of inhomogeneous and composite structures.

- In the past, the scope and depth of the experimental research in mechanics had been seriously limited by theoretical and technological factors such as:

  - a limited knowledge of the mechanism of measurement and the related theory which lead to the notion of a *"black box"*.
  - a limited knowledge of the patterns of quasi-static and dynamic responses of the *"system testing machine - specimen - measurement instruments"*.
  - a very limited accuracy and resolution of measurement systems, which lead to the acceptance of large measurement errors, e.g., in the range of 3 to 10%. Only the first-order instruments yielded results with higher accuracy.

- At present, the situation is entirely different. There are no any longer measurement problems in engineering experimentation. The theory of measurement system is well developed (Doeblin, 1983). The patterns of the flow of the information-energy through the system *testing machine-specimen* are well understood and the possibility of misunderstanding of the indicated responses is minimized (Pindera, 1981, 1987a&b). The measurement ranges, accuracy, and resolution of modern measurement system developed by teams consisting of physicists, chemists, photonics engineers, information engineers, and design engineers surpass the most demanding engineering requirements. The only one major issue that is still unresolved satisfactorily is how to educate a researcher to enable him to optimally utilize the tremendous progress in measurement and information sciences.

- One more problem was solved, more or less satisfactorily, during the last decade. It pertains to the *guessed quantities*, such as a *thin plate* versus a *thick plate, small* versus *large*, or *negligible* versus *essential*. Three factors contributed to this development:

- **The first factor** is related to the incredible progress in theory and techniques of measurements. As a result, the material responses which in the past had not been even noticed by the researchers, such as the load-induced thermoelastic effect described by the coupled equation of thermoelasticity, are routinely used in experimental stress analysis.

- **The second factor** is related to the rapid development of so-called *smart sensors*, and *smart materials and* structures. This development requires educational programs above the level of typical engineering courses in mechanical and civil engineering.

- **The third factor** is a direct consequence of the rapidly increasing societal notion of the *acceptable risk*, or/and *hazard*. This factor allows to quantify societally acceptable losses caused by unexpected failures, and, thus, allows to tell more precisely a small, acceptable simplification from a large, unacceptable simplification (*see* Pindera, 1998b).

- **The notion of the stress-at-a point**: As stated above, all the notions related to the stress states in solids are of a major importance to a reliable stress analysis which is a major component of a modern designing process. The basic notion is the notion of *stress*, as a local quantity which could be defined, *as dealt with in Chapter 2*, as a *"stress at a point"*. Thus, in the experimental determination of the stress/strain state, the range of validity of such a notion of stress is of major importance. This problem was considered not to be important when the procedures of theory of elasticity had been applied to mechanical design using metals. At that time, because of the relatively low resolution of the measurement instruments, the grain size of a few hundredth of a millimetre was considered very small and of no consequence. Situations changed when the relations of the theory of elasticity, or of engineering mechanics, were applied to such polymeric, natural composite materials as natural wood, or to such a manmade composite material as rod or wire reinforced concrete. A complete alteration of the situation occurred when the contemporary instruments allowed, for the first time in history, to record positions of single atoms, or to move the selected atoms across the face of a specimen.

- Stress is a derived quantity, which could not be measured directly. Only strain can be directly determined, as a normalized quantity. The most common definition of strain, *as dealt with in Chapter 3*, is the ratio of an alteration of the distance between two points of a solid body over the original distance which is called *"base of measurement"*. Such a definition implies that the stress and deformation states within the tested region are homogeneous. However, such excellent structural composite materials as wood, reinforced concrete, or fibre reinforced solids, are strongly heterogeneous. Thus, the measurement base should be long enough when compared to average local deviations; e.g., Dantu, 1958, and Müller, 1964. It is common to present the measurement base in the term of a dimensionless reduced base which represents a ratio of the actual base over the maximal grain size. It was shown that the evaluation error in strain determination in concrete, is less than 3% when the reduced base is longer than 30 dimensions of the maximal grain. The notion of a stress at a point collapses when the measurement distance is taken as small as about size of 5 atoms, or 5 molecules, or 5 crystallites in metallic materials. Such distances could be easily determined, with high accuracy, by means of modern instruments. At such distances, *the notion of stress* should be replaced by *the notion of a force field* (see, e. g., Axelrad and Haddad,1998, and Haddad, 1998&1995) . Thus, the notion of stress in heterogeneous materials, e.g. polycrystals and composite systems, should be applied cautiously, including the interpretation of stress values predicted by analytical solutions and evaluated experimentally.

- Great care is required when the actual stresses in heterogenous materials, e.g., composites, are determined either by means of analytical relations, or by using experimental or structural models which are designed to simulate the actual stress state. The same pertains to the derived quantities such as the *stress concentration*, or the *"stress intensity factor"*.

5.4. Problems

1. State two physical restrictions imposed on the form of the constitutive equations for a class of materials.

2. State the restrictions imposed by thermodynamics on the formulation of a constitutive equation for an engineering material.

3. Explain briefly the following terms:
   - Admissability
   - Equipresence
   - Fading memory

4. What is a *"simple material"*?

5. Differentiate between the response of an *"elastic material"* and a *"material with*

memory" from a thermodynamical point of view.

5.5. References

Axelrad, D. R. and Haddad, Y. M. (1998) On the behaviour of materials with binary microstructures, in *Multilayered and Fibre-Reinforced Composites: Problems and Prospects*, NATO ARW, Kiev, June 2-6, 1997, edited by Y. M. Haddad, Kluwer, Dordrecht, The Netherlands, pp. 163-72.

Biot, M.A. (1958) Linear thermodynamics and the mechanics of solids, Proc. 3rd U.S. Nat. Congr. Appl. Mech., pp. 1-18.

Biot, M.A. (1973) Nonlinear thermoeleasticity, irreversible thermodynamics and elastic instability, *Indi. Math. J.* **23**, 309-35.

Bodner, S. R. (1968) Constitutive equations for dynamic material behavior, in: *Mechanical Behavior of Materials under Dynamic Loads*, edited by U. S. Lindholm, Symposium held in San Antonio, Texas, Sept. 6-8, 1967, Springer-Verlag New York Inc., New York, pp. 176-90.

Coleman, B. D. (1964a) Thermodynamics of materials with memory, *Arch. Rat. Mech. Anal.* **17**, 1-46.

Coleman, B. D. (1964b) Thermodynamics, strain impulses and viscoelasticity, *Arch. Rat. Mech. Anal.* **17**, 230-54.

Coleman, B.D. and Gurtin, M.E. (1967) Thermodynamics with internal state variables, *J. Chem. Phys.* **47**, 597-613.

Coleman, B.D. and Mizel, V. J. (1963) Thermodynamics and departures from Fourier's law of heat conduction, *Arch. Rat. Mech. Anal.* **13**, 245-61.

Coleman, B.D. and Mizel, V. J. (1964) Existence of coloric equations of state in thermodynamics, J. Chem. Phys. **40**, 1116-25.

Coleman, B.D. and Mizel, V.J. (1967) A general theory of dissipation in materials with memory, *Arch. Ration. Mech. Anal.* **27**, 255-74.

Coleman, B.D. and Mizel, V.J. (1968) On the general theory of fading memory, *Arch. Ration. Mech. Anal.* **29**, 18-31.

Coleman, B.D. and Noll, W. (1960) An approximation theorem for functionals with applications in Continuum Mechanics, *Arch. Ration. Mech. Anal.* **6**, 355-70.

Coleman, B.D. and Noll, W. (1963) The thermodynamics of elastic materials with heat conduction and viscosity, *Arch. Ration. Mech. Anal.* **13**, 167-78.

Coleman, B.D. and Noll, W. (1964) Simple fluids with fading Memory, *Proc. Int. Sympos. Second Order Effects*, Heifa, 1962, MacMillan, New York, pp. 530-52.

Coleman, B.D. and Owen, D.R. (1970) On the thermodynamics of materials with memory, *Arch. Ration. Mech. Anal.* **36**, 245-69.

Crochet, M.J. and Naghdi, P.M. (1974) On a restricted non-isothermal theory of simple materials, *J. de Mécanique* **13**, 97-114.

Dantu, P. (1958) I - Étude des contraintes dans les milieux hétérogènes. Application au beton. II - Utilisation des réseaux pour l'étude des déformations. Laboratoire Central des Ponts et Chaussees, Publication 57-6. Annales de l'Institut Technique du Bâtiment et des Travaux Publics, Paris, pp.17-25.

Day, W. A. (1972) *The Thermodynamics of Simple Materials with Memory*, Springer-Verlag, New York.

Doeblin, E. O. (1983) *Measurement Systems. Application and Design*, McGraw-Hill Book Co., New York.

Drucker, D. C. (1968) Closing Comments by Session Chairmen, In: *Mechanical Behavior of Materials under Dynamic Loads*, edited by U. S. Lindholm, Symposium held in San Antonio, Texas, Sept. 6-8, 1967, Springer-Verlag New York Inc., New York, pp. 403-19.

Eringen, A.C. (1962) *Nonlinear Theory of Continuous Media*, McGraw-Hill, New York.

Eringen, A.C. (1967) *Mechanics of Continua*, John Wiley & Sons, New York.

Feynman, R. (1993) *The Character of Physical Law*, The MIT Press, Cambridge, Massachusetts.

156

Green, A. E. and Laws, N. (1967) On the formulation of constitutive equations in thermomechanical theories of continua, *Quart. J. Mech. Appl. Math.* **20**, 265-75.

Green, A.E. and Naghdi, P.M. (1965) A general theory of an elastic-plastic continuum, *Arch. Ration. Mech. Anal.* **18**, 251-81.

Green, A.E., Rivlin, R.S. and Spencer, A.J.M. (1959) The mechanics of nonlinear materials with memory, Part II, *Arch. Ration. Mech. Anal.* **3**, 82-90.

Gurtin, M.E. (1965) Thermodynamics and the possibility of spatial interaction in elastic materials, *Arch. Ration. Mech. Anal.* **19**, 339-52.

Gurtin, M.E. and Williams, W.O. (1966) On the inclusion of the complete symmetry group in unimodular group, *Arch. Ration. Mech. Anal.* **23**, 163-72.

Gurtin, M.E. and Williams, W.O. (1967) *An axiomatic foundation for continuum thermodynamics*, Arch. Rational Mech. Anal. **26**, 83-117.

Haddad, Y. M. (1995) *Viscoelasticity of Engineering Materials*, Kluwer, Dordrecht.

Haddad, Y. M. (1998) On the stochastic micromechanical approach to the response behaviour of engineering materials, in *Multilayered and Fibre-Reinforced Composites: Problems and Prospects*, NATO ARW, Kiev, June 2-6, 1997, edited by Y. M. Haddad, *Kluwer*, Dordrecht, pp. 99-110.

Hunter, S.C. (1976) *Mechanics of Continuous Media*, Ellis Hordwood, Chister, England.

Kac, M. (1969) Some Mathematical Models in Science, *Science* **166**, 695-9.

Kestin, J. (1966) *A Course in Thermodynamics, Volumes I and II*, Hemisphere Publishing Corp. and McGraw-Hill Book Company, New.York.

Ladevèse, P. (ed.) (1985) *Local Effects in the Analysis of Structures*, Elsevier, NewYork.

Laws, N. (1967) On the thermodynamics of certain materials with memory, *Int. J. Eng. Science* **5**, 427-34.

Meixner, J. (1969) Processes in simple thermodynamic materials, *Arch. Ration. Mech. Anal.* **33**, 33-53.

Müller, R. (1964) Des Einfluss der Messlänge auf die Ergebnisse bei Dehnmessungen an Beton, beton 14, 205-8.

Müller, I. (1967) On the entropy inequality, *Arch. Ration. Mech. Anal.* **26**, 118-41.

Noll, W. (1958) A mathematical theory of the mechanical behaviour of continuous media, *Arch. Ration. Mech. Anal.* **2**, 197-226.

Owen, D.R. (1968) Thermodynamics of materials with elastic range, *Arch. Ration. Mech. Anal.* **31**, 91-112.

Owen, D.R. (1970) A mechanical theory of materials with elastic range, *Arch. Ration. Mech. Anal.* **37**, 85-110.

Pindera, J. T. (1981) Foundations of Experimental Mechanics: Principles of Modelling, Observation and Experimentation, in J. T. Pindera (ed.), *New Physical Trends in Experimental Mechanics*, International Centre for Mechanical Sciences, Udine, and Springer-Verlag, Wien, pp. 188-236.

Pindera, J. T. (ed.) (1984) *Modelling problems in Crack Tip Mechanics*, Martinus Nijhoff Publishers, Dordrecht.

Pindera, J. T. (1987a) Advanced Experimental Mechanics in Modern Engineering Science and Technology, *Transactions of the CSME* **11**, 125-38.

Pindera, J. T. (1987b) Advanced Experimental Mechanics and its Components: Theoretical, Physical, Analytical and Social Aspects, in: *Developments in Engineering Mechanics*, A. P. S. Selvadurai (ed.), Elsevier, New York, pp. 367-414.

Pindera, J. T. (1988) Local Effects - A Major Problem of Contemporary Stress/Strength Analysis of Homogeneous and Composite Structures, in: *Analytical and Testing Methodologies for Design with Advanced Materials*, G. C. Sih, J. T. Pindera and S. V. Hoa (eds.), North-Holland, Amsterdam, pp. 9-55.

Pindera, J. T. (1989) Local effects and defect criticality in homogeneous and laminated structures, *Trans. ASME, J. Pressure Vessel Technology* **111**, 136-50.

Pindera, J. T. (1992) Actual three-dimensional stresses and related dynamic fractures in some adhesively bonded structures, in: *Composite Structures and Materials*, S. V. Hoa and R. Gauvin (eds.), Elsevier Applied Science, London and New York, pp. 332-40.

Pindera, J. T. (1998a) Actual three-dimensional stresses in composite structures and in-local effects in homogeneous structures. Case studies, in: *Multilayered and Fibre-Reinforced Composites: Problems and Prospects*, NATO ARW, Kiev, June 2-6, 1997, edited by Y. M. Haddad, *Kluwer*, Dordrecht, pp. 57-84.

Pindera, J. T. (1998b) Principles and approaches of advanced experimental mechanics in service of modern technology, in: *Multilayered and Fibre-Reinforced Composites: Problems and Prospects*, NATO ARW, Kiev, June 2-6, 1997, edited by Y. M. Haddad, Kluwer, Dordrecht, pp. 13-56.

Pindera, J. T. and Krasnowski, B. R. (1982) Determination of Stress Intensity Factors in Thin and Thick Plate Using Isodyne Photoelasticity, in: *Fracture Problems and Solutions in the Energy Industry*, L. A. Simpson (ed.), Pergamon Press, pp. 147-56.

Pindera, J. T. and Liu, X. (1992) On the actual three-dimensional stresses in notches and cracks, *Composites Engineering* 1, 281-301.

Pindera, J. T. and Pindera, M.-J. (1989) *Isodyne Stress Analysis*, Kluwer, Dordrecht.

Popper, K. R. (1968) *The Logic of Scientific Discovery*, Harper and Row, New York.

Prager, W. (1968) Closing comments by session chairmen, in: *Mechanical Behavior of Materials under Dynamic Loads*, edited by U. S. Lindholm, Symposium held in San Antonio, Texas, Sept. 6-8, 1967, Springer-Verlag New York Inc., New York, pp. 403-19.

Rivlin, R.S. (1975) The thermodynamics of materials with fading memory, in: *Theoretical Rheology*, J.F. Hutton, J.R.A. Pearson & K. Walters (Eds.), Applied Science Publishers, London, 83-103.

Schapery, R. A. (1964) Application of thermodynamics to thermomechanical, fracture, and birefringent phenomena in viscoelastic media, *J. appl. Physics* 35(5), 1451-65.

Sokolnikoff, I. S. (1956) *Mathematical Theory of Elasticity*, McGraw-Hill Book Co., New York.

Timoshenko, S. P. and Goodier, J. N. (1970) *Theory of Elasticity*, McGraw-Hill Book Company, New York.

Truesdell, C. (1951) A new definition of a fluid. II. The Maxwellian fluid, *J. Math. Pures Appl.* 30, 111-58.

Truesdell, C. and Toupin, R.A. (1960) Classical field theories, in: *Handbuch der Physik*, Ed. S. Flügge, Vol. III/1. Springer, Berlin, 226.

Wang, C.C. and Bowen, R.M. (1966) On the thermodynamics of non-linear materials with quasi-elastic response, *Arch. Ration. Mech. Anal.* 22, 79-99.

Wilson, A.H. (1957) *Thermodynamics and Statistical Mechanics*, Cambridge Univ. Press, Cambridge.

Życzkowski, M. (1981) *Combined Loading in the Theory of Plasticity*, PWN - Polish Scientific Publishers, Warszawa.

5.6. Further Reading

Coleman, B.D. (1963) The thermodynamics of elastic materials with heat conduction and viscosity, *Arch. Ration. Mech. Anal.* 13, 167-78.

Coleman, B.D. and Gurtin, M.E. (1967) Equipresence and constitutive equations for rigid heat conductors, *Z. Angew Math. Phys.* 18, 199-208.

Crochet, M.J. and Naghdi, P.M. (1969) A class of simple solids with fading memory, *Int. J. Eng. Sci.* 7, 1173-98.

Crochet, M.J. and Naghdi, P.M. (1979) On 'thermo-rheologically simple' solids, *Proc. IUTAM Symp., Thermoelasticity, Springer*, New York, 59-86.

Fung, Y.C. (1965). *Foundations of Solid Mechanics*, Prentice-Hall, Englewood Cliffs, N.J., pp. 377-446.

Green, A.E. and Laws, N. (1967) On the formulation of constitutive equations in thermomechanical theories of continua, *Quart. J. Mech. Appl. Math.* 20, 265-75.

Landel, R.F. and Peng, S.T.T. (1986) Equations of state and constitutive equations, *J. Rheology* 30(4), 741-65.

Schapery, R.A. (1969) On a thermodynamic constitutive theory and its application to various nonlinear materials, *Proc. IUTAM Symp. Thermoinelasticity*, Springer, Berlin, pp. 259-85.

CHAPTER 6

ELASTIC RESPONSE BEHAVIOUR

6.1. Introduction

In the foregoing chapters, several equations have been developed which hold for the deformation or motion of thermomechanical continua under any system of externally applied loads. These equations may be grouped as follows:

Continuity: $\dfrac{\partial \rho}{\partial t} + (\rho v_i)_{,i} = 0$

Motion: $\sigma_{ji,j} + \chi_i = \rho \dot{v}_i$

Moment of momentum: $\sigma_{ij} = \sigma_{ji}$ $\qquad\qquad$ (6.1)

Energy: $\rho \dot{e} = \rho h + \sigma_{ij} d_{ij} - q_{i,i}$

Clausius-Duhem inequality: $\rho - \dfrac{\rho h}{T} + \dfrac{q_i}{T_{,i}} > 0$

The system of response equations above amounts to eight independent equations. If one assumes that the body forces χ_i and the local heat production h are prescribed, then the number of the unknowns in these equations is, *in general*, nineteen, namely, $\sigma_{ij}, q_i, v_i, \rho, e, t$ and T. Consequently, eleven additional equations would be required to make the *"response problem"* being dealt with determinate. Consequently, the constitutional character of the material is brought into the formulation through appropriate constitutive relations for the particular material under consideration. For the majority of applications of engineering materials, within the context of the *"phenomenological approach"*, these eleven equations comprise six constitutive relations which characterize the mechanical behaviour of the particular class of material of interest, three other constitutive relations in the form of temperature-heat conduction relations, and the remaining required two equations are taken as thermodynamic equations of state (*see Chapter 5*). As emphasized in Chapter 5, since real materials respond in a complex manner under various loading conditions, the constitutive equations would only reflect the response behaviour of *"idealized"* materials or material models and, thus, would have restricted range of applicability.

6.2. Nonlinear Elasticity

Constitutive equations for elastic materials, *in general*, may be written in the following *general* forms (e.g., Sokolnikoff, 1956 and Provan, 1973, 1974)::

$$\sigma_{ij} = X_{K,i} X_{L,j} \Pi_{KL} \mid C_{KL}, T, X_I \mid \tag{6.2}$$

$$q_i = X_{K,i} Q_K \mid C_{KL}, T, X_I \mid \tag{6.3}$$

$$e = e \mid C_{KL}, T, X_I \mid \tag{6.4}$$

$$\psi = \psi \mid C_{KL}, T, X_I \mid \tag{6.5}$$

where $\Pi_{KL} = x_{i,K} x_{i,L} \sigma_{ij}$, and $Q_K = x_{k,K} q_k$.

Consider now the Clausius-Duhem inequality as per the form shown by (6.1); namely,

$$\rho - \frac{\rho h}{T} + \frac{q_i}{T_{,i}} > 0$$

and the energy equation, from (6.1),

$$\rho \dot{e} - \rho h - \sigma_{ij} d_{ij} + q_{i,i} = 0$$

The above two equations can be combined to read:

$$\rho T - \rho \dot{e} + \sigma_{ij} d_{ij} - \frac{q_i T_{,i}}{T} > 0 \tag{6.6}$$

which is an alternative form of the Clausius-Duhem inequality. Substituting in (6.6), the expression for the material derivative of internal energy, i.e.,

$$\dot{e} = \frac{\partial e}{\partial C_{KL}} \dot{C}_{KL} + \frac{\partial e}{\partial T} \dot{T}$$

$$= (\frac{\partial}{\partial C_{KL}} \dot{C}_{KL} + \frac{\partial}{\partial T} \dot{T}) \, e \tag{6.7}$$

it follows that

$$\rho \left| \left(T \frac{\partial}{\partial C_{KL}} - \frac{\partial e}{\partial C_{KL}} \right) \dot{C}_{KL} + \left(T \frac{\partial}{\partial T} - \frac{\partial e}{\partial T} \right) \dot{T} \right| + \sigma_{ij} d_{ij} - \frac{q_i T_{,i}}{T} > 0 \qquad (6.8)$$

Recalling, *from Chapter 3*, that

$$\dot{C}_{KL} = 2 d_{ij} x_{i,K} x_{j,L}$$

where d_{ij} is the rate of deformation tensor.

Thus, the Clausius-Duhem inequality, Eqn. (6.8), may be expressed as

$$\left| 2\rho \left(T \frac{\partial}{\partial C_{KL}} - \frac{\partial e}{\partial C_{KL}} \right) x_{i,K} x_{j,L} + \sigma_{ij} \right| d_{ij}$$

$$+ \rho \left(T \frac{\partial}{\partial T} - \frac{\partial e}{\partial T} \right) \dot{T} - \frac{q_i T_{,i}}{T} > 0 \qquad (6.9)$$

In the case of a *"thermodynamic admissible process"*, one may consider that $\dot{T}$ and d_{kl} as completely independent terms. Hence, for a homogeneous temperature distribution, i.e., $T_{,i} = 0$, one has

$$\left| 2\rho \left(T \frac{\partial}{\partial C_{KL}} - \frac{\partial e}{\partial C_{KL}} \right) x_{i,K} x_{j,L} + \sigma_{ij} \right| d_{ij}$$

$$+ \rho \left(T \frac{\partial}{\partial T} - \frac{\partial e}{\partial T} \right) \dot{T} > 0 \qquad (6.10)$$

for all arbitrary d_{ij} and $\dot{T}$. Since strict equality is admissible and d_{ij} and $\dot{T}$ are independent, the coefficients of these two terms, in (6.10) above, must vanish. Thus, it follows that

$$\sigma_{ij} = 2\rho \left(\frac{\partial e}{\partial C_{KL}} - \frac{\partial}{\partial C_{KL}} \right) x_{i,K} x_{j,L} \,,$$

$$T \frac{\partial}{\partial T} - \frac{\partial e}{\partial T} = 0 \qquad (6.11)$$

Substituting the above result in Eqn. (6.9), then,

$$q_i \, T_{,i} > 0$$

A more condensed form of these constitutive equations may be deduced by introducing the specific Helmholtz free energy ψ by employing the transformation:

$$\sigma_{ij} = 2 \rho \, \frac{\partial \psi}{\partial C_{KL}} \, x_{i,K} \, x_{j,L}$$

$$= - \frac{\partial \psi}{\partial T}$$

$$- q_k \, T_{,k} > 0$$

Meanwhile, since $\Pi_{KL} = x_{i,K} \, x_{i,L} \, \sigma_{ij}$, it follows that

$$\Pi_{KL} = 2 \rho \, \frac{\partial \psi}{\partial C_{IJ}} \, C_{KI} \, C_{LJ} \tag{6.12}$$

6.3. Linear Elasticity

In this section, we introduce the equations governing the linear elastic response behavior of a homogeneous isotropic medium under the influence of external loading. First, we deal with the *"linearity approximation"* and with the implications of the axiom of *"material invariance"*. Second, we present a brief summary of the basic field equations. Specific solved problems are, then, dealt with in Section 6.6.

6.3.1. LINEARITY AND MATERIAL INVARIANCE

We reconsider the constitutive equation

$$\sigma_{ij} = 2 \rho \, \frac{\partial \psi}{\partial C_{KL}} \, x_{i,K} \, x_{j,L} \tag{6.13}$$

and make the following modifications (*see Chapter 3*)

$$C_{KL} = 2 \, E_{KL} + \delta_{KL} \, ,$$

and

$$\psi = \psi\,(E_{KL,}\,T,\,X_K)\,,$$

where ρ_o is the initial density of the material. The constitutive equation (6.13) then becomes:

$$\sigma_{ij} = \rho\,\frac{\partial\psi}{\partial E_{KL}}\,x_{i,K}\,x_{j,L} \tag{6.14}$$

Expanding the modified Helmholtz free energy ψ in terms of the strain components $E_{KL,}$ then:

$$\psi = \psi_o + \psi_{KL}\,E_{KL} + \frac{1}{2}\,\psi_{KLMN}\,E_{MN} + \ldots.$$

with the understanding that, in general, the coefficients are functions of the absolute temperature T and the coordinates X_K. One may, also, note the following symmetry conditions on these coefficients brought about by the symmetric nature of the Lagrangian strain tensor and the scalar nature of ψ:

$$\psi_{KL} = \psi_{LK}\,,$$

$$\psi_{KLMN} = \psi_{LKMN} = \psi_{KLNM} = \psi_{LKNM}$$

$$= \psi_{MNKL} = \psi_{NMKL} = \psi_{MNLK} = \psi_{NMLK}\,.$$

An approximate linear theory is obtained when one makes the following substitutions and approximations:

i)   All terms in the polynomial expansion of ψ which are of higher order than the quadratic are neglected.

ii) $\dfrac{\rho}{\rho_o} \doteq 1 - II_1$

iii) $x_{j,L} = (\,\delta_{ML} + E_{ML} + R_{ML}\,)\,\delta_{Mj}$

iv)   The neutral state of stress, i.e., $\sigma_{ij} = 0$ when $E_{KL} = 0$ implies that $\psi_{KL} = 0$.

v)   Without loss of generality we may choose $\psi_o = 0$.

vi)   All terms higher than the linear ones are to be neglected.

163

Applying these seven conditions to the stress constitutive relation (6.13), the latter may be approximated by

$$\sigma_{ij} = \psi_{LMNK} E_{NK} \delta_{Li} \delta_{Mj}$$

where ψ_{LMNK} satisfies the previously mentioned symmetry conditions. In this context, ψ_{KLMN} has 21 independent terms which may be displayed as follows

$$\psi_{KLMN} \sim \begin{vmatrix} \psi_{1111} & \psi_{1122} & \psi_{1133} & \psi_{1123} & \psi_{1113} & \psi_{1112} \\ & \psi_{2222} & \psi_{2233} & \psi_{2223} & \psi_{2213} & \psi_{2212} \\ & & \psi_{3333} & \psi_{3323} & \psi_{3313} & \psi_{3312} \\ & & & \psi_{2323} & \psi_{2313} & \psi_{2312} \\ & & & & \psi_{1313} & \psi_{1312} \\ & & & & & \psi_{1212} \end{vmatrix}$$

where the unwritten terms are identical to those located symmetrically with respect to the diagonal. Such a material is termed *"general anisotropic"*; e.g., Love (1944).

Following the *"axiom of material invariance"*, the material coefficients for a homogeneous solid must remain invariant under the transformation

$$X_J' = A_{IJ} X_I,$$

where,

$$A_{IJ} A_{KJ} = A_{JI} A_{JK} = \delta_{IK}; \quad |A_{IJ}| = \pm 1 .$$

Accordingly, the Helmholtz free energy must satisfy the relationship

$$\psi = \frac{1}{2} \psi_{KLMN}' E_{KL}' E_{MN}'$$

On the other hand, one has

$$E_{KL} = A_{PK} A_{QL} E_{PQ}$$

which shows that ψ_{KLMN} must transform, following Cartesian tensor rules, according to

$$\psi'_{PQRS} = A_{KP} A_{LQ} A_{MR} A_{NS} \psi_{KLMN}$$

In order to illustrate briefly some of the consequences of various material symmetries, we consider the following situations:

i)   *Symmetry with respect to a plane*
     Assuming that the material is symmetric with respect to the plane $X_3 = 0$, for example, then the material coefficients must remain invariant when we replace:

$$X'_1 = X_1 \quad ; \quad X'_2 = X_2 \quad ; \quad X'_3 = - X_3$$

This implies that

$$A_{KL} = \begin{vmatrix} 1 & 0 & 0 \\ 0 & 1 & 0 \\ 0 & 0 & -1 \end{vmatrix}.$$

Under this particular transformation, all coefficients of ψ_{KLMN} in which the index 3 occurs an odd number of times are zero. Accordingly, under the condition of symmetry with respect to one plane, only 13 material constants would remain.

ii)   *Symmetry with respect to two orthogonal planes (Orthotropic)*
     If, in addition to the above symmetry condition, we also have symmetry about, say, the $X_1 = 0$ plane, then all coefficients, of the thirteen remaining ones, in wich the index 1 occurs an odd number of times are zero. Then, nine independent coefficients would remain. In reviewing these nine constants, it is observed that the index 2 does not occur an odd number of times and, hence, the material is also symmetric with respect to the $X_2 = 0$ plane.

iii)   *Hexagonal Symmetry*
     A material is said to have hexagonal symmetry. If upon an arbitrary rotation about one of the axes, X_3, ψ remains invariant. For this particular example of rotation about the X_3 - axis, one has:

$$A_{IJ} = \begin{vmatrix} \cos \alpha & \sin \alpha & 0 \\ -\sin \alpha & \cos \alpha & 0 \\ 0 & 0 & 1 \end{vmatrix}.$$

Under this transformation it may be shown that:

$$\psi_{1133} = \psi_{2233} \quad ; \quad \psi_{1111} = \psi_{2222} \quad ; \quad \psi_{1313} = \psi_{2323}$$

and,

$$\psi_{1212} = \frac{1}{2} \left(\psi_{1111} - \psi_{1122} \right)$$

which bring the number of independent coefficients to 5.

iv) *Isotropic Symmetry*
An isotropic material possesses no preferred direction which implies that ψ must remain invariant with respect to the full orthogonal group A_{IJ}. This may be obtained by superimposing on the above notation a further rotation about another orthogonal axis. Under this symmetry condition, one obtains

$$\psi_{1111} = \psi_{3333} \quad ; \quad \psi_{1313} = \psi_{1212} = \mu,$$

and:

$$\psi_{1133} = \psi_{1122} = \lambda$$

Hence, we may write the components for the completely isotropic ψ_{KLMN} in the form:

$$\psi_{KLMN} = \begin{vmatrix} \lambda + 2\mu & \lambda & \lambda & 0 & 0 & 0 \\ & \lambda + 2\mu & \lambda & 0 & 0 & 0 \\ & & \lambda + 2\mu & 0 & 0 & 0 \\ & & & \mu & 0 & 0 \\ & & & & \mu & 0 \\ & & & & & \mu \end{vmatrix}$$

where λ and μ are the "*Lamé constants*". Hence, we may write

$$\psi_{KLMN} = \lambda \, \delta_{KL} \, \delta_{MN} + \mu \left(\delta_{KM} \, \delta_{LN} + \delta_{KN} \, \delta_{LM} \right).$$

Further, one recalls that

$$E_{MN} = \varepsilon_{mn} \, \delta_{Mm} \, \delta_{Nn},$$

and, hence, we may set,

$$\sigma_{ij} = a_{ijk\ell} \, \varepsilon_{k\ell}$$

where,

$$a_{ijk\ell} = \Psi_{KLMN} \, \delta_{Ki} \, \delta_{Lj} \, \delta_{Mk} \, \delta_{N\ell} \, .$$

For a "*completely isotropic linear elastic*" material, one also has

$$a_{ijk\ell} = \lambda \, \delta_{ij} \, \delta_{k\ell} + \mu \, (\, \delta_{i\ell} \, \delta_{jk} + \delta_{ik} \, \delta_{j\ell} \,)$$

Thus, the constitutive relation is expressed as

$$\sigma_{ij} = \lambda \, \varepsilon_{kk} \, \delta_{ij} + 2 \, \mu \, \varepsilon_{ij} \tag{6.15a}$$

This relation can easily be inverted by substituting

$$\sigma_{ii} = (\, 3 \, \lambda + 2 \, \mu \,) \, \varepsilon_{ii} \, ,$$

$$\varepsilon_{kk} = \frac{1}{3 \, \lambda + 2 \, \mu} \, \sigma_{kk}$$

in equation (6.15a). Thus,

$$\varepsilon_{ij} = \frac{1}{2 \mu} \, \sigma_{ij} - \frac{\lambda}{2 \, \mu \, (\, 3 \, \lambda + 2 \, \mu \,)} \, \sigma_{kk} \, \delta_{ij}$$

In linear elasticity, the following relations between pertaining material parameters are common,

$$2 \, \mu = \frac{E}{1 + \nu} \quad ; \, \lambda = \frac{E \nu}{(\, 1 + \nu \,) (\, 1 - 2 \nu \,)}$$

$$E = \frac{\mu \, (\, 3 \, \lambda + 2 \, \mu \,)}{\lambda + \mu} \quad ; \, \nu = \frac{\lambda}{2 \, (\, \lambda + \mu \,)}$$

whereas E is the Young's modulus and ν is Poisson's ratio. Thus, the constitutive relation (6.15a) may be, also, expressed in the form

$$\varepsilon_{ij} = \frac{1 + \nu}{E} \, \sigma_{ij} - \frac{\nu}{E} \, \sigma_{kk} \, \delta_{ij} \qquad (6.15b)$$

6.3.2. BASIC EQUATIONS OF ELASTICITY

Following our earlier presentation, the basic field equations governing the linear deformation of an *isotropic elastic* medium are presented below.

We first considered in Chapter 2 the definitions of the *"surface force vector"* and the *"body force vector"*. These, from a continuum mechanics point of view, were presented, respectively, by

$$T(n) = \lim_{\Delta S \to 0} \frac{\Delta f}{\Delta S} \qquad (6.16)$$

and,

$$\chi = \lim_{\Delta V \to 0} \frac{\Delta B}{\Delta V} \qquad (6.17)$$

Also, we arrived at the following relation between the stress vector and stress tensor:

$$T_i = \sigma_{ji} \, n_j \qquad (6.18)$$

Further, the equations of motion is expressed as

$$\sigma_{ji,j} + \chi_i = \rho \, \dot{v}_i \qquad (6.19)$$

which is called the *"First Cauchy equation of motion"*. Meantime, the *"Second Cauchy equation of motion"*, being derived (*Chapter 2*) by applying the principle of conservation of angular momentum to an arbitrary volume V, is written as

$$\sigma_{ij} = \sigma_{ji} \qquad (6.20)$$

which, as previously mentioned in Chapter 2, is referred-to as the *"stress boundary conditions"* equation.

A stress field which satisfies the equilibrium form of (6.19) and the boundary conditions (6.20) is referred to as *"statically admissible"* or *"statically consistent"*.

In case of an infinitesimal deformation, the small or linear strain tensor is derived in terms of the displacement-gradient components (*Chapter 3*) as

$$\varepsilon_{ij} = \frac{1}{2}\left(u_{i,j} + u_{j,i}\right) \tag{6.21}$$

Meantime, recalling (6.15), the constitutive equations governing the response of a homogeneous, isotropic linear elastic material can be written as

$$\sigma_{ij} = \lambda\,\varepsilon_{kk}\,\delta_{ij} + 2\mu\,\varepsilon_{ij}\,, \tag{6.22a}$$

and,

$$\varepsilon_{ij} = \frac{1+\nu}{E}\,\sigma_{ij} - \frac{\nu}{E}\,\sigma_{kk}\,\delta_{ij}\,. \tag{6.22b}$$

whereby one is the inverse of the other. As identified earlier, λ and μ, in (6.22), are the *lamé constants* whereas E is *Young's modulus* and ν is *Poisson's ratio*. As noted latter in this section, The *shear modulus* G, and the *bulk modulus* K, can be expressed in terms of these four coefficients.

The stress-strain relationship (6.22) can be expressed in component form as:

$$\sigma_{11} = \lambda\,\Delta + 2\mu\,\varepsilon_{11}\,,\ \sigma_{22} = \lambda\,\Delta + 2\,\mu\,\varepsilon_{22}\,,\ \sigma_{33} = \lambda\,\Delta + 2\,\mu\,\varepsilon_{33}\,,$$

$$\sigma_{23} = \mu\,\varepsilon_{23}\,,\ \sigma_{31} = \mu\,\varepsilon_{31}\,,\ \sigma_{12} = \mu\,\varepsilon_{12} \tag{6.22c}$$

where $\Delta = \varepsilon_{11} + \varepsilon_{22} + \varepsilon_{33}$ is the "*dilatation*" wich represents, as noted in Chapter 3, the change in volume of unit cube.

Within the framework of the general theory of elasticity, four elastic constants are normally used. These are: *Young's modulus* E, *Poisson's ratio* ν, the *bulk modulus* K, and the *rigidity (sheer) modulus* which is identical with Lamé constant μ:

E, *Young's modulus*. It is defined as the ratio between the applied stress and the fractional extension, when a cylindrical or prismatic specimen is subjected to a uniform stress over its plane ends, while its lateral surfaces are free form constraint, e.g., $E = \sigma_{11} / \varepsilon_{11}$.

ν, *Poisson's ratio*. It is defined as the "*absolute*" numerical value of the ratio between the lateral contraction and the longitudinal extension of the specimen, with the lateral surfaces again are free, i.e., $\nu = \left|\varepsilon_{22} / \varepsilon_{11}\right|$.

K, *bulk modulus*. It is defined as the ratio between the applied pressure, p, and the fractional change in volume, Δ, when the solid is subjected to uniform hydrostatic compression. Thus,

$$\sigma_{11} = \sigma_{22} = \sigma_{33} = -p$$

and,

$$\sigma_{23} = \sigma_{31} = \sigma_{12} = 0$$

The fractional change in volume is

$$-(\varepsilon_{11} + \varepsilon_{22} + \varepsilon_{33}) = -\Delta$$

so that

$$K = \frac{p}{\Delta} = \lambda + \frac{2\mu}{3}$$

μ, *rigidity (sheer) modulus*. It corresponds to the ratio between the shear stress and the shear strain.

Using equations (6.22c) E, μ and K may be expressed in terms of λ and μ. In this context, the following relations between the constants can be proved:

$$E = \frac{\mu(3\lambda + 2\mu)}{\lambda + \mu}$$

$$\mu = \frac{\lambda}{2(\lambda + \mu)} \qquad\qquad (6.23)$$

$$K = \lambda + \frac{2\mu}{3}$$

The six constitutive equations (6.22c) expressing the generalized Hooke's law for an isotropic linearly elastic material can be expressed explicitly in terms of the strain as follows

$$\varepsilon_{11} = \frac{\sigma_{11}}{E} - \frac{\nu}{E}(\sigma_{22} + \sigma_{33})$$

$$\varepsilon_{22} = \frac{\sigma_{22}}{E} - \frac{\nu}{E}(\sigma_{11} + \sigma_{33})$$

$$\qquad\qquad (6.24)$$

$$\varepsilon_{33} = \frac{\sigma_{33}}{E} - \frac{\nu}{E}(\sigma_{11} + \sigma_{22})$$

$$\varepsilon_{12} = \frac{2\mu_{12}}{\mu}, \; \varepsilon_{23} = \frac{2\sigma_{23}}{\mu}, \; \varepsilon_{31} = 2\frac{\sigma_{31}}{\mu}$$

Some remarks on the constitutive equations (6.24)

– The six constitutive equations (6.24) can be solved simultaneously to express the stress components explicitly in terms of strains.
– Compression normal stresses should be substituted with negative signs in the above equations.
– The positive sense of a shear strain component corresponds to the positive direction of the corresponding shear stress component.
– The three elastic constants, E, ν, and μ are not independent of each other, and that for an isotropic, linear elastic material, there are only *two* constants.
– From the definitions of both *"plane stress"* and '*plane strain"* states, a plane strain response would not correspond to a plane stress input. Similarly, a plane stress response would not result from a plane strain input, in the case of an isotropic linear elastic material, as demonstrated in Table 6.1.

TABLE 6.1. Definitions of "*Plane Stress*" and "*Plane Strain*"

	Plane Stress	Plane Strain
INPUT	$\begin{pmatrix} \sigma_{11} & \sigma_{12} & 0 \\ \sigma_{21} & \sigma_{22} & 0 \\ 0 & 0 & 0 \end{pmatrix}$	$\begin{pmatrix} \varepsilon_{11} & \varepsilon_{12} & 0 \\ \varepsilon_{21} & \varepsilon_{22} & 0 \\ 0 & 0 & 0 \end{pmatrix}$
OUTPUT	$\begin{pmatrix} \varepsilon_{11} & \varepsilon_{12} & 0 \\ \varepsilon_{21} & \varepsilon_{22} & 0 \\ 0 & 0 & \varepsilon_{33} \end{pmatrix}$	$\begin{pmatrix} \sigma_{11} & \sigma_{12} & 0 \\ \sigma_{21} & \sigma_{22} & 0 \\ 0 & 0 & \sigma_{33} \end{pmatrix}$

6.4. Problems

1. Given that:

E	is Young's Modulus	K	is the Bulk Modulus
ν	is Poisson's Ratio	$\left.\begin{array}{l}\lambda\\\mu\end{array}\right\}$	are the Lamé coefficients
G	is the Shear Modulus		

$$\mu = G = \frac{E}{2(1+v)} \quad ; \quad \lambda = \frac{Ev}{(1+v)(1-2v)} \quad ; \quad K = \lambda + \frac{2}{3}\mu$$

show, in any order, that:

$$\text{i) } \lambda = \frac{2\mu v}{1-2v} \quad \text{ii) } G = \frac{3KE}{9K-E} \quad \text{iii) } v = \frac{\lambda}{3K-\lambda}$$

$$\text{iv) } E = \frac{\mu(3\lambda+2\mu)}{\lambda+\mu} \quad \text{v) } K = \frac{E}{3(1-2v)}$$

2. From the basic constitutive relations for isotropic linear elasticity show that the stress invariant $I_1 = \sigma_{ii}$ is related to the strain invariant $II_1 = \varepsilon_{ii}$ by the relation:

$$I_1 = 3 K \, II_1$$

6.5. The Elastic Boundary Value Problem

In classical elasticity where the response behaviour of the material is time-independent, boundary value problems are conventionally classified as **static** or **dynamic** in view of the time-dependency of the boundary conditions. Static problems of elasticity are often classified into the following two categories (e.g., see Fung, 1965, Gakhof, 1966, Fichera, 1972 and Gladwell, 1980):

(i)   Uniform boundary conditions: In this category, either the external loading (stress vector) or, alternatively, the external displacement are specified everywhere on the external boundary.

(ii) Mixed boundary conditions: In this class of problems, the external loading is specified over a part of the boundary while the external displacement is specified over the recurring part.

Due to the time-independency characteristic of the static elastic problem, the boundary conditions as classified by the above two categories are fixed, i.e., they are time-invariant.. In dynamic problems, however, a set of initial conditions on both the components of the external loading and the displacement must be specified in the volume V and over the surface S of the continuous body B.

6.5.1. FORMULATION OF THE ELASTIC BOUNDARY VALUE PROBLEM

In compliance with the principles of continuum mechanics, the motion of a an elastic (continuum) body is generally governed by the laws of conservation of mass and momentum, the stress-strain constitutive relations, the boundary conditions and the initial conditions. As demonstrated in the remainder of this section, the formulation of this set of governing conditions is determined by the type of the boundary value problem considered.

Isothermal, Linear Elastic Boundary Value Problem

In this class of boundary value problem, all the geometrical assumptions of infinitesimal elasticity theory are implied. These would usually include the assumptions of small deformations and small strains, the boundary conditions applied to undisturbed surfaces and the neglect of any convective terms in the acceleration. The governing set of conditions for an isothermal, linear elastic boundary value problem are as follows:

(i) Initial conditions. We assume that the body is initially undisturbed. In other words, it is initially stress free and in mechanical equilibrium. Thus, the initial conditions are,

$$u_i(t) = 0, \quad \varepsilon_{ij}(t) = 0, \quad \sigma_{ij}(t) = 0; \quad -\infty < t < 0 \tag{6.I}$$

where u_i designate the components of the displacement vector in a rectangular Cartesian coordinate system.

(ii) Boundary conditions. The boundary B of the body is considered to be composed of two parts B_σ and B_u. That is

$$B = B_\sigma + B_u$$

where B_σ denotes the part of the boundary of the body over which the components of the stress σ are prescribed; and B_u indicates the remaining part of the boundary over which the components of the displacement u are specified. The boundary conditions may be assigned in the form of:

- traction vector components T_i over B_σ such that

$$\sigma_{ij}(x) n_j = T_i(x); \quad x \quad on \quad B_\sigma \tag{6.IIa}$$

where n_j are the components of the outward unit normal to B_σ, and

- displacement vector components U_i over B_u as

$$u_i(x) = U_i(x); \quad x \quad on \quad B_u \tag{6.IIb}$$

The boundary conditions (6.IIa&b) are assumed to be fixed, that is, both the traction vector components T_i and the displacement vector components U_i are considered to be prescribed for all t.

(iii) Balance of linear momentum. One of the following two situations may be considered:

- *A static problem.* In this case, the equilibrium equation is

$$\sigma_{ij,j} + \chi_i = 0 \qquad (6.III)$$

- *A dynamic problem.* In this, the equation of motion is

$$\sigma_{ij,j} + \chi_i = \rho \, \frac{\partial^2 u_i}{\partial t^2} \qquad (6.IV)$$

where, in (6.III) and (6.IV), χ_i are the body force components per unit volume.

(iv) Linear strain-displacement relations.

$$\varepsilon_{ij} = \frac{1}{2}\left(u_{i,j} + u_{j,i}\right) \qquad (6.V)$$

in which a comma indicates partial differentiation with respect to the coordinates x_i of the material particle.

(v) Stress-strain relations. For the case of isotropic, linear elastic solid, the constitutive relations are:

$$\sigma_{ij} = \lambda \, \varepsilon_{kk} \, \delta_{ij} + 2\mu \, \varepsilon_{ij} \,, \qquad (6.VIa)$$

or, alternatively,

$$\varepsilon_{ij} = \frac{1 + \nu}{E} \, \sigma_{ij} - \frac{\nu}{E} \, \sigma_{kk} \, \delta_{ij} \,. \qquad (6.VIb)$$

whereby one is the inverse of the other. As identified earlier, λ and μ, in (6.VI), are the *lamé constants* whereas E is the *Young's modulus* and ν is *Poisson's ratio*.

6.5.2. UNIQUENESS OF SOLUTION

An important question concerning the solution of a boundary value problem in continuum mechanics is whether the formulated problem has a solution and whether the solution is unique or not (Fung, 1965). On physical grounds, this question may be dealt with by reference to the thermodynamics of the problem involved. On mathematical grounds, however, this question must be answered by the theory of partial differential equations. A satisfactory solution to the problem in hand must comply with both the laws of physics and principles of mathematics. In solving boundary value problems of static equilibrium within classical elasticity, for example, one may proceed in the following sequence: (i) one solves the equations of equilibrium for the stresses σ_{ij}; equation (6.III). (ii) the constitutive response equations (6.VI) are then solved for the strains ϵ_{ij} by using the stress components σ_{ij} obtained from (i). Here, an infinite set of solutions may be found. However, the unique solution would be singled out by employing, for instance, the conditions of compatibility (*Chapter 3*).

6.6. Solved Problems in Linear Elasticity

The number of "exact" solutions to the basic field equations of linear elasticity are few. In this section, we discuss some of these applications. In this context, the definitions of stress vector, body force vector, and symmetry of stress tensor, *as introduced earlier in Chapter 2*, are assumed to apply. In the presentation of the following examples, we follow closely Provan (1973&1974).

6.6.1. DETERMINATION OF THE STRESS DISTRIBUTION IN ROTATING DISCS

The stress distribution in rotating circular discs is of practical importance. If the thickness of the disc is small in comparison to its radius, the variation of the radial and tangential stresses over the thickness may, as a first approximation, be neglected. For a disc rotating at a constant angular velocity ω, the centrifugal force may be considered as a body force acting radially at each point of the disc. Using polar cylindrical coordinates, we set

$$\rho \chi_r = \rho \omega^2 r \quad ; \quad \rho \chi_\theta = 0 \quad ; \quad \rho \chi_z = 0 \tag{6.25}$$

where ρ is the mass density, such that the equilibrium equation for this axisymmetric problem reduces to

$$\frac{d\sigma_{rr}}{dr} + \frac{1}{r}(\sigma_{rr} - \sigma_{\theta\theta}) + \rho \omega^2 r = 0$$

or $\qquad \dfrac{d}{dr}(r\sigma_{rr}) - \sigma_{\theta\theta} + \rho \omega^2 r^2 = 0 \tag{6.26}$

Meantime, the strains in polar cylindrical coordinates reduce to

$$\varepsilon_{ij} = \begin{pmatrix} \dfrac{du_r}{dr} & 0 & 0 \\[3mm] 0 & \dfrac{u_r}{r} & 0 \\[3mm] 0 & 0 & 0 \end{pmatrix} \qquad (6.27)$$

while the constitutive equations become

$$\begin{aligned}
\varepsilon_{\theta\theta} &= -\frac{\nu}{E}(\sigma_{rr} + \sigma_{\theta\theta}) + \frac{(1+\nu)}{E}\sigma_{\theta\theta} \\[2mm]
&= \frac{1}{E}(\sigma_{\theta\theta} - \nu\sigma_{rr})
\end{aligned} \qquad (6.28)$$

In a similar manner:

$$\varepsilon_{rr} = \frac{1}{E}(\sigma_{rr} - \nu\sigma_{\theta\theta}) \qquad (6.29)$$

Analysis
We consider the equilibrium equation (6.26). If we set:

$$r\,\sigma_{rr} = F(r) \quad \text{and} \quad \sigma_{\theta\theta} = \frac{dF(r)}{dr} + \rho\,\omega^2\,r^2 \qquad (6.30)$$

then, the equilibrium equation is identically satisfied. Furthermore, upon considering the strain-displacement field (6.27), the u_r displacement component may be eliminated by setting:

$$\frac{d\varepsilon_{\theta\theta}}{dr} = \frac{d}{dr}\left(\frac{u_r}{r}\right) = \frac{1}{r}\varepsilon_{rr} + \left(-\frac{1}{r^2}\right)u_r = \frac{1}{r}(\varepsilon_{rr} - \varepsilon_{\theta\theta}) \qquad (6.31)$$

Combining (6.28) and (6.31), then,

$$\frac{d}{dr}\left[\frac{1}{E}(\sigma_{\theta\theta} - v\,\sigma_{rr})\right] = \frac{1}{r}\left[\frac{1}{E}(\sigma_{rr}\,v\,\sigma_{\theta\theta} - \sigma_{\theta\theta} + v\,\sigma_{rr})\right]$$

$$\text{thus,}\quad \frac{d\sigma_{\theta\theta}}{dr} - v\frac{d\sigma_{rr}}{dr} = \frac{1}{r}(1 + v)(\sigma_{rr} - \sigma_{\theta\theta})$$

(6.32)

Also, combining equations (6.30) and (6.32), it follows that:

$$\frac{d^2F(r)}{dr^2} + 2\rho\,\omega^2\,r - \frac{v}{r}\frac{dF}{dr} + \frac{vF}{r^2} = \frac{(1 + v)}{r}\left(\frac{F}{r} - \frac{dF}{dr} - \rho\,\omega^2\,r^2\right)$$

$$\text{thus,}\quad r^2\frac{d^2F(r)}{dr^2} + r\frac{dF}{dr} - F = -(3 + v)\rho\,\omega^2\,r^3$$

(6.33)

This is a nonhomogeneous, 2nd order linear ordinary differential equation with variable coefficients.

We consider the homogeneous part of (6.33) and change the variable

$$r = e^t \quad \rightarrow \quad t = \ell n\,r \quad \rightarrow \quad \frac{dt}{dr} = \frac{1}{r}$$

(6.34)

Hence,

$$\frac{dF_h}{dr} = \frac{dF_h}{dt}\frac{dt}{dr} = \frac{1}{r}\frac{dF_h}{dt}$$

(6.35a)

and,

$$\frac{d^2F_h}{dr^2} = \frac{d}{dr}\left(\frac{1}{r}\frac{dF_h}{dt}\right) = -\frac{1}{r^2}\frac{dF_h}{dt} + \frac{1}{r}\frac{d^2F_h}{dt^2}\cdot\frac{1}{r}$$

$$= \frac{1}{r^2}\frac{d^2F_h}{dt^2} - \frac{1}{r^2}\frac{dF_h}{dt}$$

(6.35b)

Inserting (6.35) into the homogeneous part of the differential equation (6.33), it follows that

$$\frac{d^2 F_h}{d t^2} - F_h = 0 \tag{6.36}$$

which is a 2nd order linear ordinary differential equation with constant coefficients.

Let:

$$F_h = e^{mt} \rightarrow \frac{d^2 F_h}{dt^2} = m^2 e^{mt} \tag{6.37}$$

$$e^{mt} (m^2 - 1) = 0 \rightarrow m = \pm 1 \tag{6.38}$$

Then, by combining (6.36) and (6.37), it follows that

$$F_h = A_1 e^t + A_2 e^{-t} \tag{6.39}$$

Recalling the substitution (6.34), namely $t = \ln r$, then (6.39) can be expressed again as

$$F_h = A_1 r + A_2 \frac{1}{r} \tag{6.40}$$

Using the technique of *"variation of parameters"* to obtain the particular integral of the differential equation (6.33), one may suppose that the particular integral to be of the form:

$$F_p = A_1 (r) r + A_2 (r) \frac{1}{r} \tag{6.41}$$

$$\frac{d F_p}{dr} = \frac{d A_1}{dr} r + A_1 + \frac{1}{r} \frac{d A_2}{dr} - \frac{1}{r^2} A_2 \tag{6.42}$$

With reference to (6.42) above, we choose the following relation as an "arbitrary equation" in A_1 and A_2,

$$\frac{dA_1}{dr} r + \frac{1}{r} \frac{dA_2}{dr} = 0 \tag{6.43}$$

then, from (6.42) and (6.43), it follows that

$$\frac{d^2 F_p}{dr^2} = \frac{dA_1}{dr} - \frac{1}{r^2} \frac{dA_2}{dr} + \frac{2}{r^3} A_2 \tag{6.44}$$

Using the substitutions of (6.42) and (6.44) in (6.33), then,

$$r^2 \frac{dA_1}{dr} - \frac{dA_2}{dr} = -(3+v)\rho \omega^2 r^3$$

$$\text{thus,} \quad r^2 \frac{dA_1}{dr} - \frac{dA_2}{dr} = -(3+v)\rho \omega^2 r^3 \tag{6.45}$$

Adding (6.43) and (6.45), then,

$$\frac{dA_1}{dr} = -\frac{(3+v)\rho \omega^2 r}{2} \tag{6.46}$$

$$\text{thus,} \quad A_1 = -\frac{(3+v)\rho \omega^2}{4} r^2 \tag{6.47}$$

Upon subtracting (6.45) from (6.33) it is noted that:

$$\frac{dA_2}{dr} = \frac{(3+v)\rho \omega^2}{2} r^3$$

$$\text{thus,} \quad A_2 = +\frac{(3+v)\rho \omega^2}{8} r^4 \tag{6.48}$$

Hence, by using (6.47) and (6.48), the particular integral (6.41) becomes

$$F_p = -\frac{(3+v)\rho \omega^2}{4} r^3 + \frac{(3+v)\rho \omega^2}{8} r^3 = -\frac{(3+v)\rho \omega^2}{8} r^3 \tag{6.49}$$

which, along with the homogeneous solution (6.40), dictates that:

$$F(r) = A_1 r + A_2 \frac{1}{r} - \frac{(3+v)\rho\omega^2}{8} r^3 \qquad (6.50)$$

Substituting for F(r) in the expressions (6.30) using (6.50), the associated with stresses may be determined as

$$\sigma_{rr} = \frac{F}{r} = A_1 + \frac{A_2}{r^2} - \frac{(3+v)}{8}\rho\omega^2 r^2 \quad ;$$

$$\sigma_{\theta\theta} = A_1 - \frac{A_2}{r^2} - \frac{(1+3v)}{8}\rho\omega^2 r^2 \qquad (6.51)$$

Having obtained these general expressions, (6.51), we now have to specify the boundary conditions. In this context, we consider the following two different boundary value problems:

I) The Solid Disc

Further to our presentation above, since σ_{rr} must be finite at the centre of the disc, A_2 must be zero. Furthermore, the stress boundary conditions at r=a impose that:

$$T_r = \sigma_{rr} n_r + \sigma_{\theta r} n_\theta + \theta_{zr} n_z = 0 \quad ; \quad n_i = (1,0,0) \qquad (6.52)$$

which upon substituting in (6.51) gives

$$\sigma_{rr} = \frac{3+v}{8}\rho\omega^2(a^2 - r^2) \quad ;$$

$$\sigma_{\theta\theta} = \frac{(3+v)}{8}\rho\omega^2 a^2 - \frac{(1+3v)}{8}\rho\omega^2 r^2 \qquad (6.53)$$

Thus, $\sigma_{rr} = 0$ at $r = a$

i.e., $A_1 = \frac{(3+v)}{8}\rho\omega^2 a^2$ and $A_2 = 0$ (6.54)

One notes that the stresses are greatest at the disc centre with

$$\sigma_{rr} = \sigma_{\theta\theta} = \frac{(3+v)}{8} \rho \omega^2 a^2 \qquad \text{at} \quad r = 0 \tag{6.55}$$

The displacement field may be determined by substituting the expressions for the stresses $\sigma_{\theta\theta}$ and σ_{rr}, namely (6.53), back into the constitutive equations (6.28), for example:

$$
\begin{aligned}
u_r = r \, e_{\theta\theta} &= \frac{r}{E} (\sigma_{\theta\theta} - v \, \sigma_{rr}) \\
&= \frac{r \rho \omega^2}{8E} [(3+v) a^2 - (1+3v) r^2 - v(3+v) a^2 \\
&\quad + v(3+v) r^2] \\
&= r \frac{\rho \omega^2}{8E} [(3 - 2v - v^2) a^2 + (v^2 - 1) r^2]
\end{aligned}
\tag{6.56}
$$

Again, one notes that the displacement tends to zero as the radius tends to zero.

II) A Disc with a Stress Free Circular Hole

The stress boundary conditions in this case may be formulated as follows

At the outside $r = a$:

$T_i \sim (0,0,0) \quad n_i \sim (1,0,0)$

then, $\sigma_{rr} = 0$

also, $A_1 = -\dfrac{A_2}{a^2} + \dfrac{3+v}{8} \rho \omega^2 a^2$

At the inside $r = b$:

$T_i \sim (0,0,0) \quad n_i \sim (-1,0,0)$

then $\sigma_{rr} = 0$

also, $A_1 = -\dfrac{A_2}{b^2} + \dfrac{3+v}{8} \rho \omega^2 b^2$

Thus, $\quad A_2 \left(\dfrac{1}{b^2} - \dfrac{1}{a^2} \right) = -\dfrac{3+v}{8} \rho \omega^2 (a^2 - b^2)$

i.e., $\quad A_2 = -\dfrac{3+v}{8} \rho \omega^2 a^2 b^2$

$$\tag{6.57a}$$

$$\text{Accordingly,} \quad A_1 = \frac{3 + v}{8} \rho \omega^2 (a^2 + b^2) \tag{6.57b}$$

Hence, the stresses (6.51) become

$$\sigma_{rr} = \frac{3 + v}{8} \rho \omega^2 \left(a^2 + b^2 - \frac{a^2 b^2}{r^2} - r^2 \right) \quad ;$$

$$\sigma_{\theta\theta} = \frac{3 + v}{8} \rho \omega^2 \left(a^2 + b^2 + \frac{a^2 b^2}{r^2} - \frac{(1 + 3v)}{(3 + v)} r^2 \right) \tag{6.58}$$

The position of maximum "*radial*" stress can be shown to be at $r = \sqrt{ab}$ while the maximum "*hoop*" stress is at $r = b$. Furthermore, the stresses at these positions may easily be shown to be:

$$\sigma_{rr_{max}} = \frac{3 + v}{8} \rho \omega^2 (a - b)^2 \quad ; \quad \sigma_{\theta\theta_{max}} = \frac{3 + v}{8} \rho \omega^2$$

$$\left(2 a^2 + 2 \frac{(1 - v)}{(3 + v)} b^2 \right) \tag{6.59}$$

Thus, as $b \to 0$, it is immediately observed upon comparing the maximum hoop stresses described by (6.58) and (6.59) that

$$\sigma_{\theta\theta} \big|_{hole} = 2 \sigma_{\theta\theta} \big|_{solid} \tag{6.60}$$

The value of 2, appearing in (6.60) above, is referred to as the "*stress concentration factor*" for this particular problem.

6.6.2. THE DEFORMATION OF A LONG ROD STANDING VERTICALLY IN A GRAVITATIONAL FIELD

With reference to Figure 6.1, Let us consider the Cartesian coordinate frame to be:

$$x_i \sim (x_1, x_2, x_3)$$

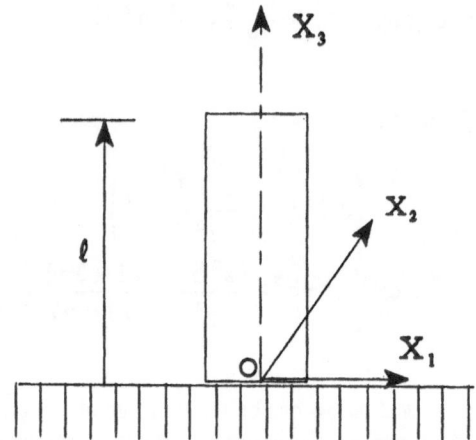

Figure 6.1. A long rod standing vertically in a gravitational field.

The Problem

(i) *Stress Boundary Conditions.* $\qquad T_i = \sigma_{ji} n_j$

- On all the lateral surfaces;

$$T_i \sim (0,0,0) \quad ; \quad n_i \sim (n_1, n_2, 0) \tag{6.61}$$

$$\therefore \ \sigma_{11} n_1 + \sigma_{21} n_2 = 0 \quad ; \quad \sigma_{12} n_1 + \sigma_{22} n_2 = 0 \quad ; \quad \sigma_{13} n_1 + \sigma_{23} n_2 = 0 \tag{6.62}$$

- On the top surface; $x_3 = \ell$

$$T_i \sim (0,0,0) \quad ; \quad n_i \sim (0,0,1)$$

$$\text{Accordingly,} \quad \sigma_{31} = \sigma_{32} = \sigma_{33} = 0 \tag{6.63}$$

- On the bottom surface; $x_3 = 0$

$$T_i \sim (0, 0, +\rho g \ell) \quad ; \quad n_i \sim (0, 0, -1) \tag{6.64}$$

$$\text{then,} \quad \sigma_{31} = \sigma_{32} = 0 \quad ; \quad \sigma_{33} = -\rho g \ell \tag{6.65}$$

(ii)   *Equations of Equilibrium or Motion.* $\sigma_{ji,j} + \rho\,\chi_i = 0$

$$\text{Accordingly,} \quad \begin{aligned} \sigma_{11,1} + \sigma_{21,2} + \sigma_{31,3} &= 0 \\ \sigma_{12,1} + \sigma_{22,2} + \sigma_{32,3} &= 0 \\ \sigma_{13,1} + \sigma_{23,2} + \sigma_{33,3} &= \rho g \end{aligned} \tag{6.66}$$

Thus, the static equilibrium is satisfied.

(iii)   *Strain-Displacement relations.*
We consider the expression of infinitesimal strain, i.e.,

$$\varepsilon_{ij} = \frac{1}{2}\,(u_{i,j} + u_{j,i})$$

$$\tag{6.67}$$

Here, in order to eliminate the possibility of any rigid body motion, we select the condition that:

$$\text{at} \quad x_i = 0 \quad u_i \sim (0,0,0) \tag{6.68}$$

(iv)   *Constitutive Relations.*
Since the stress field will be determined first, the following constitutive equations are utilized;

$$\varepsilon_{ij} = -\frac{\nu}{E}\,\sigma_{kk}\,\delta_{ij} + \frac{1+\nu}{E}\,\sigma_{ij}$$

$$\left.\begin{aligned}
\varepsilon_{11} &= \frac{1}{E}\sigma_{11} - \frac{\nu}{E}(\sigma_{22} + \sigma_{33}) \quad ; \quad & \varepsilon_{12} &= \frac{1+\nu}{E}\sigma_{12} = \frac{1}{2\mu}\sigma_{12} \\
\varepsilon_{22} &= \frac{1}{E}\sigma_{22} - \frac{\nu}{E}(\sigma_{11} + \sigma_{33}) \quad ; \quad & \varepsilon_{13} &= \frac{1+\nu}{E}\sigma_{13} = \frac{1}{2\mu}\sigma_{13} \\
\varepsilon_{33} &= \frac{1}{E}\sigma_{33} - \frac{\nu}{E}(\sigma_{11} + \sigma_{22}) \quad ; \quad & \varepsilon_{23} &= \frac{1+\nu}{E}\sigma_{23} = \frac{1}{2\mu}\sigma_{23}
\end{aligned}\right\} \tag{6.69}$$

Analysis of the Problem
One can easily verify that the following stress field satisfies the equilibrium equations (6.66) and the boundary conditions (6.63) and (6.64):

$$\sigma_{ij} \sim \begin{pmatrix} 0 & 0 & 0 \\ 0 & 0 & 0 \\ 0 & 0 & -\rho g(\ell - x_3) \end{pmatrix} \qquad (6.70)$$

Substituting the stress field (6.70) into the constitutive equations (6.69), it follows that

$$\varepsilon_{ij} \sim \begin{pmatrix} \dfrac{\nu \rho g(\ell - x_3)}{E} & 0 & 0 \\ 0 & \dfrac{\nu \rho g(\ell - x_3)}{E} & 0 \\ 0 & 0 & -\dfrac{\nu \rho g(\ell - x_3)}{E} \end{pmatrix} \qquad (6.71)$$

Substituting the strain field (6.71) into the strain-displacement relations (6.67), then,

$$u_{1,1} = \frac{\nu \rho g(\ell - x_3)}{E} \quad ; \quad u_{2,2} = \frac{\nu \rho g(\ell - x_3)}{E} \quad ; \quad u_{3,3} = -\frac{\nu \rho g(\ell -}{E}$$
$$u_{1,2} + u_{2,1} = 0 \quad ; \quad u_{1,3} + u_{3,1} = 0 \quad ; \quad u_{2,3} + u_{3,2} = \qquad (6.72a\text{-}f)$$

Integrating the first three relations of (6.72), it follows that

$$u_1 = \frac{\nu \rho g}{E}(\ell - x_3) x_1 + f_1(x_2, x_3) \quad ; \quad u_2 = \frac{\nu \rho g}{E}(\ell - x_3) x_2 + f_2(x_1, x_3)$$
$$u_3 = -\frac{\rho g}{E}\left(\ell x_3 - \frac{1}{2} x_3^2\right) + f_3(x_1, x_2) \qquad (6.73)$$

From (6.72d):

$$f_{1,2} + f_{2,1} = 0 \quad \Rightarrow \quad f_{1,2} = 0 \quad ; \quad f_{2,1} = 0$$
$$\text{then,} \quad f_1(x_3) \quad \text{and} \quad f_2(x_3) \qquad (6.74)$$

From (6.72e):

$$-\frac{v\rho g}{E}x_1 + f_{1,3} + f_{3,1} = 0 \rightarrow f_{1,3} = 0 \; ; \; f_{3,1} = \frac{v\rho g}{E}x$$

(6.75a-e)

$$\text{thus,} \quad f_1 = 0; \quad f_3 = \frac{v\rho g}{2E}x_1^2 + h_1(x_2)$$

From (6.72f):

$$-\frac{v\rho g}{E}x_2 + f_{2,3} + f_{3,2} = 0 \rightarrow \quad f_{2,3} = 0 \; ; \; f_{3,2} = \frac{v\rho g}{E}x_2$$

(6.76a-e)

$$\text{i.e.,} \quad f_2 = 0; \quad f_3 = \frac{v\rho g}{2E}x_2^2 + h_2(x_1)$$

Combining (6.75b) with (6.76b), then the general expression for f_3 may be taken as:

$$f_3 = \frac{v\rho g}{2E}(x_1^2 + x_2^2) + C$$

(6.77)

The constant C is evaluated from (6.68) and becomes:

$$C = 0$$

(6.78)

Hence, from (6.73) and (6.75) to (6.78), the displacement field is:

$$u_1 = \frac{v\rho g}{E}(\ell - x_3)x_1$$

$$u_2 = \frac{v\rho g}{E}(\ell - x_3)x_2$$

(6.79)

$$u_3 = -\frac{\rho g}{2E}\left[\ell^2 - (\ell - x_3)^2 - v(x_1^2 + x_2^2)\right]$$

6.6.3. SAINT-VENANT TORSION

The solution of the problem of torsion of prismatic bars by mechanical couples applied at the ends of the bar was first given by Saint-Venant in 1855. In his solution, Saint-Venant began with particular assumptions as to the deformation of the twisted bar and, then, proceeded to illustrate that with these assumptions the equations of equilibrium and the boundary conditions were satisfied. This method is now known as the "*Semi-inverse Method*". The referred-to method asserts that, within the framework of linear, isotropic, homogenous, isothermal, 3-dimensional elasticity, the assumed displacement field is the "exact solution" to the torsion problem; e.g., Sokolnikoff (1956).

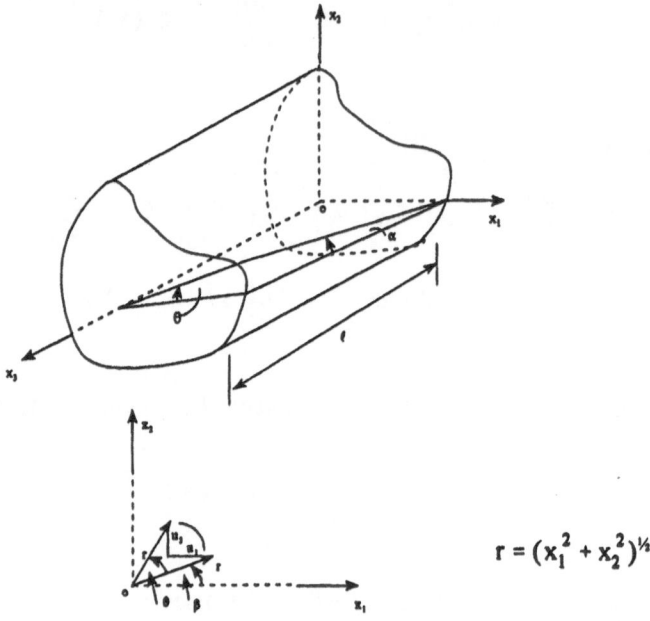

$$r = (x_1^2 + x_2^2)^{1/2}$$

Figure 6.2. Torsion of a prismatic bar.

With reference to Figure 6.2, the displacement components in the plane x_1, x_2 may be chosen as

$$u_1 = r \cos(\beta + \theta) - r \cos \beta \quad ; \quad u_2 = r \sin(\beta + \theta) - r \sin \beta \tag{6.80}$$

As the angle θ is very small, it follows that

$$x_1 \approx r \cos \beta \quad , \quad x_2 \approx r \sin \beta$$
$$\cos \theta \approx 1 \quad , \quad \sin \theta \approx \theta \tag{6.81}$$

and the two displacement components of (6.80) may be written, respectively, as

$$u_1 = -x_2 \theta \quad , \quad u_2 = x_1 \theta \tag{6.82}$$

This is with the understanding that θ is a function of x_3. Hence the torsion angle, α, per unit length is expressed as

$$\alpha = \frac{d\theta}{dx_3} \quad \text{or} \quad \theta = \alpha x_3 \tag{6.83}$$

where $u_\alpha = 0$ at $x_3 = 0$. The above displacement components may be, thus, written as

$$u_1 = -\alpha x_2 x_3 \quad ; \quad u_2 = \alpha x_1 x_3 \tag{6.84}$$

Meanwhile, the displacement component in the x_3 direction may be chosen as

$$u_3 = \alpha \varphi (x_1, x_2) \tag{6.85}$$

where φ is known as the *"warping function"*. Hence, the displacement field may be chosen, Saint-Venant (1855), as:

$$u_i \sim (-\alpha x_2 x_3, \ \alpha x_1 x_3, \ \alpha \varphi) \tag{6.86}$$

and, thus, it remains to illustrate, by using the *semi-inverse method*, that the assumed displacement field (6.86) is the correct one.

Adopting the linear form of the strain-displacement relations, the strain tensor corresponding to the displacements (6.86) is expressed as

$$\varepsilon_{ij} = \frac{1}{2}(u_{i,j} + u_{j,i}) \sim \begin{pmatrix} 0 & 0 & \frac{1}{2}(-\alpha_2 x_2 + \alpha \varphi_{,1}) \\ \cdot & 0 & \frac{1}{2}(\alpha x_1 + \alpha \varphi_{,2}) \\ \cdot & \cdot & 0 \end{pmatrix} \tag{6.87}$$

Substituting these strains into the constitutive relations (6.6.VIa), then,

$$\sigma_{ij} \sim \begin{pmatrix} 0 & 0 & \mu \alpha (\varphi_{,1} - x_2) \\ \cdot & 0 & \mu \alpha (\varphi_{,2} + x_1) \\ \cdot & \cdot & 0 \end{pmatrix} \tag{6.88}$$

Does the obtained stress field (6.88) satisfies the equilibrium equation (6.III)? Since the body and acceleration forces are being neglected, the equilibrium equations, $\sigma_{ij,j} = 0$, becomes:

$$\sigma_{11,1} + \sigma_{12,2} + \sigma_{13,3} = 0$$
$$\sigma_{21,1} + \sigma_{22,2} + \sigma_{23,3} = 0$$
$$\sigma_{31,1} + \sigma_{32,2} + \sigma_{33,3} = 0$$
$$\mu\,\alpha\,(\varphi_{,11} + \varphi_{,22}) = 0$$

i.e., identically satisfied.
i.e., identically satisfied.
becomes:

(6.89)

Accordingly, $\quad \nabla'^2\,\varphi = 0 \quad$ in R $\qquad$ (6.90)

Hence, one reaches the conclusion that the warping function must be harmonic if the displacement field (6.88) is to be the correct one.

The only remaining basic relation is that concerning the stress boundary conditions, namely $T_i = \sigma_{ij}\,n_j$, with $n_j \sim (n_1, n_2, 0)$ everywhere on the lateral surfaces. Hence, on these lateral surfaces:

$$T_1 = \sigma_{11}\,n_1 + \sigma_{12}\,n_2 + \sigma_{13}\,n_3 = 0 \quad ; \quad T_2 = \sigma_{21}\,n_1 + \sigma_{22}\,n_2 + \sigma_{23}\,n_3 = 0 \qquad (6.91)$$

$$T_3 = \sigma_{31}\,n_1 + \sigma_{32}\,n_2 + \sigma_{33}\,n_3 = 0$$
$$= \mu\,\alpha\,(\varphi_{,1} - x_2)\,n_1 + \mu\,\alpha\,(\varphi_{,2} + x_1)\,n_2 = 0 \qquad (6.92)$$

Accordingly, $\quad \varphi_{,1}\,n_1 + \varphi_{,2}\,n_2 = x_2\,n_1 - x_1\,n_2 \quad$ on C. $\qquad$ (6.93)

This set of boundary condition on φ may be further simplified as follows: Letting dn and dt be elemental lengths in the directions of n and t, respectively (*see Figure 6.3*), then,

either:

$$n_1 = \lim_{\Delta n \to 0} \frac{\Delta x_1}{\Delta n} = \frac{dx_1}{dn} = \cos \delta \quad \text{(i)}$$

$$n_2 = \lim_{\Delta n \to 0} \frac{\Delta x_2}{\Delta n} = \frac{dx_2}{dn} = \sin \delta \quad \text{(ii)}$$

or

$$n_1 = \lim_{\Delta t \to 0} \frac{\Delta x_2}{\Delta t} = \frac{dx_2}{dt} = \cos \delta \quad \text{(iii)}$$

$$n_2 = \lim_{\Delta t \to 0} \frac{\Delta x_1}{\Delta t} = \frac{- dx_1}{dt} = - \sin \delta \quad \text{(iv)}$$

$$(6.94)$$

Combining (6.93) and (6.94), it follows that

$$\frac{\partial \varphi}{\partial x_1} \frac{dx_1}{dn} + \frac{\partial \varphi}{\partial x_2} \frac{dx_2}{dn} = x_2 \frac{dx_2}{dt} + x_1 \frac{dx_1}{dt}$$

$$(6.95)$$

$$\text{then,} \quad \frac{d \varphi}{dn} = \frac{1}{2} \frac{dr^2}{dt} \quad \text{on} \quad C$$

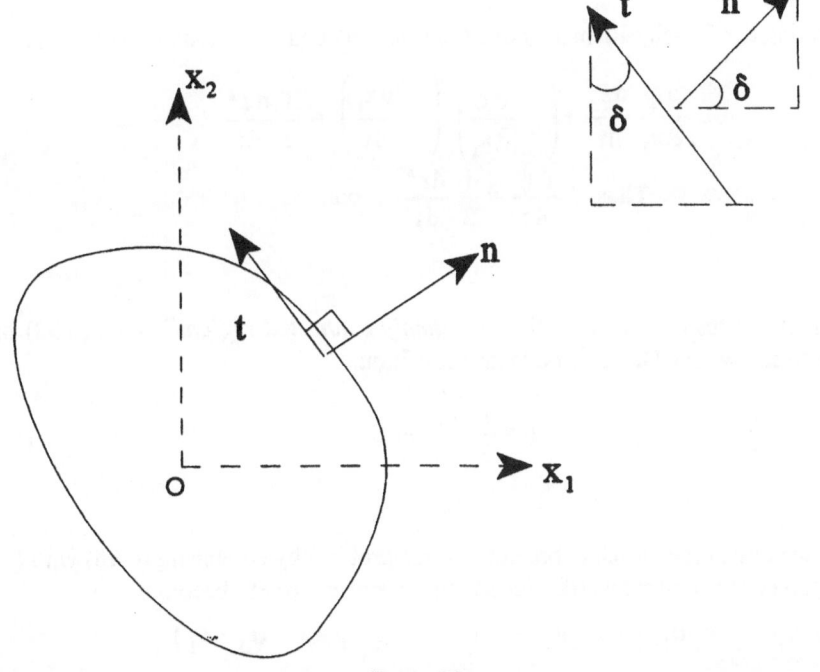

Figure 6.3

Hence the boundary value problem for the determination of the warping function $\varphi(x_1, x_2)$ is expressed by (6.90) and (6.95). Because of the inconvenientboundary conditions expressed in (6.95), however, the problem is transformed into a *"Dirichlet"* type of boundary value problem. A *"Dirichlet Problem"* is one in terms of a harmonic function ψ with ψ itself prescribed on the boundary. This transformation is brought about by introducing the conjugate harmonic function to φ, i.e.:

$$z = \varphi + i\,\psi \tag{6.96}$$

The *"Cauchy-Riemann"* conditions automatically apply:

$$\left.\begin{array}{l} \varphi_{,1} = \psi_{,2} \\ \varphi_{,2} = -\psi_{,1} \end{array}\right\} \quad \varphi_{,\alpha} = \varepsilon_{\alpha\beta}\,\psi_{,\beta} \tag{6.97}$$

where $\varepsilon_{\alpha\beta}$ is the two-dimensional skew-symmetric tensor.

Utilizing these conditions, it is easy to show that the harmonic function ψ satisfies, also, Laplace's equation:

$$\nabla'^2\,\psi = 0 \quad \text{in} \quad R \tag{6.98}$$

Substituting the Cauchy-Riemann conditions into the boundary condition (6.93) results in

$$\frac{\partial\psi}{\partial x_2}\frac{dx_2}{dt} + \left(-\frac{\partial\psi}{\partial x_1}\right)\left(-\frac{dx_1}{dt}\right) = \frac{1}{2}\frac{dr^2}{dt}$$

$$\text{Thus,} \quad \frac{d\psi}{dt} = \frac{1}{2}\frac{dr^2}{dt} \quad \text{on} \quad C \tag{6.99}$$

Further, If we restrict our attention to *"simply connected regions"*, then (6.99) may be integrated to give the Dirichlet boundary condition:

$$\psi = \frac{1}{2}r^2 \quad \text{on} \quad C \tag{6.100}$$

Hence, we obtain the Dirichlet boundary value problem by combining (6.98) with (6.100). It is observed that under this transformation the stresses (6.88) become:

$$\sigma_{13} = \mu\,\alpha\,(\psi_{,2} - x_2) \quad ; \quad \sigma_{23} = \mu\,\alpha\,(-\psi_{,1} + x_1)$$

This boundary value problem is still not in its simplest and most useful form. The most appropriate boundary value problem is the "*Poisson Problem*" obtained by making the substitution:

$$\Upsilon = \psi - \frac{1}{2}(x_1^2 + x_2^2)$$ (6.101)

Accordingly, the boundary conditions (6.99) on C become :

$$\frac{d\Upsilon}{dt} = 0 \quad \text{on} \quad C$$ (6.102)

while the Laplacian operator on ψ dictates that:

$$\nabla'^2\Upsilon = -2 \quad \text{on} \quad R$$ (6.103)

Furthermore, using the substitution (6.101), the stresses become, from (6.88):

$$\sigma_{ij} \sim \begin{pmatrix} 0 & 0 & \mu\,\alpha\,\Upsilon_{,2} \\ \cdot & 0 & -\mu\,\alpha\,\Upsilon_{,1} \\ \cdot & \cdot & 0 \end{pmatrix}$$ (6.104)

The above analysis may be carried out further to include the treatment of, e.g., non-simply connected regions, determination of torsional rigidity and total applied torque, ... We discuss, below, a few illustrative problems.

Torsion of a Circular Cross-section

The governing equations are expressed by the above Poisson boundary value problem. Since we are dealing with a simply connected region of radius a, the boundary conditions (6.102) can be expressed as

$$\Upsilon = 0 \quad \text{on} \quad r = a$$ (6.105)

Consequently, one may choose Υ to be:

$$\Upsilon = k\,(x_1^2 + x_2^2 - a^2)$$ (6.106)

thus; $\Upsilon = 0$ on C and, also, $\nabla'^2\Upsilon = 4k$. This results in:

$$k = -\frac{1}{2}$$

$$\text{then,} \quad \Upsilon = -\frac{1}{2}(x_1^2 + x_2^2) + \frac{1}{2}a \qquad \text{(6.107)}$$

Having found Υ for this boundary value problem, one needs to reverse the two transformations introduced in forming the Poisson problem in order to determine an expression for the warping function φ. First, from (6.101), one can write that

$$\psi = \Upsilon + \frac{1}{2}(x_1^2 + x_2^2) = \frac{1}{2}a^2 = \text{constant} \qquad \text{(6.108)}$$

Second, if a conjugate harmonic is a constant, then, the real term, from the Cauchy Riemann conditions, is also a constant. Accordingly,

$$\varphi = \text{constant} \qquad \text{(6.109)}$$

which indicates that for a circular body, the cross-section does not warp.

From (6.104) and (6.107), the stress tensor can be displayed as:

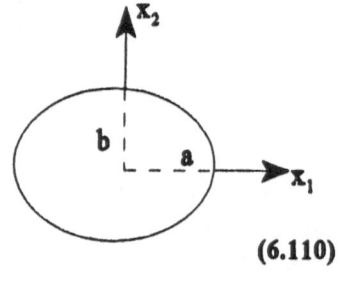

$$\sigma_{ij} \sim \begin{pmatrix} 0 & 0 & -\mu\,\alpha\,x_2 \\ \cdot & 0 & +\mu\,\alpha\,x_1 \\ \cdot & \cdot & 0 \end{pmatrix} \qquad \text{(6.110)}$$

Torsion of an Elliptical Cross-section

We consider an elliptical cross-section with radii a (major) and b (minor); *see sketch below.*

In this case, the governing equations are

$$\nabla'^2 \Upsilon = -2 \quad ; \quad \Upsilon = 0 \quad \text{on C}$$

The equation of an elliptical boundary is:

$$\frac{x_1^2}{a^2} + \frac{x_2^2}{b^2} = 1 \tag{6.111}$$

Hence, we choose,

$$\Upsilon = k\left(\frac{x_1^2}{a^2} + \frac{x_2^2}{b^2} - 1 \right) \tag{6.112}$$

Accordingly, $\Upsilon = 0$ on C, and

$$\nabla'^2 \Upsilon = 2k\left(\frac{1}{a^2} + \frac{1}{b^2} \right) = -2 \quad , \text{ then } \quad k = -\frac{a^2 b^2}{b^2 + a^2} \tag{6.113}$$

$$\text{also,} \quad \Upsilon = \frac{a^2 b^2}{b^2 + a^2}\left(1 - \frac{x_1^2}{a^2} - \frac{x_2^2}{b^2} \right) \tag{6.114}$$

Thus, from (6.104), the stresses are

$$\sigma_{13} = -\frac{2\mu\alpha a^2}{b^2 + a^2} x_2 \quad ; \quad \sigma_{23} = +\frac{2\mu\alpha b^2}{b^2 + a^2} x_1 \tag{6.115}$$

Meanwhile,

$$\begin{aligned}
\psi &= \Upsilon + \frac{1}{2}(x_1^2 + x_2^2) \\
&= \frac{a^2 b^2}{a^2 + b^2}\left(1 - \frac{x_1^2}{a^2} - \frac{x_2^2}{b^2} \right) + \frac{1}{2}(x_1^2 + x_2^2) \\
&= \frac{a^2 b^2}{a^2 + b^2} + \left[\frac{a^2 - b^2}{2(a^2 + b^2)} \right]x_1^2 + \left[\frac{b^2 - a^2}{2(a^2 + b^2)} \right]x_2^2
\end{aligned} \tag{6.116}$$

Further, the warping function can be obtained via the Cauchy-Riemann conditions (6.97), i.e.,

$$\varphi_{,1} = \psi_{,2} = \frac{b^2 - a^2}{a^2 + b^2} x_2$$

$$\text{Then,} \quad \varphi = \frac{b^2 - a^2}{a^2 + b^2} x_1 x_2 + f(x_2) \tag{6.117}$$

However, $\varphi_{,2} = -\psi_{,1}$

$$\text{then,} \quad \frac{b^2 - a^2}{a^2 + b^2} x_1 + f_{,2} = + \frac{b^2 - a^2}{a^2 + b^2} x_1 \tag{6.118}$$

$$\text{thus,} \quad f_{,2} = 0 \quad \rightarrow \quad f = C$$

$$\text{Accordingly,} \quad \varphi = \frac{b^2 - a^2}{a^2 + b^2} x_1 x_2 + C \tag{6.119}$$

Hence, from (6.85), the displacement, apart from a nonessential constant, in the x_3-direction is given by

$$u_3 = \alpha \, \varphi \, (x_1, x_2) = \frac{b^2 - a^2}{a^2 + b^2} \alpha \, x_1 x_2 \tag{6.120}$$

Torsion of an Equilateral Triangle
The boundary line equations shown in the diagram, Figure 6.4, are

$$x_1 - a = 0$$
$$x_1 - \sqrt{3} \, x_2 + 2a = 0 \tag{6.121}$$
$$x_1 + \sqrt{3} \, x_2 + 2a = 0$$

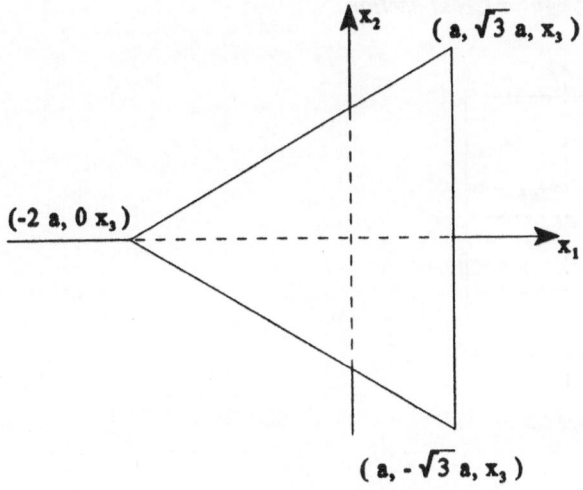

Figure 6.4

Thus, the following expression is valid on any boundary point:

$$(x_1 - a)(x_1 - \sqrt{3}\, x_2 + 2a)(x_1 + \sqrt{3}\, x_2 + 2a) = 0 \qquad (6.122)$$

The student can choose, by analogy to the previous examples:

$$\Upsilon = k\,(x_1^3 - 3\,x_1\,x_2^2 + 3\,a\,[\,x_1^2 + x_2^2\,] - 4a^3) \qquad (6.123)$$

and continue to complete the solution.

196

Torsion of a Rectangular Cross-section

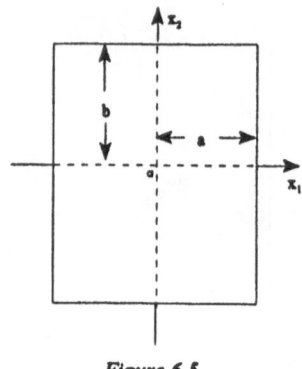

Figure 6.5

The governing equations in this example are:

$$\nabla'^2 \Upsilon = -2 \qquad \text{in} \begin{cases} -a \le x_1 \le a \\ -b \le x_2 \le b \end{cases} \qquad (6.124)$$

$$\Upsilon = 0 \qquad \text{on} \begin{cases} x_1 = \pm a \\ -x_2 = \pm b \end{cases} \qquad (6.125)$$

In this problem, one introduces a slightly different stress function F which is zero on two sides of the rectangle and non-zero on the remaining two, e.g.,

$$F = \Upsilon + x_2^2 - b^2 \qquad (6.126)$$

Thus, one obtains from (6.124) and (6.125) the expression

$$\nabla'^2 F = \nabla'^2 \Upsilon + 2 = 0 \qquad \text{in} \quad R \qquad (6.127)$$

$$\left. \begin{array}{l} F = 0 \quad \text{on} \quad x_2 = \pm b \\ F = x_2^2 - b^2 \quad \text{on} \quad x_1 = \pm a \end{array} \right\} \quad \text{on} \quad C \qquad (6.128)$$

Now, one may solve the homogeneous governing equation (6.127) by the method of *"separation of variables"*. In this context, one may choose:

$$F = X(x_1)\, Y(x_2) \tag{6.129}$$

then, $\nabla^2 F = X'' Y + X Y'' = 0$

thus, $\dfrac{X''}{X} + \dfrac{Y''}{Y} = 0$ $\tag{6.130}$

Thus, since X and Y are functions, only, of x_1 and x_2, respectively, one may set:

$$\frac{X''}{X} = \lambda^2 \qquad \text{and} \qquad \frac{Y''}{Y} = -\lambda^2$$

$$Y'' + \lambda^2 Y = 0 \tag{6.131}$$

thus, $X'' = \lambda^2 X$

$X \sim \{1, x_1, \cosh \lambda x_1, \sinh \lambda x_1\}$ $\qquad X \sim \{1, x_2, \sin \lambda x_2, \cos \lambda x$

Since the torsion of a rectangular section is symmetrical with respect to both x_1 and x_2, one may discard the odd functions of the above sets of solutions to obtain:

$$X = A \cosh \lambda x_1 \qquad Y = B \cos \lambda x_2 \tag{6.132}$$

Substituting these values into (6.129), it follows that

$$F = C \cosh \lambda x_1 \cos \lambda x_2 \tag{6.133}$$

Applying the first boundary condition of (6.128), one arrives from (6.133) at

$$C \cosh \lambda x_1 \cos \lambda b = 0$$

A trivial solution would result in either C or $\cosh \lambda x_1$ being zero. Thus, one is obliged to set:

$$\cos \lambda b = 0 = \cos \frac{(2n-1)\pi}{2} \qquad n = 1, \dots \qquad (6.134)$$

$$\therefore \lambda_n = \frac{(2n-1)\pi}{2b} \qquad (6.135)$$

Thus, a general solution for F may be expressed as:

$$F = \sum_{n=1}^{\infty} C_n \cosh \lambda_n x_1 \cos \lambda_n x_2 \qquad (6.136)$$

Meantime, the second boundary condition in (6.128) dictates that:

$$x_2^2 - b^2 = \sum_{n=1}^{\infty} C_n \cosh \lambda_n a \cos \lambda_n x_2 \qquad (6.137)$$

Applying certain properties of orthogonal functions, namely $\cos \lambda_n x_2$, the constants $C_n \cosh \lambda_n a$, may be found by integrating both sides of the previous equation from $x_2 = 0$ to $x_2 = b$ after multiplying by $\cos \lambda_m x_2$. From a knowledge of Fourier series:

$$\frac{2}{b} \int_0^b (x_2^2 - b^2) \cos \lambda_n x_2 \, dx_2 = \frac{2}{b} C_n \cosh \lambda_n a \int_0^b \cos^2 \lambda_n x_2 \, dx_2$$

$$\text{thus,} \quad \int_0^b x_2^2 \cos \lambda_n x_2 \, dx_2 - b^2 \int_0^b \cos \lambda_n x_2 \, dx_2 = C_n \cosh(\lambda_n a)(b/2) \qquad (6.138)$$

The solution of (6.138) will lead to:

$$\frac{16 b^3 (-1)^n}{\pi^3 (2n-1)^3} = C_n \cosh(\lambda_n a) \cdot \frac{b}{2}$$

$$\text{Thus,} \quad C_n = \frac{32 b^2 (-1)^n}{\pi^3 (2n-1)^3} \cdot \frac{1}{\cosh \lambda_n a} \qquad (6.139)$$

In order to determine the warping function φ from the expression (6.140) of F, one first employs (6.126), i.e.,

$$\Upsilon = F - x_2^2 + b^2$$

i.e.,
$$F = \frac{32b^2}{\pi^3} \sum_{n=1}^{\infty} \frac{(-1)^n}{(2n-1)^3} \cdot \frac{\cosh \lambda_n x_1}{\cosh \lambda_n a} \cos \lambda_n x_2 \qquad (6.140)$$

and, then, find ψ from (6.101), i.e.,

$$\psi = \Upsilon + \frac{1}{2}(x_1^2 + x_2^2) = b^2 + \frac{1}{2}(x_1^2 - x_2^2) + F \qquad (6.141)$$

One is then able to determine φ from ψ, expressed in a combination of (6.141) and (6.140), by utilizing the standard Cauchy-Riemann relations, as

$$\varphi_{,1} = \Upsilon_{,2} = -x_2 - \frac{16b}{\pi^2} \sum_{n=1}^{\infty} \frac{(-1)^n}{(2n-1)^2} \cdot \frac{\cosh \lambda_n x_1}{\cosh \lambda_n a} \sin \lambda_n x_2$$

$$\therefore \varphi = -x_1 x_2 - \frac{32b^2}{\pi^3} \sum_{n=1}^{\infty} \frac{(-1)^n}{(2n-1)^3} \cdot \frac{\sinh \lambda_n x_1}{\cosh \lambda_n a} \sin \lambda_n x_2 + f(x_2) \qquad (6.142)$$

then,
$$\varphi_{,2} = -x_1 - \frac{16b}{\pi^2} \sum_{n=1}^{\infty} \frac{(-1)^n}{(2n-1)^2} \frac{\sinh \lambda_n x_1}{\cosh \lambda_n a} \cos \lambda_n x_2 + f_{,2} \qquad (6.143)$$

Meanwhile, from (6.141) and (6.140), one has

$$\varphi_{,2} = \Upsilon_{,1} = -x_1 - \frac{16b}{\pi^2} \sum_{n=1}^{\infty} \frac{(-1)^n}{(2n-1)^2} \cdot \frac{\sinh \lambda_n x_1}{\cosh \lambda_n a} \cos \lambda_n x_2 \qquad (6.144)$$

then,
$$f_{,2} = 0 \quad \rightarrow \quad f = C. \qquad (6.145)$$

Accordingly, it follows, from (6.142) and (6.145), that

$$\varphi = -x_1 x_2 - \frac{32 b^2}{\pi^3} \sum_{n=1}^{\infty} \frac{(-1)^n}{(2n-1)^3} \cdot \frac{\sinh \lambda_n x_1}{\cosh \lambda_n a} \sin \lambda_n x_2 + C. \tag{6.146}$$

Thus, the axial component of displacement $u_3 = \alpha \varphi$ can be determined from (6.85). Meanwhile, the stresses would be obtained from any of (6.88), or (6.104). Accordingly, the only non-zero components of the stress tensor σ_{ij} are:

$$\sigma_{13} = \sigma_{31} = -\mu \alpha \left\{ 2 x_2 + \frac{16b}{\pi^2} \sum_{n=1}^{\infty} \frac{(-1)^n}{(2n-1)^2} \cdot \frac{\cosh \lambda_n x_1}{\cosh \lambda_n a} \sin \lambda_n x_2 \right\} \tag{6.147}$$

$$\sigma_{23} = \sigma_{32} = -\frac{16 b \mu d}{\pi^2} \sum_{n=1}^{\infty} \frac{(-1)^n}{(2n-1)^2} \cdot \frac{\sinh \lambda_n x_1}{\cosh \lambda_n a} \cos \lambda_n x_2$$

$$\tag{6.148}$$

6.6.4. LAMÉ SPHERICAL PROBLEM

In this section, we deal with the Lamé spherical problem. In this context, we utilize the *Navier displacement equation of equilibrium*. In the absence of body forces, this equation can be expressed in a vectorial form as follows:

$$\mu \nabla^2 u + (\lambda + \mu) \nabla (\nabla \cdot u) = 0 \tag{6.149}$$

If we consider the internal and external radii of a pressurized spherical cavity to be $r_1 = b$ and $r_2 = a$, respectively, and the internal and external pressures to be p_i and p_e respectively, then the displacement vector u is everywhere radial and is a function of the radius r only.

Recalling the identity $\nabla \times (\nabla \times u) = -\nabla^2 u + \nabla (\nabla \cdot u)$ and since $\nabla \times u = 0$, it follows from (6.149) that:

$$\nabla^2 u = \nabla (\nabla \cdot u) = 0 \tag{6.150}$$

One can, thus, write that:

$$\nabla \cdot u = \text{Constant} = 3A \tag{6.151}$$

In spherical coordinates, since there is only dependence on r, div. u reduces to:

$$\frac{1}{r^2}\frac{d}{dr}(r^2 u_r) = 3 A$$

$$\text{then,} \quad r^2 u_r = A r^3 + B \qquad (6.152)$$

$$\text{i.e.,} \quad u_r = A r + \frac{B}{r^2}$$

where A and B are constants to be determined.

Accordingly, the strain tensor components can be displayed, in spherical coordinates, as:

$$\varepsilon_{ij} = \begin{pmatrix} \dfrac{du_r}{dr} = A - \dfrac{2B}{r^3} & , & 0 , & 0 \\[2mm] \cdot & \dfrac{u_r}{r} = A + \dfrac{B}{r^3} & , & 0 \\[2mm] \cdot & \cdot & \dfrac{u_r}{r} = A + \dfrac{B}{r^3} \end{pmatrix} \qquad (6.153)$$

Meanwhile, the corresponding stress components are

$$\sigma_{ij} = 3\lambda A \delta_{ij} + 2\mu \varepsilon_{ij} = \begin{pmatrix} (3\lambda A + 2\mu)A - \dfrac{\lambda B}{r^3} , & 0 , & 0 , \\[2mm] \cdot , & (3\lambda + 2\mu)A + \dfrac{2\mu B}{r^3} , & 0 , \\[2mm] \cdot , & \cdot , & (3\lambda + 2\mu)A + \dfrac{2\mu B}{r^3} \end{pmatrix}$$

$$(6.154)$$

Meanwhile, the constants A and B can be determined from the stress boundary conditions. That is:

At $r = b$; $T_{i|_b} = (p_i, 0, 0)$; $n_{i|_b} = (-1, 0, 0)$.

From the relation $T_i = \sigma_{ji} n_j$, it can be shown that.

$$0\sigma_{rr}|_b = -p_i = (3\lambda + 2\mu)A - \frac{4\mu B}{b^3} \tag{6.155}$$

At $r = a$; $\quad T_i|_a = (-p_o, 0, 0)$; $\quad n_i|_a = (1, 0, 0)$.

$$\therefore \sigma_{rr}|_a = -p_o = (3\lambda + 2\mu)A - \frac{4\mu B}{a^3} \tag{6.156}$$

Meantime, subtracting (6.155) from (6.156), it follows that

$$p_i - p_o = 4\mu B\left(\frac{1}{b^3} - \frac{1}{a^3}\right)$$

i.e., $\quad B = \frac{b^3 a^3}{a^3 - b^3} \frac{p_i - p_o}{4\mu}$ $\tag{6.157}$

Combining (6.157) and (6.155) [or (6.156)], it follows that:

$$-p_i = (3\lambda + 2\mu)A - \frac{a^3}{a^3 - b^3}(p_i - p_o)$$

thus, $\quad A = \frac{p_i b^3 - p_o a^3}{(a^3 - b^3)(3\lambda + 2\mu)}$ $\tag{6.158}$

Meanwhile, substituting A and B from (6.158) and (6.157) into (6.152), the radial displacement u_r can be expressed as

$$u_r = \frac{(p_i b^3 - p_o a^3)}{(a^3 - b^3)(3\lambda + 2\mu)} r + \frac{b^3 a^3 (p_i - p_o)}{(a^3 - b^3) 4\mu r^2} \tag{6.159}$$

Now, substituting (6.157) and (6.158) into (6.154), the following stress field is determined:

$$\sigma_{rr} = \frac{p_i b^3 - p_o a^3}{a^3 - b^3} - \frac{b^3 a^3 (p_i - p_o)}{(a^3 - b^3) r^3} \tag{6.160}$$

$$\sigma_{\varphi\varphi} = \sigma_{\theta\theta} = \frac{p_i b^3 - p_o a^3}{a^3 - b^3} + \frac{b^3 a^3 (p_i - p_o)}{2(a^3 - b^3) r^3} \tag{6.161}$$

Special Cases.

i) $p_i = p$, $p_o = 0$;

then, $\sigma_{rr} = \dfrac{pb^3}{a^3 - b^3}\left(1 - \dfrac{a^3}{r^3}\right)$; $\sigma_{\theta\theta} = \sigma_{\varphi\varphi} = \dfrac{pb^3}{a^3 - b^3}\left(1 + \dfrac{a^3}{2\,r^3}\right)$ (6.162)

ii) $h = a - b \ll r$;

it follows that $u_r = \dfrac{pr^2}{h}\dfrac{(4\mu + 1)}{(3\lambda + 2\mu)\,4\mu}$; $\hat{\sigma}_{rr} = \dfrac{1}{2}p$; $\sigma_{\theta\theta} = \sigma_{\varphi\varphi} = \dfrac{pr}{2h}$ (6.163)

where $\hat{\sigma}_{rr}$ is the mean value of the radial stress over the thickness of the shell.

iii) $p_o = p$ at $a = \infty$;

In this case, one arrives at the stress distribution in an infinite elastic medium with a spherical cavity of radius b subjected to a hydrostatic compression $p_o = p$ at $a = \infty$. That is

$$\sigma_{rr} = -p\left(1 - \dfrac{b^3}{r^3}\right) \quad ; \quad \sigma_{\theta\theta} = \theta_{\varphi\varphi} = -p\left(1 + \dfrac{b^3}{2\,r^3}\right) \qquad (6.164)$$

6.7. References

Fichera, G. (1972) Boundary value problems of elasticity with unilateral constraints, in: *Encyclopedia of Physics*, Vol. VI a/2: Mechanics of Solids II, Ed. C. Truesdell, Springer-Verlag, Berlin, pp. 391-423.

Fung, Y.C. (1965) *Foundations of Solid Mechanics*, Prentice-Hall, Englewood Cliffs, New Jersey.

Gakhof, F.D. (1966) *Boundary Value Problems*, Pergamon, Oxford.

Gladwwell, G.M.L. (1980) *Contact Problems in the Classical Theory of Elasticity*, Sijthoff and Noordhoff, Alphen aan den Riju..

Love, A. E. H. (1944) *The Mathematical Theory of Elasticity*, Dover, New York.

Provan, J. W. (1973, 1974) *A Continuum Mechanics Course*, McGill University, Montreal (*unpublished*).

Sokolnikoff, I. S. (1956) *Mathematical Theory of Elasticity*, McGraw-Hill Book Co., New York.

6.8. Further Reading

Bellet, D. (1990) *Problémes d'Élasticité*, Cepadues-Éditions, Toulouse, France.

Bowen, R. M. (1989) *Introduction to Continuum Mechanics for Engineers*, Plenum Press, New York.

Brull, M.A. (1953) A structural theory incorporating the effect of time-dependent elasticity, *Proc.*

1ⁿ Mid- Western Conf. Solid Mechanics, 141- 7.

Chung, T.J. (1988) *Continuum Mechanics*, Prentice-Hall, Englewood Cliffs, New Jersey.

Fredrick, D. and Chang, T.S. (1965) *Continuum Mechanics*, Allyn and Bacon, Boston.

Fung, Y.C. (1965) *Foundations of Solid Mechanics, Prentice-Hall*, Englewood Cliffs, N.J.

Gurtin, M.E. (1981) *An Introduction to Continuum Mechanics*, Academic Press, New York.

Gurtin, M.E. and Williams, W.O.(1967) *An axiomatic foundation for continuum thermodynamics*, Arch. *Rational Mech. Anal.* 26, 83-117.

Hetenyi, M. (1939, 1940) Some applications of photoelasticity in turbine generator design, *Trans. A.S.M.E* 61, 1939, A151; 62, 1940, A80.

Hunter, S.C. (1968) The motion of a rigid sphere embedded in an adhering elastic or viscoelastic medium, in: *Proceedings, Edinburgh Mathematical Society* 16(Series II), Part I, pp. 55-69.

Hunter, S.C. (1976) *Mechanics of Continuous Media*, Ellis Hordwood, Chister, England.

Malvern, L.E. (1969) *Introduction to the Mechanics of a Continuous Medium*, Prentice-Hall, Englewood Cliffs, N.J.

Mandel, J. (1978) Propriétés Méchaniques des Matériaux, Eyrolles, 1978.

Newton, R. E. (1940) Photoelastic study of stresses in rotating discs, *Trans. A.S.M.E* 62, A57& A174.

Nisida, M. (1963) New photoelastic methods for torsion problems, *Proceedings International Symposium on Photoelasticity*, edited by M. M. Frocht, Pergamon, New York, pp. 109-121.

Prager, W. (1961) *Introduction to Mechanics of Continua*, Ginn, Boston, Mass.

Reismann, H. and Pawlik, P.S. (1980) *Elasticity-theory and applications*, Wiley, New York.

Rivlin, R. S. and Ericksen, J. L. (1955) Stress-deformation relations for isotropic materials, *J Ratl. Mech. Anal.* 4, 323-5.

Sneddon, I.N. and Hill, R. (Eds.), (1960-1963) *Progress in Solid Mechanics*, Vol 1 (1960), Vol. 2(1961), Vol. 3(1963), Vol. 4 (1963), North-Holland Pub. Co., Amsterdam.

Sommerfeld, A. (1950) *Mechanics of Deformable Bodies*, Academic Press, New York.

Stackgold, I. (1967) *Boundary Value Problems of Mathematical Physics*,Vol. 1, McMillan, New York.

Timoshenko, S. P. and Goodier, J. N. (1970) *Theory of Elasticity*, McGraw-Hill Book Company, New York.

CHAPTER 7

ELASTIC-PLASTIC BEHAVIOUR

7.1. Introduction

7.1.1. ELASTIC vs. INELASTIC DEFORMATION

Elastic Deformation
It depends only upon the stress level in the material, meanwhile it is not strain- or time-history dependent. Further, an elastic deformation process is described, from a thermodynamical point of view, *as we dealt with in Chapter 5*, to be a *reversible* process. Thus, upon the removal of the load, a complete recovery to the undeformed configuration should take place. An elastic response of an engineering material is formulated within the realm of classical elasticity. Such an elastic response could be linear or nonlinear pending on the form of the constitutive law that is used in its description (*Chapter 6*).

Inelastic Deformation
It depends, as explained below, on both the stress level and the strain-history of the material. A particular illustration of inelastic deformation is the plastic deformation in metals, which occurs after a metal has reached its yield point, i.e., the yield stress. The magnitude of the latter is denoted in this text, for the simple case of uniaxial tension, by the symbol σ_y. As we recall, plastic deformation in metals results from microstructural sources, e.g., dislocations and various mechanisms of slip within the grains and/or along the grain boundaries. Such microstructural mechanisms are not fully recoverable upon the removal of the load and, thus, resulting in a permanent portion of the strain after full removal of the external load. Hence, an inelastic deformation process is seen from a thermodynamical point of view as an *irreversible* process. In uniaxial tension, the yield point is not always well defined. In some cases, it is considered to be coinciding with the *"proportionality limit"*, e. g., point P in the stress-strain diagram of Figure 7.1. The latter represents the upper end point of the linear portion *"OP"* whereby the proportionality between the stress and the strain for the linear elastic range ceases to exist. In other situations, various offset methods are used to determine the yield point. In this context, the stipulation that the yield point corresponds to the stress at 0.20 % permanent strain is frequently used.

7.1.2. LOADING *vs.* UNLOADING

Within the elastic range of the material, an increase of the load will result in an increase of the magnitude of the incurred deformation, thus causes the stress-strain state point on the stress-strain curve to move upward along the path *OP*, Figure 7.1. Meanwhile, on unloading, the stress-strain state point will move downward along the same path.

In the plastic range of the material, unloading from a point such as *A*, Figure 7.1, would result in the state point following the unloading path *AB*. The latter is essentially parallel with the linear elastic portion *OP* of the curve. When complete unloading occurs and the stress magnitude becomes zero as the unloading line reaches the point *B*, a permanent plastic strain ϵ^P remains. Meantime, at this state, the recoverable elastic strain is ϵ^E.

A reloading from *B* back to *A* would not follow exactly the path *BA*, but forming, as shown in Figure 7.1, a *"hysteresis loop"*, due to the energy loss in the unloading-reloading cycle.

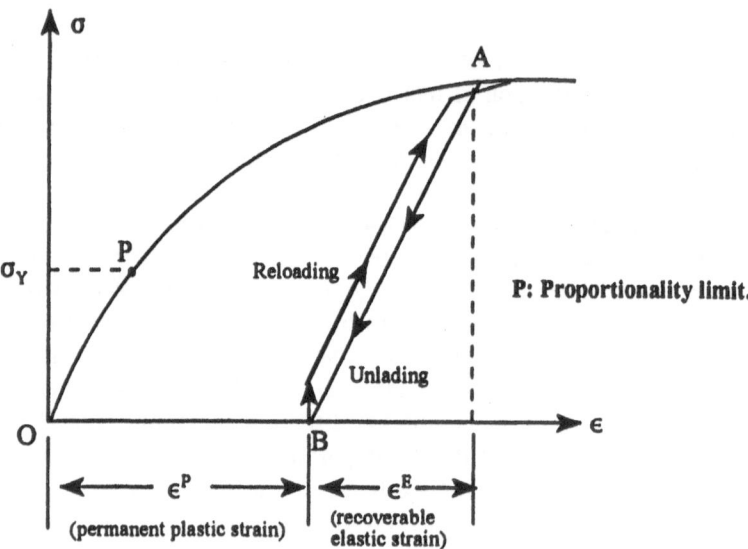

Figure 7.1. A hypothetical stress-strain diagram for an elastic-plastic material.

As illustrated in Figure 7.1, upon full reloading to point A, a load increase would be further required to cause additional deformation. The latter phenomenon is referred-to as *"strain-hardening"*, or, alternatively, *"work-hardening"*.

Thus, in the plastic range of a material, the stress depends on the present magnitude of the stress as well as its entire history in the material. Consequently, the plastic response is described to be "strain-history" dependent.

In the "conventional" theory of plasticity, the following two assumptions are often made:

(i)   the assumption of isothermal deformation.
(ii)   the assumption of rate-independency. Under this assumption, the effect of rate of loading is neglected and, hence, plastic deformation is seen to be time-independent.

7.1.3. IDEALIZED MODELLING OF PLASTIC BEHAVIOUR

Much of the three-dimensional theory for analysing plastic behaviour may be seen as generalization of certain idealizations of the actual one-dimensional behaviour of the particular class of material under consideration. Thus, we present below a number of the most popular idealized models pertaining to the one-dimensional elastic-plastic behaviour:

Rigid-perfectly plastic behaviour
In this model, neither elastic response nor work-hardening is assumed to exist. This behaviour is shown in Figure 7.2a.

Elastic-perfectly plastic
In this model, elastic behaviour prior to yielding is included, although strain-hardening is assumed not to exist (Figure 7.2b).

Rigid-linear work hardening
Here, the elastic response is disregarded, and a "linear" work hardening is assumed to exist (Figure 7.2c).

Elastic-linear work hardening
In this model, both linear elastic and linear work hardening are assumed to exist. This is illustrated in Figure 7.2d.

7.2. Elastic-Plastic Behaviour under Static Loading

7.2.1 YIELD CONDITIONS

A yield condition is the mathematical relationship that must exist among the components of the stress tensor at a point in the body in order for yielding (the onset of plastic deformation) to occur at this point.

The yield condition may be generally expressed by the relationship

$$f(\sigma_{ij}) = C_y \qquad (7.1)$$

where σ_{ij} is the *Cauchy's* stress tensor and C_y is referred to as the *"yield constant"*. Alternatively, the yield condition may be expressed in a more compact form as

$$\Upsilon(\sigma_{ij}) = 0 \qquad (7.2)$$

where $\Upsilon(\sigma_{ij})$ is known as the *"yield function"*.

In the case of an assumed isotropy of the body with respect to the components of the stress tensor, the yield condition will be, by consequence, direction-independent. In such situation, the yield condition may be expressed as a function of the stress invariants. Alternatively, the yield condition may be represented as a symmetric function of the principal stresses. In this case, the yield condition (7.1) would take the form

$$f_1(\sigma_1, \sigma_2, \sigma_3) = C_y \qquad (7.3)$$

In this case, the yield condition format (7.2) may be re-written in the following alternative form:

$$\Upsilon_1(\sigma_1, \sigma_2, \sigma_3) = 0 \qquad (7.4)$$

A large number of yield conditions have been proposed in the literature for different classes of materials. In case of isotropic materials, both the *Tresca* and *von Mises* yield conditions have proven to be useful. We introduce these two yield conditions below.

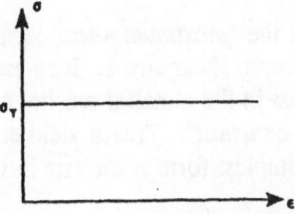

(a) Rigid-perfectly plastic behaviour: Neither elastic response nor work-hardening is assumed to exist.

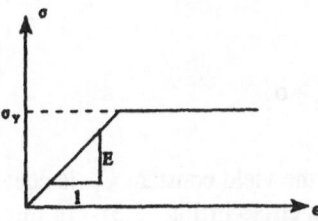

(b) Elastic-perfectly plastic behaviour: Elastic behaviour prior to yielding is included, but strain-hardening is assumed not to exist.

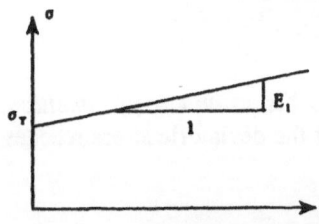

(c) Rigid-linear work hardening behaviour: The elastic response is disregarded and a "linear" work-hardening is assumed to exist.

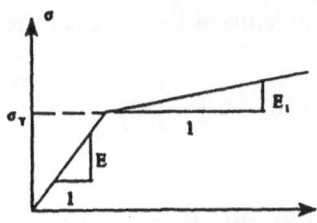

(d) Elastic-linear work-hardening behaviour: Both linear elastic and linear work-hardening are assumed to exist.

Figure 7.2. Idealized models for elastic-plastic behaviour.

Tresca Yield Condition

This yield condition is often referred to as the *"maximum shear yield condition"* as it expresses the yield condition in terms of maximum shear stress. It asserts that yielding in the body occurs when the maximum shear stress in the material reaches a prescribed value C_y. The latter is often referred to as the *"yield constant"*. Tresca yield condition may take different mathematical forms, although its simplest form is the one in terms of principal stresses, i.e.

$$\frac{1}{2}(\sigma_1 - \sigma_3) = C_y \tag{7.5}$$

This is with the assumption that

$$\sigma_1 > \sigma_2 > \sigma_3 \tag{7.6}$$

In case of pure shear, it can be shown that the yield constant C_y is equal to the so-called pure shear yield point value k (*see* Mohr's circle in Fig. 7.3). In this case, the Tresca yield condition is represented by the form

$$\sigma_1 - \sigma_3 = 2k \tag{7.7}$$

von Mises Yield Condition

This yield condition is often referred to as the *"distortion energy condition"*. It asserts that yielding occurs when the second invariant of the deviatoric stress reaches a certain value, i.e.,

$$-I_2' = C_y \tag{7.8}$$

which may be, also, expressed (*Chapter 2*) in terms of the principal stresses as

$$(\sigma_1 - \sigma_2)^2 + (\sigma_2 - \sigma_3)^2 + (\sigma_3 - \sigma_1)^2 = 6C_y \tag{7.9}$$

The von Mises yield condition (7.8), or (7.9), may be also expressed in terms of the magnitude of the yield stress in tension in the following form

$$(\sigma_1 - \sigma_2)^2 + (\sigma_2 - \sigma_3)^2 + (\sigma_3 - \sigma_1)^2 = 2\sigma_y^2 \tag{7.10}$$

or, alternatively, in terms of the yield stress in pure shear as

$$(\sigma_1 - \sigma_2)^2 + (\sigma_2 - \sigma_3)^2 + (\sigma_2 - \sigma_1)^2 = 6k^2 \tag{7.11}$$

where k, as mentioned earlier, is the pure shear yield value.

7.3. Yield Surfaces

Consider the stress space (Fig. 7.4) whose coordinate-axes are associated with the three principal stresses σ_1, σ_2 and σ_3. In this stress space, one may consider the position vector to an arbitrary point P $(\sigma_1, \sigma_2, \sigma_3)$. This position vector may be resolved into a component OA along the line OL, which makes equal angles with the coordinate axes, and a component OB in the plane that is perpendicular to OL and passes by the origin O. This plane is often referred to as the "$\prod$-plane". In view of Fig. 7.4, the equation of the $\prod$-plane is expressed as

$$\sigma_1 + \sigma_2 + \sigma_3 = 0 \tag{7.12}$$

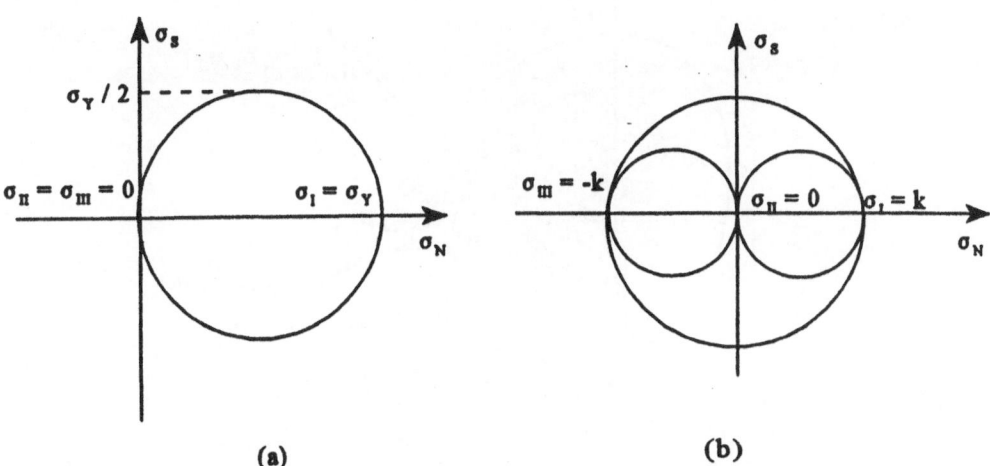

(a)   (b)

Figure 7.3. Mohr's circle; (a) Simple tension, and (b) Pure shear.

212

Thus, the position vector component OA represents the state of stress $\sigma_1 = \sigma_2 = \sigma_3$, i.e., a state of *"hydrostatic stress"*. Meantime, the component *OB* (in the $\prod$-plane) represents the deviator portion of the considered stress state at the point *O*.

In the stress space presented in Fig. 7.4, the yield condition

$$f_1\,(\sigma_1, \sigma_2, \sigma_3) = C_y \tag{7.13}$$

represents a surface which is referred to as the *'yield surface'*.

> *Thus, under the assumption that yielding is independent of hydrostatic stress, the yield surface would be presented by a general cylinder formed by a generator parallel to OL. Accordingly, coordinate points that locate stress-states inside the mentioned cylindrical yield surface represent elastic stress-states, meantime, those locate stress-states on the yield surface represent incipient plastic stress-states.*

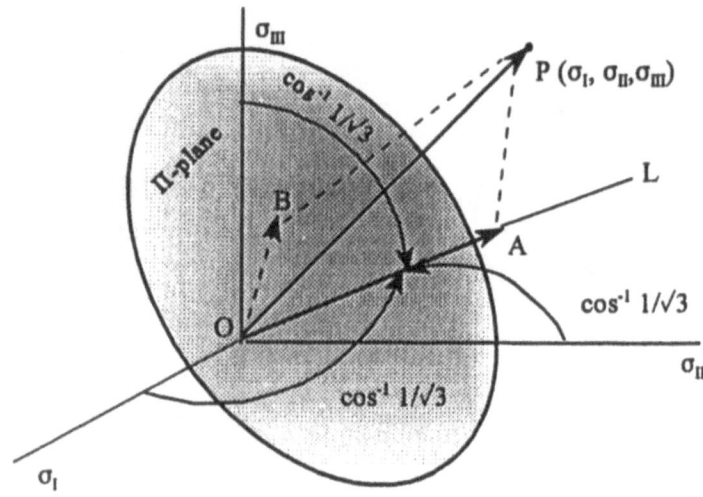

Figure 7.4. The concept of *"yield surface"*.

7.3.1. YIELD CURVE

A yield curve is defined as the intersection of the yield surface with the $\prod$-plane. In this context, as illustrated in Fig. 7.4, looking to the $\prod$-plane, along LO, the principal stress axes, emanating from the origin, appear symmetrically placed 120° apart. In view of expressions (7.5) and (7.10), the respective Tresca and von Mises yield curves, as based on the yield stress in pure tension, are represented as shown in Fig. 7.5(a). As illustrated in the latter figure, the von Mises circle of radius $\sqrt{2/3}\ \sigma_y$ circumscribes the Tresca's regular hexagon.

The corresponding two yield curves, as based on the yield stress in pure shear, equations (7.7) and (7.11), respectively, are represented in Fig. 7.5(b). In this case, as illustrated in Fig. 7.5(b), the *von Mises circle* is inscribed in *Tresca's hexagon*.

With reference to Fig. 7.4, the projection on the $\prod$-plane of an arbitrary point $P(\sigma_1,\sigma_2,\sigma_3)$ is easily determined as each of the principal stress axes makes an angle of a magnitude $\cos^{-1}\sqrt{2/3}$ with the $\prod$-plane.

Accordingly, the projected deviatoric components are $\left(\sqrt{\dfrac{2}{3}}\,\sigma_1, \sqrt{\dfrac{2}{3}}\,\sigma_2, \sqrt{\dfrac{2}{3}}\,\sigma_3 \right)$.

The reader, however, should be aware that the inverse problem of determining the stress components that will correspond with an arbitrary point in the $\prod$-plane is not unique since the hydrostatic stress components may have any value.

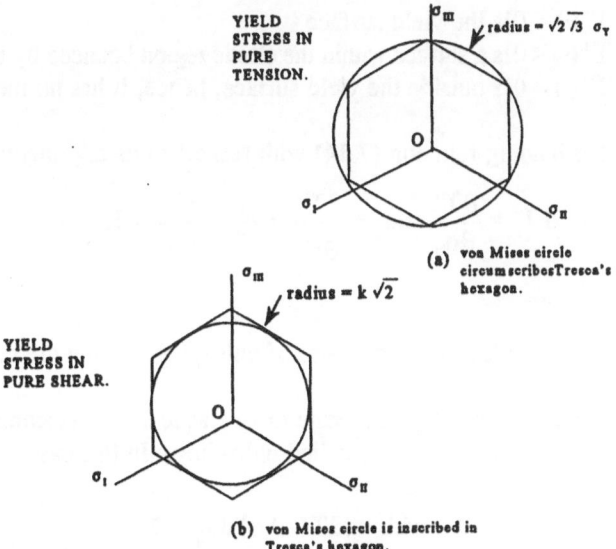

Figure 7.5. Tresca and von Mises yield curves; as based on pure tension (a), and pure shear (b).

7.4. Post-Yield Behaviour: Changes in the Yield Surface

If the specimen is continued to be loaded beyond initial yielding, plastic deformation will occur. In the case of *"perfectly-plastic"* material (Section 7.1.3), the yield surface does not change during plastic deformation, that is, the initial yield condition for this material remains the same during post-yielding deformation. In other classes of strain-history dependent materials, however, post-yielding deformation may bring changes in the geometry of the yield surface. Such changes may involve modifications in both the shape and the spatial location of the yield surface. In this case, the material is referred to as *"strain-hardening material"*.

For a strain-hardening material, the yield function $\Upsilon(\sigma_{ij}) = 0$, Eqn. (7.2), must be modified to account for subsequent changes in the yield surface due to post-yielding. In this case, a *"loading function"* is often considered. The choice of this loading function would depend on the stain-hardening characteristics of the material. Such loading function may be expressed, in a general manner, as

Loading function: $\quad \Upsilon^* (\sigma_{ij}, \varepsilon_{ij}^P, K) = 0$ $\hfill$ (7.14)

In the expression (7.14), σ_{ij} is the Cauchy stress, ε_{ij}^P is the plastic strain and K is referred to as the *"work-hardening parameter"*.

The loading function (7.14) determines a loading surface defined as follows:

(i) $\quad \Upsilon^*(\cdot) = 0$ is the yield surface
(iii) $\quad \Upsilon^*(\cdot) < 0$ is a surface within the elastic region bounded by the yield surface.
(iii) $\quad \Upsilon^*(\cdot) > 0$ is outside the yield surface, hence, it has no meaning.

Differentiating the loading function (7.14) with respect to its arguments, it follows that

$$d\Upsilon^* = \frac{\partial \Upsilon^*}{\partial \sigma_{ij}} \, d\sigma_{ij} + \frac{\partial \Upsilon^*}{\partial \varepsilon_{ij}^P} \, d\varepsilon_{ij}^P + \frac{\partial \Upsilon^*}{\partial K} \, dK \qquad (7.15)$$

In view of (7.15), the following three states of unloading/loading are often categorized:

(i) Unloading. A change from a plastic state to an elastic state unaccompanied by plastic strain (the plastic strain rate is zero) is called unloading. In this case,

$$\Upsilon^* = 0 \quad \text{with} \quad (\partial \Upsilon^*/\partial \sigma_{ij}) \, d\sigma_{ij} < 0 \qquad (7.16a)$$

(ii) Neutral loading. When a change in σ_{ij} from one plastic state to another plastic state is not accompanied by a change in plastic strain, the process is called neutral loading. In this case,

$$\Upsilon^* = 0 \quad \text{with} \quad (\partial \Upsilon^* / \partial \sigma_{ij}) \, d\sigma_{ij} = 0 \qquad (7.16b)$$

(iii) Loading. A change from one plastic to another accompanied by plastic strain constitutes loading. This process occurs when

$$\Upsilon^* = 0 \quad \text{with} \quad (\partial \Upsilon^* / \partial \sigma_{ij}) \, d\sigma_{ij} > 0 \qquad (7.16c)$$

7.4.1. HARDENING RULES

On loading, the manner in which the plastic strain ε_{ij}^P affects the loading function (7.14) is described by the so-called *"hardening rules"*. Two specially simple hardening rules are considered here, mainly *"isotropic hardening"* and *"kinematic hardening"*.

(I) Isotropic hardening
This postulates that, on loading, the yield surface increases in size but maintains its original shape. Thus, under this postulate, the yield curves for Tresca and von Mises conditions are, respectively, concentric regular hexagons and circles; as illustrated in Figure 7.6 below.

(II) Kinematic hardening
Here, the assumption is that, on loading, the initial yield surface does not change in size or shape, but is relocated to a new position in the stress space. Let α_{ij} be the coordinates of the new centre of the relocated yield surface, then the new yield surface is defined, with reference to Fig. 7.7 below, by

$$\Upsilon (\sigma_{ij} - \alpha_{ij}) = 0 \qquad (7.17)$$

replacing the initial yield condition (7.2). As a special case of kinematic hardening is *"linear hardening"*. In this case the following linear relation is applied

$$\dot{\alpha}_{ij} = a \, \dot{\varepsilon}_{ij}^P \qquad (7.18)$$

where a is a constant and the over-dot indicates time-derivative. The case of linear hardening, as described by (7.18), is illustrated in Figure 7.7 below.

216

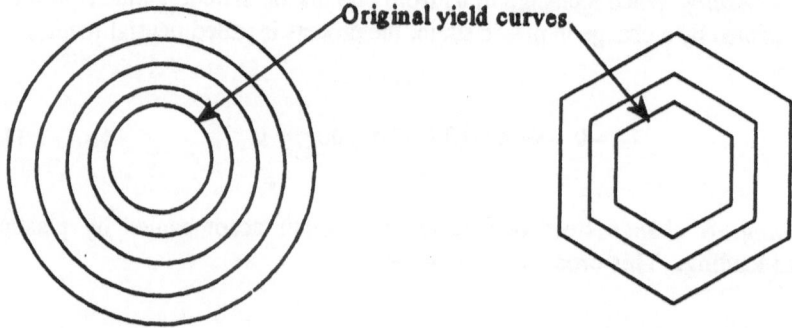

(a) Mises Circles   (b) Tresca Hexagons

Figure 7.6. Tresca and von Mises yield curves under the condition of isotropic hardening.

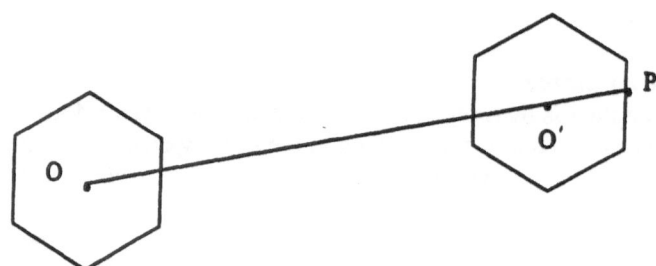

Figure 7.7. Tresca yield curve under the condition of kinematic hardening.

7.5. Constitutive Relations

As plastic strain depends upon the entire loading history of the material, plastic constitutive equations are often expressed in terms of *"strain increments"*. Hence, the pertaining constitutive relations are formulated within so-called *"incremental theories"*. Two idealized response situations are considered below, namely, *"rigid-perfectly plastic"* material and *"elastic perfectly plastic"* material.

(I) Flow rule for rigid-perfectly plastic material. Levy-Mises equations.
In this case, we neglect, by definition, the elastic portion of the response and we assume, further, that the principal axes of strain-increments coincide with the axes of the principal stresses. In other words, it is assumed that no dissipation is involved during the strain increment. In this case, the flow rule is expressed in terms of the total strain-increment in relation with the deviatoric stress via the constitutive relation

$$d\,\varepsilon_{ij} = \sigma'_{ij}\,d\lambda \qquad (7.19)$$

In the *"Levy-Mises"* response equation above, $d\lambda$ is referred to as *"proportionality factor"*. This equation appears in a differential form in order to reflect the situation that incremental strains are being related to finite deviatoric-stress components. It should be emphasized here that $d\lambda$ may change during the loading process and it is, thus, a scalar multiplier and not a fixed constant. It should be noted that the Levy-Mises response equation (7.19) specifies the total strain increment, i.e., containing both the pertaining elastic and plastic increments.

(II) Flow rule for an elastic-perfectly plastic material. Prandtl-Reuss equations.
In this case, the strain increment is seen to be composed of both elastic and plastic portions as

$$d\,\varepsilon_{ij} = d\,\varepsilon^E_{ij} + d\,\varepsilon^P_{ij} \qquad (7.20)$$

Thus, under the flow rule for an elastic-perfectly plastic material, the plastic strain increment is assumed to be related to the stress deviator by the constitutive relation

$$d\,\varepsilon^P_{ij} = \sigma'_{ij}\,d\lambda \qquad (7.21)$$

which is known as the *"Prandtl-Reuss"* equation. It is noted that this response equation does not specify the total strain increment.
The parameter $d\lambda$ appearing in the Prandtl-Reuss equation (7.21) is often expressed in

terms of the so-called *"equivalent stress"*, σ_{EQ} ,alternatively called *"effective stress"*, in conjunction with the corresponding *"equivalent or effective plastic strain increment"*, $d\,\varepsilon_{EQ}^{p}$ as

$$d\lambda = \frac{3}{2}\ \frac{d\,\varepsilon_{EQ}^{p}}{\sigma_{EQ}} \tag{7.22}$$

where σ_{EQ} can be expressed in terms of the stress tensor components as

$$\sigma_{EQ} = \frac{1}{\sqrt{2}}\left\{[(\sigma_{11} - \sigma_{22})^2 + (\sigma_{22} - \sigma_{33})^2 + (\sigma_{33} - \sigma_{11})^2]\right.$$
$$\left. + 6(\sigma_{12}^2 + \sigma_{23}^2 + \sigma_{31}^2)\right\}^{1/2} \tag{7.23}$$
$$= \sqrt{3\,\sigma_{ij}'\,\sigma_{ij}'/2} \ = \ \sqrt{3\,I_2'}$$

in which I_2' is the second invariant of the stress deviator.

In a similar manner, the equivalent strain-increment, appearing in (7.22), is expressed in terms of the components of incremental strain tensor as

$$d\varepsilon_{EQ}^{p} = \left\{\frac{2}{9}\left[(d\varepsilon_{11}^{p} - d\varepsilon_{22}^{p})^2 + (d\varepsilon_{22}^{p} - d\varepsilon_{32}^{p})^2 + (d\varepsilon_{33}^{p} - d\varepsilon_{11}^{p})^2\right]\right.$$
$$\left. + \frac{4}{3}\left[(d\varepsilon_{12}^{p})^2 + (d\varepsilon_{23}^{p})^2 + (d\varepsilon_{31}^{p})^2\right]\right\}^{1/2} \tag{7.24}$$
$$= \sqrt{\frac{2}{3}\ d\varepsilon_{ij}^{p}\ d\varepsilon_{ij}^{p}}$$

The introduction of both the equivalent stress σ_{EQ} and the equivalent plastic strain increment $d\,\varepsilon_{EQ}^{p}$ by (7.23) and (7.24), respectively, are particularly useful in the mathematical formulation of pertaining strain hardening rules.

7.5.1. PLASTIC POTENTIAL FUNCTION. PLASTIC WORK.

For a so-called *"stable plastic material"*, a plastic potential function $\Gamma(\sigma_{ij})$ may exist such that

$$d\varepsilon_{ij}^{P} = \frac{\partial \Gamma(\sigma_{ij})}{\partial \sigma_{ij}} \qquad (7.25)$$

Meantime, the work increment, due to stress, per unit volume is generally expressed by

$$dW = \sigma_{ij} d\varepsilon_{ij}$$

which can be written in terms of the elastic and plastic strain increments, in view of (7.20), as

$$\begin{aligned} dW &= (\sigma_{ij} d\varepsilon_{ij}^{E}) + (\sigma_{ij} d\varepsilon_{ij}^{P}) \\ &= dW^{E} + dW^{P} \end{aligned} \qquad (7.26)$$

Meantime for a *"plastically incompressible material"*, the plastic work increment is expressed as

$$dW^{P} = \sigma_{ij} d\varepsilon_{ij}^{P} = \sigma_{ij}' d\varepsilon_{ij}^{P} \qquad (7.27)$$

In case of an elastic-perfectly plastic material, the material may be assumed to obey the Prandtl-Reuss constitutive equation (7.21) and, for this case, the plastic work increment is expressed as

$$dW^{P} = \sigma_{EQ} d\varepsilon_{EQ}^{P} \qquad (7.28)$$

In this case, the Prandtl-Reuss constitutive equation (7.21) can be written in terms of plastic work increment, in view of equations (7.22) to (7.24) and (7.28), as

$$d\varepsilon_{ij}^{P} = \frac{3}{2} \frac{dW^{P}}{\sigma_{EQ}^{2}} \sigma_{ij}' \qquad (7.29)$$

7.5.2. WORK HARDENING. STRAIN HARDENING.

Under the assumption of *"isotropic strain-hardening plastic flow"*, the yield criterion may be expressed using one of the following two hypotheses:

(I) Work-Hardening Hypothesis

Under this hypothesis, the yield criterion is expressed as a functional of the total plastic work by

$$\Upsilon(\sigma_{ij}) = F(W^P) \tag{7.30}$$

whereby the precise functional form $F(\cdot)$ is to be determined experimentally.

Meantime, the total plastic work is expressed as

$$W^P = \int \sigma_{ij}\, d\varepsilon_{ij}^P$$

(II) Strain-Hardening Hypothesis

Here, it is assumed that the extent of hardening is a functional of the total equivalent plastic strain. In this case, the associated hardening rule is formulated as

$$\Upsilon(\sigma_{ij}) = H(\varepsilon_{EQ}^P) \tag{7.31}$$

whereby the appropriate functional form $H(\cdot)$ is determined by experiment. In expression (7.31) above, ε_{EQ}^P is the total equivalent plastic strain determined as

$$\varepsilon_{EQ}^P = \int d\varepsilon_{EQ}^P$$

7.5.3. HENCKY'S TOTAL DEFORMATION THEORY

This theory expresses the stresses as a function of the total strain in the following format

$$\varepsilon_{ij}' = \left(\gamma + \frac{G}{2}\right)\sigma_{ij}'$$

$$\varepsilon_{ii} = (1 - 2v)\frac{\sigma_{ii}}{E} \tag{7.32}$$

where γ is a rational parameter expressed as

$$\gamma = \frac{3}{2}\frac{\varepsilon_{EQ}^P}{\sigma_{EQ}} \tag{7.33}$$

Plane Plastic Strain. Elementary Slip Line Theory (Slip Line Field and Velocity Field)
In some engineering applications where unrestricted plastic flow may occur, e.g. metal forming, it may be feasible to neglect elastic deformation and assume the response of the material to be rigid perfectly plastic. Further, if the response of the material is assumed to be a case of plane strain, then, the occurring velocity field may be studied using *"slip line theory"*. In this case, the plastic strain-rate tensor components may be expressed as

$$\dot{\varepsilon}_{ij}^{P} = \begin{pmatrix} \dot{\varepsilon}_{11} & \dot{\varepsilon}_{12} & 0 \\ \dot{\varepsilon}_{12} & \dot{\varepsilon}_{22} & 0 \\ 0 & 0 & 0 \end{pmatrix} \tag{7.34}$$

Meantime, the corresponding stress tensor components are

$$\sigma_{ij} = \begin{pmatrix} \sigma_{11} & \sigma_{12} & 0 \\ \sigma_{12} & \sigma_{22} & 0 \\ 0 & 0 & \sigma_{33} \end{pmatrix} \tag{7.35}$$

Assuming, for simplification, the linear definition of strain to be applicable, the time rate of the strain tensor is expressed in terms of the gradient of the velocity components as

$$\dot{\varepsilon}_{ij} = \frac{1}{2}\left(v_{i,j} + v_{j,i}\right) \tag{7.36}$$

Following the standard slip-line notation, e.g.,

$$\sqrt{(\sigma_{11} - \sigma_{22})^2/4 + (\sigma_{12})^2} \;=\; k \quad \text{and} \quad \sigma_{33} = -p$$

then, it can be demonstrated that the principal stress values, corresponding to the stress-state (7.35), can be expressed by

$$\sigma_1 = -p + k$$
$$\sigma_2 = -p$$
$$\sigma_3 = -p - k$$

222

Meantime, the principal stress directions are given with respect to the x_1, x_2 axes, as illustrated in Figure 7.8, with

$$\tan 2\theta = 2\,\sigma_{12}\,/\,(\sigma_{11} - \sigma_{22}) \qquad (7.37)$$

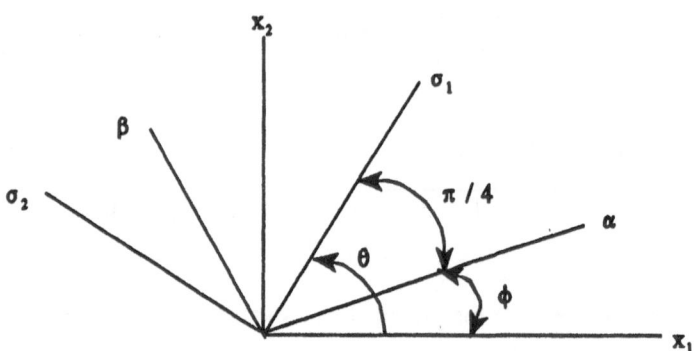

Figure 7.8. Maximum shear directions (α, β) are at 45° with respect to the principal stress directions.

As shown in Figure 7.8, the maximum shear directions, designated as α & β are at 45° with respect to the principal axes directions. In Figure 7.8, it is clear that $\theta = \dfrac{\pi}{4} + \phi$. Thus,

$$\tan (2\phi) = -\frac{1}{\tan (2\theta)} \qquad (7.38)$$

Further, for a given stress field in a plastic flow, two families of curves exist along the directions of maximum shear at every point. These curves are referred to as *"shear lines"*, or *"slip lines"*.

Consider a small curvilinear element bounded by two pairs of slip lines as shown in Fig.7.9. The components of stress are expressed as

$$\sigma_{11} = -p - k \sin (2\phi)$$
$$\sigma_{22} = -p + k \sin (2\phi) \qquad (7.39)$$
$$\sigma_{12} = k \cos (2\phi)$$

Thus, from the equilibrium of the element, it can be shown that

$$p + 2k\phi = C_1 \text{ (a constant along an } \alpha\text{-line)}$$
$$p - 2k\phi = C_2 \text{ (a constant along an } \beta\text{-line)} \qquad (7.40)$$

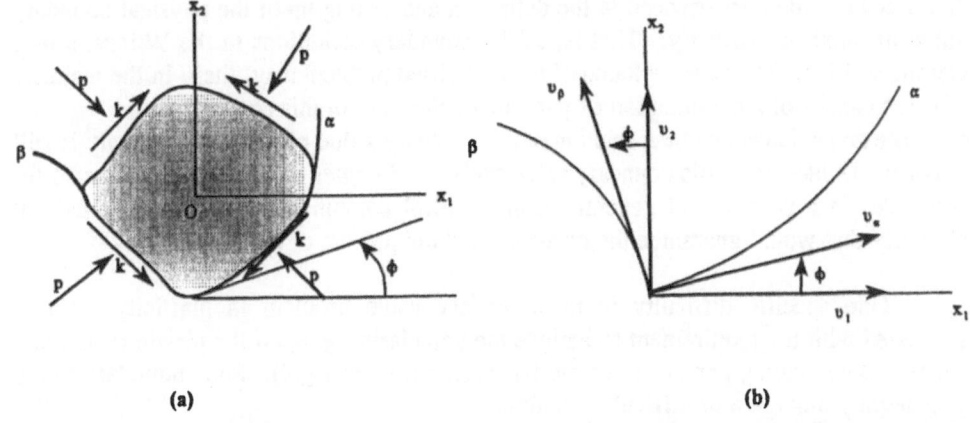

(a)   (b)

Figure 7.9. Small curvilinear element. Slip-line field (a), and velocity components (b).

Meantime, the velocity components are expressed by

$$v_1 = v_\alpha \cos \phi - v_\beta \sin (\phi)$$
$$v_2 = v_\alpha \sin \phi + v_\beta \cos (\phi)$$

(7.41)

Further, if x_1 and x_2 are slip-line directions, then, with reference to (7.41),

$$\left\{ \frac{\partial}{\partial x_1} (v_\alpha \cos \phi - v_\beta \sin \phi) \right\}_{\phi + 0} = 0$$

$$\left\{ \frac{\partial}{\partial x_2} (v_\alpha \sin \phi - v_\beta \cos \phi) \right\}_{\phi = 0} = 0$$

(7.42)

which bring us to the expressions

$$d v_1 - v_2\, d\phi = 0 \quad (\text{on } \alpha \text{ lines})$$
$$d v_2 - v_1\, d\phi = 0 \quad (\text{on } \beta \text{ lines})$$

(7.43)

For statically determinate problems, the slip line field may be determined from (7.40). Thus, using the resulting slip line field, the velocity field may be determined from (7.43).

7.6. The Boundary Value Problem in Plasticity

A difficulty is often experienced in the definition and setting up of the physical boundary value problem in plasticity. That is, of the boundary conditions in the Wildest sense, Geiringer, 1953. The actual solution of the formulated problem may, then, in the majority of cases consist of a combination of particular solutions. In this context, the existence and uniqueness of the solution of the boundary value problem in plasticity is still in doubt. Unlike the elastic boundary value problem (*Chapter 6*), for instance, one so-far concludes in a precise and general manner typical combinations of boundary data in plasticity that would guarantee the existence and uniqueness of a solution.

One specific difficulty in the boundary value problem in plasticity is often associated with the requirement to include the boundaries between the plastic region and the other surrounding parts of the material (e.g., elastic or rigid). Such boundaries may be generally unknown or difficult to define.

7.6.1. THE "THREE-DIMENSIONAL PROBLEM". QUADRATIC YIELD CONDITION.

We present in this section the three-dimensional theory as formulated by von Mises (1913, 1949); *see* Geiringer (1953). In this context, reference is made to the original work, concerning the first complete formulation of the three-dimensional problem, of Lévy (1870, 1871), De St. Venant (1870, 1871) and Tresca (1872). The three-dimensional problem consists of the following eleven equations to be solved for eleven unknown functions.

−   **Three equations of motion:**

$$\rho \, \frac{d v}{d t} = \chi + \text{div } \sigma \qquad (7.44)$$

where ρ is the density, v is the deformation rate vector and χ is the resultant vector of body (and/or external) forces.

Alternatively, we may consider the corresponding equilibrium problem in the absence of acceleration and of body forces. In this case, equation (7.44) will reduce to the following vector equation

$$\text{div } \sigma = 0 \qquad (7.45)$$

−   **The yield condition equation:**
In the von Mises theory, one has, in terms of the principal stresses, the form given earlier by (7.11), i.e.,

$$(\sigma_1 - \sigma_2)^2 + (\sigma_2 - \sigma_3)^2 + (\sigma_3 - \sigma_1)^2 = 6k^2 \qquad (7.46)$$

or, in terms of the principal shear stresses τ_i where $2\,\tau_1 = (\sigma_2 - \sigma_3)$, etc., the expression

$$\tau_1^2 + \tau_2^2 + \tau_3^2 = 3/2\, k^2 \qquad (7.47)$$

−   **The constitutive equation:**
Here, it is assumed that the strain rate tensor $\dot{\varepsilon}$ and the deviatoric stress σ' are related via the constitutive expression

$$\sigma'_{ij} + \delta_{ij} \, p = \alpha \, \dot{\varepsilon}_{ij} \tag{7.48}$$

where δ_{ij} is the Kronecher delta, p is the hydrostatic pressure, and α ( ) is a nonnegative (unknown) function of proportionality (Geiringer,1953). Equation (7.48) has *six* independent components.

—   **The continuity equation:**

$$\text{div } v = 0$$

$$\text{or,} \quad \dot{\varepsilon}_{11} + \dot{\varepsilon}_{22} + \dot{\varepsilon}_{33} = 0 \tag{7.49}$$

The continuity equation (7.49) expresses also the assumed incompressibility of the medium.

The above are *eleven* equations to be solved for the eleven unknowns: six independent components of the stress tensor σ_{ij} , three components of the velocity vector v_k , and the value of the proportionality function α and that of the hydrostatic pressure p.

7.7. Derivation of the "Plane Problem" form the "Three-Dimensional Problem". Quadratic Yield Condition.

In this section, we follow the approach of Geiringer (1953) to derive a "plane problem" from the "three-dimensional problem" introduced above in Section 7.5. In this, one considers the two particular solutions (i.e., of "plane strain" and of "plane stress") of the more general "three-dimensional problem".

7.7.1. THE PROBLEM OF "PLANE STRAIN". QUADRATIC YIELD CONDITION.

Following Geiringer (1953), we assume that one of the principal directions of the stress tensor σ_{ij} (i.e., also of the strain-rate tensor $\dot{\varepsilon}_{kl}$, in view of eqn. (7.48)) is parallel to a fixed direction (say the x_3 - axis). It is further assumed that

(i)   all components of σ and $\dot{\varepsilon}$ are independent of x_3.
(ii)   the strain rate component $\dot{\varepsilon}_{33}$ vanishes everywhere.

Thus, if one combines the above assumptions with the constitutive equation (7.48) and the continuity equation (7.49), one can write the following set of relations

$$\dot{\varepsilon}_{11} + \dot{\varepsilon}_{22} = 0, \qquad \sigma_{11} + \sigma_{22} + 2\,p = 0$$
$$\sigma_{23} + \sigma_{13} = 0, \qquad \sigma_{11} + p = 0 \tag{7.50}$$

Meantime, the von Mises's quadratic expression is written as

$$(\sigma_{11} - \sigma_{22})^2 + 4\,\sigma_{12} = \frac{16}{3}\,k^2 = \frac{4}{3}\,\sigma_y^2 \tag{7.51}$$

Also, the set of six constitutive equations (7.48) reduces to three equations, of which only two are independent; since $\dot{\varepsilon}_{11} + \dot{\varepsilon}_{22} = 0$, equation (7.50).

In view of the above assumptions and results, one arrives to the following set of governing equations for the *"plane strain problem"*:

$$\frac{\partial \sigma_{11}}{\partial x_1} + \frac{\partial \sigma_{12}}{\partial x_2} = 0, \qquad \frac{\partial \sigma_{12}}{\partial x_1} + \frac{\partial \sigma_{22}}{\partial x_2} = 0 \tag{7.A}$$

$$(\sigma_{11} - \sigma_{22})^2 + 4\,\sigma_{12}^2 = \frac{16}{3}\,k^2 = \frac{4}{3}\,\sigma_y^2 \tag{7.B}$$

$$\dot{\varepsilon}_{11} + \dot{\varepsilon}_{22} = 0, \qquad \frac{\dot{\varepsilon}_{22} - \dot{\varepsilon}_{11}}{\dot{\varepsilon}_{12}} = \frac{\sigma_{22} - \sigma_{11}}{\sigma_{12}} \tag{7.C}$$

The set of equations (7.A-C) are to be solved for the five unknowns σ_{11}, σ_{22}, σ_{12}, v_1 and v_2.

7.7.2. THE PROBLEM OF "PLANE STRESS". QUADRATIC YIELD CONDITION.

In the formulation of the complete problem of *"plane stress"*. We again assume that one of the principal directions of the stress tensor σ and, hence, also, of the strain-rate tensor $\dot{\varepsilon}$ are parallel to a fixed direction (*say* the x_3 - axis). Further, similar to the above analysis concerning the problem of *"plane strain"*, we consider that

$$\sigma_{23} = \sigma_{13} = 0 \quad \text{and} \quad \sigma_{33} = 0 = \sigma_3 \qquad (7.52)$$

Meantime the stress remaining components σ_{11}, σ_{22}, σ_{12} are seen to be independent of x_3. In this case, the von Mises's yield condition (7.46) reduces to

$$\sigma_1^2 + \sigma_2^2 - \sigma_1 \sigma_2 = 4 k^2 \qquad (7.D)$$

or, alternatively,

$$\sigma_{11}^2 + \sigma_{22}^2 - \sigma_{11} \sigma_{22} + 3 \sigma_{12}^2 = 4 k^2 \qquad (7.D)$$

Also, the equilibrium equations (7.45) reduce to the two equations (7.A), i.e.,

$$\frac{\partial \sigma_{11}}{\partial x_1} + \frac{\partial \sigma_{12}}{\partial x_2} = 0, \quad \frac{\partial \sigma_{12}}{\partial x_1} + \frac{\partial \sigma_{22}}{\partial x_2} = 0 \qquad (7.A)$$

Further, from the constitutive equation (7.48) and the continuity equation (7.49), one arrives at

$$\sigma_{11} + \sigma_{22} + 3p = 0 \qquad (7.53)$$

Thus, by substituting into the stress-strain rate relations (7.48), one has

$$\left. \begin{array}{l} 2\sigma_{11} - \sigma_{22} = \ \ 3\alpha \dot{e}_{11} \\[2mm] 2\sigma_{22} - \sigma_{11} = \ \ 3\alpha \dot{e}_{22} \\[2mm] 2\sigma_{12} = 3\alpha \dot{e}_{12} \end{array} \right\} \qquad (7.E)$$

Equations (7.A) and (7.E) constitute a system of three equations for σ_{11}, σ_{22} and σ_{12}. Corresponding to each solution of this system, there exists solution for v_1, v_2, $\alpha \geq 0$ of the three equations (7.E). Meantime, there remain three equations of the constitutive relation (7.48), i.e.,

$$\left.\begin{array}{c} \dfrac{\partial v_1}{\partial x_3} + \dfrac{\partial v_3}{\partial x_1} = 0, \qquad \dfrac{\partial v_2}{\partial x_3} + \dfrac{\partial v_3}{\partial x_2} = 0 \\[3mm] \dfrac{\partial v_3}{\partial x_3} + \dfrac{\sigma_{11} + \sigma_{22}}{3k} = 0 \end{array}\right\}$$ (7.54)

which can be solved for v_3, $\dfrac{\partial v_1}{\partial x_3}$ and $\dfrac{\partial v_2}{\partial x_3}$.

One notices at this point that the set of equations (7.48) of the deviatoric stress-strain rate relations and the von Mises's yield condition (7.46) are not independent of each other.

7.8. The Three-Dimensional Problem under General Yield Function

In order to arrive to a *"complete plane problem"* with a more general yield condition, Geiringer (1953) assumed the yield condition to be represented by a general function of the form

$$g(\sigma) = 0$$ (7.55)

in replacement of the von Mises quadratic yield condition (7.46).

The new "general" yield condition (7.55) is subjected to the following restrictions which, by consequence, influence the corresponding yield condition.

(a)   For an isotropic medium, the yield function g (·) must be independent of the coordinate system. Thus, g (·) depends only on the three invariants I_1, I_2 and I_3 of the stress tensor, or, alternatively, on the three principal stresses σ_1, σ_2 and σ_3 and in a symmetric manner.

(b)   It is assumed that g (·) remains unchanged if the stress tensor σ is replaced by the deviatoric stress σ' as an argument of g (·).

In view of the restrictions (a) and (b) above, it follows that

$$\begin{aligned} g(\sigma) = g(\sigma') &= G(\sigma_1, \sigma_2, \sigma_3) \\ &= G(\sigma_1', \sigma_2', \sigma_3') \end{aligned}$$ (7.56)

This is with the understanding that $G(\cdot)$ is a symmetric function of the principal stresses. In other words, $G(\cdot)$ depends only on the differences between the principal stresses, i.e., on $(\sigma_1-\sigma_3)$, etc., or, alternatively, on the principal shear stresses $\tau_1 = 1/2\,(\sigma_2 - \sigma_3)$, $\tau_2 = 1/2\,(\sigma_3 - \sigma_1)$ and $\tau_3 = 1/2\,(\sigma_1 - \sigma_2)$. Thus, the representation (7.56) may be extended to read as

$$g\,(\sigma) = G(\sigma_1, \sigma_2, \sigma_3) = S(\tau_1, \tau_2, \tau_3) \qquad (7.57)$$

where the two functions $G(.)$ and $S(.)$ are symmetric with respect to their arguments. The form (7.57) is dealt with by Geiringer (1953) as a " *general yield condition for an isotropic perfectly plastic material*".

In view of (7.57), one has

$$\frac{\partial g(\cdot)}{\partial \sigma_{11}} + \frac{\partial g(\cdot)}{\partial \sigma_{22}} + \frac{\partial g(\cdot)}{\partial \sigma_{33}} = \frac{\partial G(\cdot)}{\partial \sigma_1} + \frac{\partial G(\cdot)}{\partial \sigma_2} + \frac{\partial G(\cdot)}{\partial \sigma_3} = 0 \qquad (7.58)$$

-   The equation of motion (7.44), or, alternatively, the equilibrium equation, in the absence of body and/or external forces, (7.45), is applicable. Thus, considering the case of equilibrium, equation (7.45), i.e.,

$$\mathrm{div}.\ \sigma = 0$$

-   Following Geiringer (1953), we assume the following stress-strain rate proportionality relationship to be applicable

$$\mathrm{Grad}\ g\,(\sigma) = \alpha\,\dot{\varepsilon} \qquad (7.59a)$$

where α is, as introduced earlier, a proportionality function $\alpha\,(\cdot)$. Expression (7.59a) has six components which can be written explicitly as

$$\left.\begin{array}{ccc} \dfrac{\partial g(\cdot)}{\partial \sigma_{11}} = \alpha\,\dot{\varepsilon}_{11}, & \dfrac{\partial g(\cdot)}{\partial \sigma_{22}} = \alpha\,\dot{\varepsilon}_{22}, & \dfrac{\partial g(\cdot)}{\partial \sigma_{33}} = \alpha\,\dot{\varepsilon}_{33} \\[3mm] \dfrac{\partial g(\cdot)}{\partial \sigma_{12}} = \alpha\,\dot{\varepsilon}_{12}, & \dfrac{\partial g(\cdot)}{\partial \sigma_{13}} = \alpha\,\dot{\varepsilon}_{13}, & \dfrac{\partial g(\cdot)}{\partial \sigma_{23}} = \alpha\,\dot{\varepsilon}_{23} \end{array}\right\} \qquad (7.59b)$$

of which five equations are independent.

The three-dimensional problem under general yield condition consists, in view of the above, of the equilibrium equations (7.45), the continuity equations (7.49), the general yield condition (7.57), and the stress-strain rate proportionality equations (7.59). In this system of ten equations, there are ten unknowns, namely, the six components of the stress tensor σ, the three components of the deformation-rate vector v, and the proportionality function value α.

7.9. The "Plane Problem" under General Yield Condition

In this section, we derive, following Geiringer(1953), the *"plane problem"* from a three dimensional one, i.e., by utilizing a *"general yield condition"*. In doing so, one aims at introducing, as stated by Geiringer (1953), "desirable generality along with necessary conditions". We deal, below, first with *"plane strain"* followed by *"plane stress"*.

7.9.1. THE PROBLEM OF "PLANE STRAIN". GENERAL YIELD CONDITION.

Plane strain is defined by

$$\dot{\varepsilon}_3 = \dot{\varepsilon}_{33} = v_{3,3} = 0$$
$$\dot{\varepsilon}_{23} = \dot{\varepsilon}_{13} = 0$$

and all components of stress and strain are independent of x_3.

Theorem: For an isotropic medium, i.e. for a plastic potential, $h(\sigma)$, that depends only on the principal stresses, the tensor *"Grad h"* is coaxial with the stress tensor. (The reader is referred to Geiringer, 1953, pgs. 206-7 for a proof of this theorem).

Recalling (7.58), and incorporating the above theorem, it follows that

$$\frac{\partial g(\cdot)}{\partial \sigma_{33}} = \frac{\partial g(\cdot)}{\partial \sigma_{23}} = \frac{\partial g(\cdot)}{\partial \sigma_{23}} = 0$$

and,

$$\frac{\partial h(.)}{\partial \sigma_{11}} + \frac{\partial h(.)}{\partial \sigma_{22}} + \frac{\partial h(.)}{\partial \sigma_{33}} = \frac{\partial G(\cdot)}{\partial \sigma_1} + \frac{\partial G(\cdot)}{\partial \sigma_2} + \frac{\partial G(\cdot)}{\partial \sigma_3} = 0 \qquad (7.60)$$

Thus, from the above definition of *"plane strain"* in conjunction with (7.60), one can write that

$$\frac{\partial G(.)}{\partial \sigma_3} = 0 \qquad (7.F)$$

Thus, one can conclude from the property of *Grad h*, as stated by the above theorem, that

$$\sigma_{23} = \sigma_{13} = 0 \quad \text{and} \quad \sigma_{33} = \sigma_3 \qquad (7.61)$$

Accordingly (7.F) can be replaced by

$$\sigma_{23} = \sigma_{13} = 0 \qquad (7.62a)$$

$$\sigma_{33} = F(\sigma_1 + \sigma_2) = \sigma_3 \qquad (7.62b)$$

where $F(\cdot)$ is a symmetric function with respect to the two principal stresses σ_1 and σ_2. In the case of quadratic yield condition, the corresponding expression to (7.62b) is $\sigma_3 = 1/2 (\sigma_1 + \sigma_2)$

Thus, the *"plane yield condition"* may be written as

$$f(\sigma_{11}, \sigma_{22}, \sigma_{12}) = F(\sigma_1, \sigma_2) = 0 \qquad (7.G)$$

with

$$\frac{\partial f(\cdot)}{\partial \sigma_{11}} + \frac{\partial f(\cdot)}{\partial \sigma_{22}} = \frac{\partial F(\cdot)}{\partial \sigma_1} + \frac{\partial F(\cdot)}{\partial \sigma_2} = 0$$

–   **The stress-strain relations:**
The stress-strain relations of the three-dimensional problem reduce in the present case of *"plane strain"* to

$$\frac{\partial f(.)}{\partial \sigma_{11}} = \alpha \dot{\varepsilon}_{11}, \quad \frac{\partial f(.)}{\partial \sigma_{22}} = \alpha \dot{\varepsilon}_{22} \quad \text{and} \quad \frac{\partial f(.)}{\partial \sigma_{12}} = \alpha \dot{\varepsilon}_{12} \qquad \text{(7.H)}$$

of which only two are independent since

$$\left. \begin{array}{l} \dot{\varepsilon}_{11} + \dot{\varepsilon}_{22} = 0 \quad \text{and} \\[2mm] \dfrac{\partial f(\cdot)}{\partial \sigma_{11}} + \dfrac{\partial f(\cdot)}{\partial \sigma_{22}} = \dfrac{\partial f(\cdot)}{\partial \sigma_1} + \dfrac{\partial f(\cdot)}{\partial \sigma_2} = 0 \end{array} \right\} \qquad \text{(7.I)}$$

Thus, the plane strain problem, under general yield condition, consists of the following set of equations

- the two equations of equilibrium (7.A)
- the yield condition (7.G) with the restriction (7.I)
- the stress-strain equations (7.H) of which, as previously mentioned, two are independent.

These are six equations to be solved for the six unknowns σ_{11}, σ_{22}, σ_{12}, v_1, v_2 and α.

7.9.2. THE PROBLEM OF "PLANE STRESS". GENERAL YIELD CONDITION.

Plane stress is defined by

$$\sigma_{33} = \sigma_{23} = \sigma_{13} = 0 \quad (\text{i.e., } \sigma_3 = 0)$$

and σ_{11}, σ_{22} and σ_{12} depend on x_1 and x_2 only.

Meantime, the equilibrium equation (7.45) reduces to the two equations (7.A); namely

$$\frac{\partial \sigma_{11}}{\partial x_1} + \frac{\partial \sigma_{12}}{\partial x_2} = 0, \quad \frac{\partial \sigma_{12}}{\partial x_1} + \frac{\partial \sigma_{22}}{\partial x_2} = 0 \qquad \text{(7.A)}$$

Also, the yield condition is expressed in a general format as

$$g(\sigma_{11}, \sigma_{22}, \sigma_{12}) = f(\sigma_{11}, \sigma_{22}, \sigma_{12}) = G(\sigma_1, \sigma_2)$$
$$= F(\sigma_1, \sigma_2) = 0 \qquad \text{(7.J)}$$

234

where F( ) is symmetric with respect to σ_1 and σ_2.

It follows, then, that,

$$\frac{\partial g(\cdot)}{\partial \sigma_{11}} = \frac{\partial f(\cdot)}{\partial \sigma_{11}} \tag{7.J'}$$

with both $g(\cdot)$ and $f(\cdot)$ are defined by (7.J) above.

Further, one arrives, as in (7.H), at the following stress-strain relations

$$\alpha \dot{\varepsilon}_{11} = \frac{\partial f(\cdot)}{\partial \sigma_{11}} \; , \quad \alpha \dot{\varepsilon}_{22} = \frac{\partial f(\cdot)}{\partial \sigma_{22}} \; , \quad \alpha \dot{\varepsilon}_{12} = \frac{\partial f(\cdot)}{\partial \sigma_{12}} \tag{7.K}$$

with, upon the elimination of α,

$$\dot{\varepsilon}_{11} : \dot{\varepsilon}_{22} : \dot{\varepsilon}_{12} = \frac{\partial f(\cdot)}{\partial \sigma_{11}} : \frac{\partial f(\cdot)}{\partial \sigma_{22}} : \frac{\partial f(\cdot)}{\partial \sigma_{12}}$$

Accordingly the problem of plane stress, derived from the general complete problem of the perfectly isotropic solid, is defined by the *six* equations (7.A), (7.J) and (7.K), or alternatively, by the five equations

$$\left.\begin{array}{c} \dfrac{\partial \sigma_{11}}{\partial x_1} + \dfrac{\partial \sigma_{12}}{\partial x_2} = 0, \quad \dfrac{\partial \sigma_{12}}{\partial x_1} + \dfrac{\partial \sigma_{22}}{\partial x_2} = 0 \\[2mm] f(\sigma_{11} \cdot \sigma_{22}, \sigma_{12}) = F(\sigma_1, \sigma_2) = 0 \\[2mm] \dfrac{\dot{\varepsilon}_{11} \mp \dot{\varepsilon}_{22}}{\dot{\varepsilon}_{12}} = \dfrac{\dfrac{\partial f}{\partial \sigma_{11}} \mp \dfrac{\partial f}{\partial \sigma_{22}}}{\dfrac{\partial f}{\partial \sigma_{12}}} \end{array}\right\} \tag{7.L}$$

Some Remarks on the Formulation of "Plane Strain" and "Plane Stress" Problems under the Assumption of "General Yield Condition":

(i)   Comparing the governing equations for the plane strain problem under general yield conditions and the set of equations (7.L) for the corresponding plane stress problem, one realizes that the mathematical difference consists only in that in (7.L) the sum

$$\frac{\partial f(.)}{\partial \sigma_{11}} + \frac{\partial f(.)}{\partial \sigma_{22}} = m \qquad (7.63)$$

(and by consequence the sum $\dot{\varepsilon}_{11} + \dot{\varepsilon}_{22}$) must be zero identically for all stress values.

(ii)   As consequence to (i) above, one may consider as a "General Plane Problem", under general yield condition, the problem defined by (7.L) with no restriction on the sum given by (7.63).

(iii)   The particular case where the problem (7.L) is subjective to the condition

$$\frac{\partial f(.)}{\partial \sigma_{11}} + \frac{\partial f(.)}{\partial \sigma_{22}} = 0 \qquad (\text{i.e., } m = 0 \text{ in Eqn. 7.63}) \qquad (7.64)$$

was referred to by Geiringer (1953) as *"the orthogonal case"*.
In this context, the most frequently considered orthogonal case where
$f = (\sigma_{11} - \sigma_{22})^2 + 4\sigma_{12}^2$ = constant (resulting from von Mises's three-dimensional condition) in case of plane strain, as defined by the set of equations (7.A), (7.B) and (7.C), is referred to by Gieringer as *the special or classical case"*. The mathematical theory of the special or classical case is dealt with extensively in the literature. In this context, the reader is referred, for instance, to the books by Nadai (1931, 1950), Sokolovsky (1947), Hill (1950), Hodge (1950) and Prager and Hodge (1951).

EXAMPLE PROBLEM 7.1

The following expression expresses the von Mises yield condition, in terms of the principal values of the stress tensor,

$$(\sigma_1 - \sigma_2)^2 + (\sigma_2 - \sigma_3)^2 + (\sigma_3 - \sigma_1)^2 = 6 k^2 \tag{1}$$

Show that this expression can be written, in terms of the principal deviatoric stresses, as follows

$$\sigma_1^2 + \sigma_2^2 + \sigma_3^2 = 2 k^2 \tag{2}$$

Solution:

One can write the relations between the principal stresses and the corresponding principal deviatoric stresses as follows

$$\sigma_1 = \sigma_1' + \sigma_M$$
$$\sigma_2 = \sigma_2' + \sigma_M$$
$$\sigma_3 = \sigma_3' + \sigma_M$$

where σ_M is the mean stress (Chapter 2). Thus, the von Mises yield condition (1) can be written as

$$(\sigma_1' - \sigma_2')^2 + (\sigma_2' - \sigma_3')^2 + (\sigma_3' - \sigma_1')^2 = 6 k^2$$

which, upon expanding and rearranging, can be expressed as

$$\sigma_1'^2 + \sigma_2'^2 + \sigma_3'^2 - (\sigma_1' + \sigma_2' + \sigma_3')^2 / 3 = 2 k^2$$

But

$$\sigma_1' + \sigma_2' + \sigma_3' = I' = 0$$

then, the required expression (2) follows.

EXAMPLE PROBLEM 7.2

Consider the state of stress:

$$\sigma_{11} = \sigma, \; \sigma_{22} = \sigma_{33} = 0 \; , \quad \sigma_{12} = \tau, \; \sigma_{23} = \sigma_{13} = 0$$

which is produced in a tension-torsion test of a thin-walled tube. Derive the yield curves on the σ-τ plane for the Tresca and von Mises conditions, with the inclusion of the yield stress in simple tension σ_y.

Solution:

For the given state of stress, the principal stress values are
Recalling the Tresca yield condition:

$$\left. \begin{aligned} \sigma_1 &= (\sigma + \sqrt{4\tau^2 + \sigma^2})/2 \\ \sigma_2 &= 0 \\ \sigma_3 &= (\sigma - \sqrt{4\tau^2 + \sigma^2})/2 \end{aligned} \right\} \tag{1}$$

$$\sigma_1 - \sigma_3 = \sigma_y \tag{2}$$

Thus, the Tresca yield curve is $\sqrt{4\tau^2 + \sigma^2} = \sigma_y$ or, alternatively,

$$\sigma^2 + 4\tau^2 = \sigma_y^2 \tag{3}$$

which represents an ellipse on the σ - τ plane.

Similarly, recalling the von Mises yield condition:

$$(\sigma_1 - \sigma_2)^2 + (\sigma_2 - \sigma_3)^2 + (\sigma_3 - \sigma_1)^2 = 2\sigma_y^2 \tag{4}$$

Combining (1) and (4), it follows that the Mises yield curve is represented by

$$\sigma^2 + 3\tau^2 = \sigma_y^2 \tag{5}$$

which is also an ellipse on the σ - τ plane.

It is left to the student to illustrate schematically the Tresca and von Mises yield curves for the considered stress state.

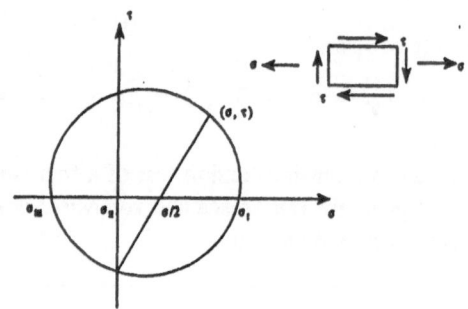

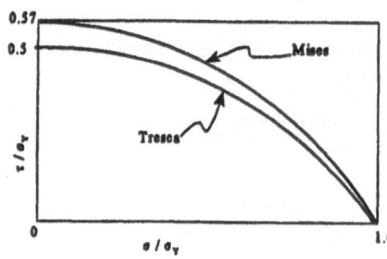

Figure 7.10

EXAMPLE PROBLEM 7.3

For a biaxial state of stress with $\sigma_{II} = 0$, determine the yield loci for the von Mises and Tresca conditions. Compare the two criteria, for this case, by a plot in the two-dimensional space:

$$\sigma_1 \Big/ \sigma_y \quad vs. \quad \sigma_3 \Big/ \sigma_y$$

Solution:

In terms of the yield stress in simple tension, the von Mises criterion is expressed as

$$(\sigma_1 - \sigma_2)^2 + (\sigma_2 - \sigma_3)^2 + (\sigma_3 - \sigma_1)^2 = 2\,\sigma_y^2$$

which with $\sigma_2 = 0$ becomes

$$\sigma_1^2 - \sigma_1\sigma_3 + \sigma_3^2 = \sigma_y^2$$

which is the ellipse:

$$(\sigma_1/\sigma_y)^2 - (\sigma_1\sigma_3/\sigma_y^2) + (\sigma_3/\sigma_y)^2 = 1$$

Similarly, recalling the Tresca yield condition as referred to the yield stress in simple tension, i.e., $\sigma_1 - \sigma_3 = \sigma_y$ and the compression equations

$$\sigma_3 - \sigma_2 = \sigma_y$$
$$\sigma_2 - \sigma_1 = \sigma_y$$

It follows, with reference to Fig. 7.11, that:

(i)   the line segments AB and ED are determined, respectively, with the equations :

$$(\sigma_1/\sigma_y) - (\sigma_3/\sigma_y) = \pm 1$$

(ii)   the line segments DC and FA are determined, respectively, with the equations $\sigma_3/\sigma_y = \pm 1$,

(iii) the line segments BC and EF are determined, respectively, with the equations $\sigma_1/\sigma_y = \mp 1$.

It is left to the student to locate the von Mises ellipse with respect to the Tresca hexagon shown on Figure 7.11.

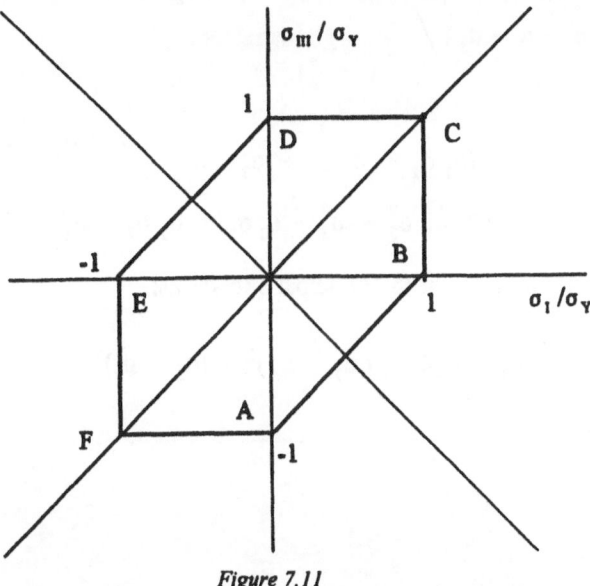

Figure 7.11

EXAMPLE PROBLEM 7.4

Express the von Mises yield condition

$$- I_2' = C_y \tag{1}$$

in terms of the principal stresses

Solution:

Recalling the characteristic equation for the deviatoric stress tensor, i.e.

$$\sigma'^3 + I_2' \, \sigma' - I_3' = 0 \tag{2a}$$

or

$$\sigma'^3 + (\sigma_1' \sigma_2' + \sigma_2' \sigma_3' + \sigma_3' \sigma_1') \, \sigma' - \sigma_1' \sigma_2' \sigma_3' = 0 \tag{2b}$$

Thus, it follows that

$$- I_2' = - (\sigma_1' \sigma_2' + \sigma_2' \sigma_3' + \sigma_3' \sigma_1')$$

Meantime recalling $\sigma_1' = \sigma_1 - \sigma_M$, $\sigma_2' = \sigma_2 - \sigma_M$ and $\sigma_3' = \sigma_3 - \sigma_M$
where $\sigma_M = (\sigma_1 + \sigma_2 + \sigma_3) / 3$
Thus

$$- I_2' = - (\sigma_1 \sigma_2 + \sigma_2 \sigma_3 + \sigma_3 \sigma_1) + \frac{(\sigma_1 + \sigma_2 + \sigma_3)^2}{3}$$

$$= 2 (\sigma_1^2 + \sigma_2^2 + \sigma_3^2 - \sigma_1 \sigma_2 - \sigma_2 \sigma_3 - \sigma_3 \sigma_1) / 6$$

and Eqn. (1) reads in terms of the principal stresses as:

$$(\sigma_1 - \sigma_2)^2 + (\sigma_2 - \sigma_3)^2 + (\sigma_3 - \sigma_1)^2 = 6 C_y$$

7.10. Problems

1. Explain briefly, and with schematics, the following terms:
    - yield surface
    - yield curve
2. Illustrate the differences between Tresca and von Mises yield conditions as based on simple tension and pure shear.
3. Show that the von Mises yield condition can be written in the form

$$(\sigma_{11} - \sigma_{22})^2 + (\sigma_{22} - \sigma_{33})^2 + (\sigma_{33} - \sigma_{11})^2 + 6(\sigma_{12}^2 + \sigma_{23}^2 + \sigma_{31}^2) = 6k^2$$

4. For the state of stress:

$$\sigma_{ij} = \begin{bmatrix} 10 & 2 & 0 \\ 2 & 10 & 0 \\ 0 & 0 & 10 \end{bmatrix}$$

determine the yield condition corresponding to von Mises and Tresca criteria.
5. Determine the yield condition for the state of stress

$$[\sigma_{ij}] = \begin{vmatrix} 6 & -3 & 0 \\ -3 & 6 & 0 \\ 0 & 0 & 6 \end{vmatrix}$$

according to Tresca's and von Mises's criteria.
6. Show that in *"plane strain"*, von Mises's three-dimensional yield condition leads to the following two-dimensional condition

$$(\sigma_1 - \sigma_2)^2 = \text{constant}$$

where σ_1, σ_2 are principal stresses.

7. Show that the Levy-Mises equations imply that the principal axes of the rate of deformation coincide with the principal axes of stress.

8. Interpret the Mises yield condition as a condition on (i) the octahedral shear stress, and (ii) the elastic distortion energy.

9. In plane plastic strain governed by the Levy-Mises equations, with $\varepsilon_{33} = 0$ and elastic strains neglected, show that $\sigma_{33} = 1/2(\sigma_{11} + \sigma_{22})$ with $\sigma_{31} = \sigma_{32} = 0$. Hence, write the Mises yield condition for plastic strain as a function of σ_{11}, σ_{22} and σ_{12}, and show it is identical to the Tresca yield condition for plane plastic strain.

7.11. Review Problems

1. (i) Expand the 3rd order tensor

$$U_{ijk} = a_{il} a_{jm} a_{kn} V_{lmn}$$

(ii) Prove the following forms
a) $\delta_{ii} = 3$
b) $\delta_{ik} e_{ikl} = 0$
c) $e_{ijk} e_{ijk} = 6$
d) $e_{ijk} e_{mjk} = 2 \delta_{im}$

2. (i) For the following state of stress, determine the stress invariants and the principal stress values. Display the matrix of the principal values of stress and determine the stress invariants in the latter case.

$$[\sigma_{ij}] = \begin{vmatrix} 6 & -3 & 0 \\ -3 & 6 & 0 \\ 0 & 0 & 8 \end{vmatrix}$$

(ii) Determine the yield condition for the state of stress shown under (i) above according to Tresca and von Mises criteria.

3. (a) For the following stress matrix, determine the principal deviatoric stress values

$$[\sigma_{ij}] = \begin{vmatrix} 12 & 4 & 0 \\ 4 & 9 & -2 \\ 0 & -2 & 3 \end{vmatrix}$$

4. Assuming that the strain tensor is given by

$$\varepsilon_{ij} = 10^{-2} \begin{vmatrix} 2 & 3 & 2 \\ 3 & 2 & 1 \\ 2 & 1 & 2.5 \end{vmatrix},$$

determine the principal strain and the corresponding principal directions.

5. Determine an expression for the first deviatoric stress invariant.

6. The second deviatoric stress invariant can be expressed as $I_2' = \dfrac{1}{2} \sigma_{ij}' \sigma_{ij}'$

where σ_{ij}' are the components of the deviatoric stress tensor. Show that this expression can be equivalently written as

$$I_2' = \sigma_{kk}^2 - I_2$$

where I_2 is the stress tensor second invariant.

7. Derive an expression for the deviatoric stress third invariant in terms of the stress tensor third invariant.

8. Given the following expression for the second deviatoric stress invariant, derive an expression for the letter in terms of the octahedral shear stress, σ_{oct}.

$$I_2' = \frac{1}{3} \sigma_{ij}' \sigma_{jk}' \sigma_{ki}'$$

9. Show that the deviatoric stress σ_{ij}' is equal to the total stress σ_{ij} if $i \neq j$.

10. Show that the elastic tensor modulus of an isotropic material may depend only on two material parameters.

11. Illustrate the differences between Tresca and von Mises yield conditions as based on simple tension and pure shear.

12. For the state of stress:

$$\sigma_{ij} = \begin{bmatrix} 10 & 2 & 0 \\ 2 & 10 & 0 \\ 0 & 0 & 10 \end{bmatrix},$$

determine the yield condition corresponding to von Mises and Tresca criteria.

13. What is meant by an "*Octahedral plane*"? Express the shear stress acting on this plane in terms of the stress tensor components. Then, derive from the letter an expression for the octahedral shear stress in terms of the principal stresses.

14. The following relation expresses the set of Prandtl-Reuss equations

$$d\,\varepsilon_{ij}^p = \sigma_{ij}'\, d\lambda$$

Show that this set of equations implies that the principal axes of plastic strain increments coincide with principal stress axes

7.12. Transition to the Creep of Metals and Alloys

7.12.1. INTRODUCTION

In the present section, we present a brief introduction to the important subject of creep of crystalline solids, such as metals and alloys, at elevated temperatures. Creep of such materials is often regarded as a continuing plastic deformation process when the material specimen is subjected to constant loading. Research efforts in recent years have contributed significantly towards understanding of different microscopic and macroscopic mechanisms that may determine the physical nature of high temperature creep in crystalline solids. From an atomistic point of view, creep of such materials at high temperatures has been explained (Mukherjee, 1974) by assuming a cooperative mechanism between the occurring dislocations in the solid and the kinetics of reaction. In this, it is often argued that creep of crystalline solids occurs as the result of thermally-activated migration of dislocations, diffusion of vacancies, grain boundary deformation, and possibly as the outcome of stochastic interaction between the foregoing mentioned effects under various conditions. Deformation diagrams are often constructed (e.g., Ashby, 1972 and Frost and Ashby, 1973&1982) which show the fields of stress and temperature in which a given mechanism may be dominant and the strain rate that it may yield.

From a microstructural point of view, one may differentiate between the various mechanisms that may be responsible for the following two aspects of creep.

(i) Low-temperature Creep
At low temperatures (above the absolute zero), dislocations can overcome only the lowest energy barriers that may obstruct their motion. Thus, during a deformation process, dislocations would progressively proceed to new and often high energy barriers which may be able to override, but often with a continuously decreasing rate. This may be primarily due to the infrequency of high intensity thermal fluctuations in energy that would be needed to activate the low energy barriers at low temperatures. Hence, low-temperature creep of crystalline solids is often characterized by a monotonously decreasing (ever diminishing) creep rate.

(ii) High-temperature Creep
At high temperatures, thermal fluctuations that may be needed to stimulate the low energy mechanisms, such as the nucleation of kink pairs, cross-slip, etc., become quite frequent such that barriers to dislocations motion would be no longer effective. When such conditions prevail, the gliding dislocations, however, may be eventually arrested by high energy barriers that could result, for instance, from long stress fields. Thus, high-temperature creep for a crystalline solid is often characterized by a larger *"apparent initial creep strain"*, when compared with low-temperature creep, under the same applied level of loading. This would be, then, followed by immeasurably small creep rates.

At temperatures about one half the melting temperature $(0.5T_m)$, creep of the crystalline solid may continue to take place despite the fact that glide dislocations at such temperatures often become arrested at high energy barriers. It is just at these high temperatures that creep of crystalline solids may continue until it is modified by auxiliary microstructural mechanisms such as grain boundary fissuring, grain cavitation and eventually necking which would ultimately lead to fracture. A number of different diffusion-controlled mechanisms are known to determine high-temperature creep rates and each mechanism may be particularly influenced by various substructural and microstructural effects.

For a review of the subject of creep of crystalline solids at high temperatures, reference is made to Kanter (1936), Finnie (1959), Johnson (1960), Dorn and Mote (1963), Garafalo (1965), Grant and Mullendore (1965), Oding (1965), Hult (1966), McLean (1966), Kachanov (1967), Sherby and Burke (1967), Rabotnov (1968),Weertman (1968), Bird *et al* (1969), Mukherjee *et al.* (1969, 1982), Grant (1971), Greenfield (1972), Lagenborg (1972), Mukherjee (1974), Leckie (1981), Odqvist (1981), Frost and Ashby (1982), Gooch and How (1986), Webster and Ainsworth (1994), among others.

7.12.2. HIGH-TEMPERATURE CREEP

Creep tests of crystalline solids produce creep curves of different forms, but they all demonstrate the essential features of the so-called "*idealized*" or "*classical*" creep curve. Classically, the creep curve for crystalline solids is seen as being divided into four regions: an "*initial*" or "*instantaneous*" response followed by characteristic three stages of creep. The instantaneous strain response occurs upon the application of the load. If the applied stress exceeds the elastic limit of the solid at the temperature concerned, the instantaneous response will be composed of both elastic and plastic components. The elastic component will be recovered if the stress is removed. The instantaneous strain response is not a characteristic stage of the creep curve, but it is the initial strain response that usually occurs upon or immediately after the application of the load (e.g., Oding, 1965).

Figure 7.12 is due to Mukherjee (1974). It illustrates several common types of high-temperature creep curves of crystalline solids. The three stages of response characteristic of creep are identified in this Figure: a "*primary*" creep period (*Stage I*), followed by a "*secondary*" stage of steady-state creep (*Stage II*) and, then, a "*tertiary*" period (*Stage III*) characterized by an increasing creep rate. The notions "*primary*", "*secondary*" and "*tertiary*" creep are often referred to Costa Andrade (1910); *see* Odqvist, 1981. With reference to Fig. 7.12, Stage II is an extended stage of a steady creep rate. It is characterized by a balance between the rate of generation of the creep-resistant substructure and the rate of its thermal recovery under the imposed stress field. The *tertiary* stage III, is a final period in the creep deformation process of the crystalline solid. It is often characterized by local plastic deformation, micro-fissuring and creep rupture. Major differences between high-temperature creep of most metals and alloys occur, however, during the *primary* creep period, Stage I, which is a transient stage characterized by a gradually decreasing rate of strain. This stage of creep often depends on the initial dislocation structure. As shown in Fig. 7.12, one may encounter the following variations concerning the *primary* creep stage I:

(i)   *An Usual Type of Creep Response (Curve Type A)*.
        This type of creep behaviour would exhibit a decelerating primary creep rate, illustrating the formation of a more creep-resistant substructure during the deformation process (*e.g.*, Hazlett and Hansen, 1954).

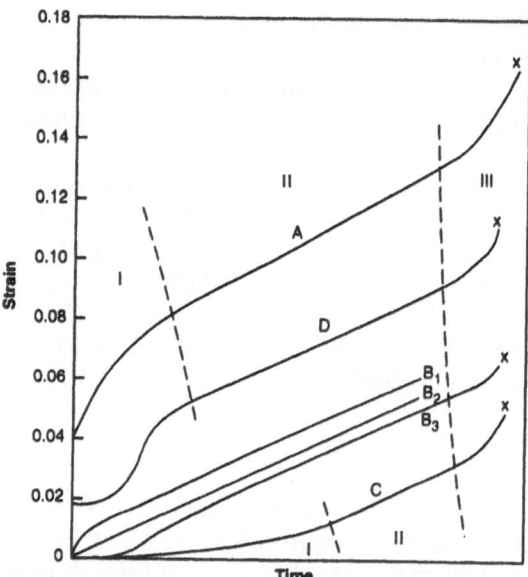

Figure 7.12. Common types of creep curves. From: Mukherjee, A.K. (1974) High-temperature creep, in *Plastic Deformation of Materials*, Treatise on Materials Science and Technology, Ed. H. Herman, Vol .6, Academic Press, New York, pp. 163-224. Reprinted with kind permission of Academic Press.

(ii)   *An Immediate Entry to the Steady-state (Curve Type B)*
Creep curves of this type would enter the steady-state (Stage II) almost immediately. This type of creep behaviour may be typical of crystalline solids in which the substructure relevant to creep remains unchanged during the period.

(iii)   *A Smooth Transition to the Secondary Creep Stage (Curve Type C)*
This type of creep response is characterized by a gradual increasing creep rate over the primary Stage I with a smooth transition to the secondary creep stage II.   In this, Raymond and Dorn (1964) have shown that metals and alloys of Type A, mentioned under (i) above, may behave according to Type C when they have previously undergone creep at higher applied stresses. Meantime, Hazlett and Hansen (1954) have argued that previous cold working may have the same effect.   Hence, the increasing creep rate over the *primary* stage I for these materials may be attributed to the recovery of the pertinent substructure to a steady-state condition.

(iv)   *A Sigmoidal Creep Response (Curve Type D)*
This mode of creep response has been observed, for instance, in certain disposed phase alloys (e.g., Webster and Piearcey, 1967). This may suggest the nucleation and spread of slip zones during the *primary* stage I.

The shape of the creep curve, for a particular material, would generally depend on both the stress and temperature levels. At very low stresses for a given temperature level, the creep curve may, for instance, consist of only two stages (Fig. 7.13). In this case, the creep stage II, which is characterized by a fixed creep rate, may last for a quite extended period of time and the subsequent creep stage III may not occur at all. In this situation (*Curve a of Fig. 7.13*), the steady-state creep rate is significantly low. On the other hand, at higher stresses, failure would occur after short periods of service and the creep stage II would be only of a short duration or non-existent, as the creep stage III may appear almost immediately after the primary creep stage I (*Curves b, c and d of Fig. 7.13*); e.g., Oding (1965).

7.12.3. CONSTITUTIVE RESPONSE EQUATIONS

Most of the published experimental data concerning creep pertains to the *secondary* stage II. By a consequence, most existing creep formulations are principally related to this stage of creep.

A large number of constitutive models have been proposed to determine the creep response of crystalline materials. The majority of these models are, however, empirical; e.g. Bailey and Roberts (1933), Kanter (1938), Fastov (1950), *see* Oding (1965), Sherby *et al.* (1954), Kennedy (1962), Rabotnov (1969) and Webster and Ainsworth (1994).

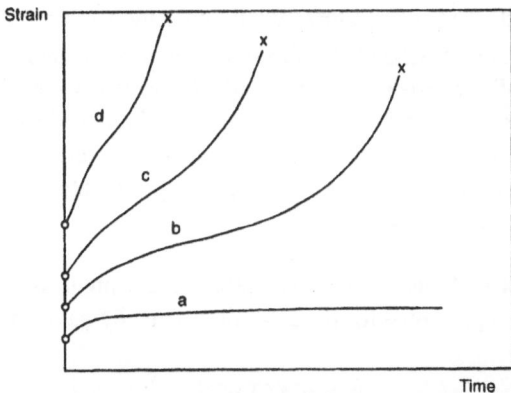

Figure 7.13. Effect of increase of stress (at constant temperature) on the shape of the creep curve for metals. Reprinted from: Oding, I.A. (1965) Creep and Stress Relaxation in Metals, translated from Russian by E. Bishop.

For the constant stress case, Bailey and Roberts (1933) and Kanter (1938) introduced, respectively, the following two empirical relationships between the temperature and the occurring rate of plastic strain (creep rate) in a polycrystalline solid

$$\dot{\varepsilon}_p = a\, e^{kt} \quad \text{(Bailey and Roberts, 1933)}$$

and

$$\dot{\varepsilon}_p = a^{-b/kT} \quad \text{(Kanter, 1938)}$$

where a, b and k are material constants. Other relationships concerning the influence of temperature on the creep rate have been suggested by, for instance, Kauzmann (1941), Dushman et al. (1944) and Nowick and Machlin (1946). Fastov (1950), see Oding (1965), expressed a relationship between creep rate and temperature with the inclusion of a limiting temperature $T_{crit.}$, below which, as he asserted, creep could not occur. The limiting temperature T_{crit} is considered to be determinable by experiment. For the creep rate, Fastov (1950) suggests the following approximate relationship

$$\ln \dot{\varepsilon}_p = a_0 - \frac{b_0}{T} + b_1 \left(1 - \frac{T_{crit.}}{T} \right) \sigma$$

which is assumed to be valid for $T > T_{crit.}$ and where a_0, b_0 and b_1 are material constants. Sherby et al. (1954), however, allow for the effect of temperature on creep by the introduction of a time-temperature influence factor t_k expressed by

$$t_k = t\, e^{-Q/RT} \tag{7.65}$$

where t indicates a time variable corresponding to the time of creep testing and Q is the activation energy for creep. Accordingly, creep curves measured at different constant stress levels, but at different temperatures would coincide to form a "master curve" when plotted as the plastic (creep) strain ε_p against the new time variable t_k. Sherby et al. (1954) demonstrated the validity of their equation (7.65) for elevated temperature creep data on aluminum, constantan, copper, gold, iron, lead, nickel, platinum and zinc. This equation holds good agreement for elevated temperature creep at test temperature exceeding $0.45T_m$. Under these conditions, the activation energy for creep would approximate the value of the activation energy for self-diffusion (e.g., Oding, 1965).

The dependence of the creep strain rate on the input stress level σ and the temperature T is often represented for the secondary stage by a power law of the form

$$\dot{\varepsilon}_p = \sigma^n f(T) \tag{7.66}$$

In the above constitutive equation, the exponent n is usually considered to be constant over a wide range of stress and temperature for a large class of crystalline solids. According to

Mukherjee (1974),

- The exponent n may vary from about 3 to about 7 for various solid solution alloys and pure metals. Pure metals often give values as $4.2 < n < 6.9$ which may suggest an average value for n to be about 5 (e.g., Sherby, 1962).

- Some alloys such as Ni-Au may exhibit a value of n about 3 (e. g., Sellers and Quarrell, 1961). However, other alloys, such as Fe-Si and Fe-Al may have values for n as high as $6 < n < 7$ (e.g., Davies, 1963 and Lawley et al., 1960).

A particular form of the constitutive equation (7.66) is Norton's Law (1929)

$$\dot{\varepsilon}_p = k \sigma^n \qquad (7.67)$$

where k and n are material constants depending on temperature only.

In the general three-dimensional case, the yield condition is $f(\sigma_{ij}) = 0$ where σ_{ij} is the Cauchy's stress tensor (*Chapter 2*). The latter converts to $\sigma = \sigma_y$ in the uniaxial loading situation where σ_y is the uniaxial yield stress, as we have discussed earlier in this chapter (e.g., Hill, 1950). On the assumption that the plastic hardening is kinematic, the yield condition in the uniaxial case becomes

$$f(\sigma - \alpha) = 0 \qquad (7.68)$$

Leckie (1981) advanced that the parameter α, appearing in the above expression, may be assumed to be linearly related to the plastic strain ε^p via the relation

$$\alpha = m \varepsilon^p \qquad (7.69)$$

where m is determinable from a kinematic hardening stress-train diagram as illustrated in Fig. 7.14. Meanwhile, the yield condition reduces to the non-hardening situation when $\alpha = 0$.

In order to determine the constitutive three-dimensional relation, Onat (1976) proposed a state space description where at least two state variables may be needed. In this approach, the selected two-state variables are the second-order tensor a_{ij} and a scalar λ with a_{ij} representing the current position of a yield surface and λ is a measure of its size. Thus, the stress tensor $(\sigma_{ij} - a_{ij})$ may be interpreted, for instance, as the "*back stress*" due to dislocation pileups or bowing (e.g., Krieg et al., 1976). Meantime, λ could be interpreted as the mechanical strength determined by existing barriers. In this model, the constitutive equation (*e.g.*, Leckie, 1981) is expressed as follows

$$\varepsilon_{ij}^p = f(a_{ij}, \lambda) \, \dot{\sigma}_{ij}$$
$$\dot{a}_{ij} = g(a_{ij}, \lambda, \dot{\varepsilon}_{ij}), \, \dot{\lambda} = h(a_{ij}, \lambda, \dot{\varepsilon}_{ij}) \qquad (7.70)$$

where $\dot{\varepsilon}_{ij}$ is the total strain rate. The formalism, presented by the constitutive equation (7.70), is based on time-independent plasticity (discussed earlier in this Chapter). In order to introduce time-effects, recovery terms need to be added. It is, however, recognized that the experimental determination of the state variables a_{ij} and λ is complicated by existing multi-axial non-proportional loading with time-dependent high-temperature existing phenomena.

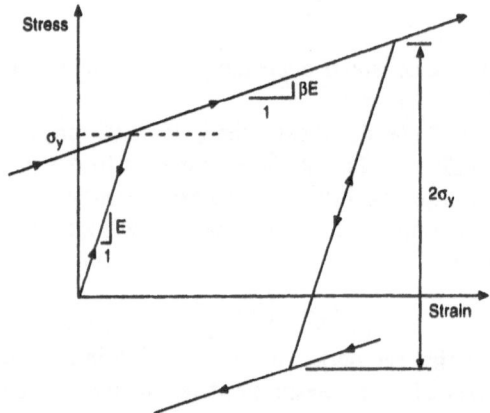

Figure 7.14. Kinematic hardening stress-strain diagram. From: Leckie, F.A. (1981) Advances in Creep Mechanics, 3rd Symposium, IUTAM, Eds. A.R.S. Ponter and D.R. Hayhurst, Leicester, UK, Sept. 8-12, 1980, Springer-Verlag, Berlin, pp. 13-47. Reprinted with kind permission of Springer-Verlag.

It is often argued in the literature that the creep rate in metals and alloys is influenced by two counteracting processes, i.e., *"strain hardening"* and *"thermal softening"*. In this, strain hardening is expected to be the dominant process at high rates of strain, whereby, plastic deformation occurs and the metal hardens. On the other hand, thermal softening is considered to be the controlling mechanism at low rates of strain; whereby the hardness of the material would be reduced. Further, if the two processes, i.e., *strain hardening and thermal softening*, are in equilibrium, one would generally expect that the creep process to be stationary, i.e., a constant creep rate is accompanying constant stress. The above argument formulates what is known by the *"Recovery Model"*. In this context, Mitra and McLean (1966) have made predictions of the stationary creep state using direct measurements of strain hardening and thermal softening in aluminum and nickel. Other applications of the recovery model have been considered by Ponter and Leckie (1976) and Leckie (1981).

In the recovery model, denoting the inelastic strain rate by $\dot{\varepsilon}$, then,

$$\dot{\varepsilon} = H(\sigma - \bar{\sigma}) \qquad (7.71)$$

where $\bar{\sigma}$ is the flow stress and H (.) Is the Heaviside step function (*Appendix B*).

Meantime, the change of the flow stress $\bar{\sigma}$ is determined by the net outcome of the two counteracting processes of strain hardening and thermal softening. This may be expressed as (e.g., Leckie, 1981)

$$\frac{d\bar{\sigma}}{dt} = \left(\frac{\partial\bar{\sigma}}{\partial\varepsilon}\right)_t \dot{\varepsilon} + \left(\frac{\partial\bar{\sigma}}{\partial t}\right)_\varepsilon$$

$$\text{where} \quad \left(\frac{\partial\bar{\sigma}}{\partial\varepsilon}\right)_t \quad \text{and} \quad \left(\frac{\partial\bar{\sigma}}{\partial t}\right)_\varepsilon \qquad (7.72)$$

indicate, respectively, the hardening rate at fixed time and the softening rate at constant strain.

With reference to Figure 7.15, three different stages of deformation may be interpreted from the above model:

(i)   *Accumulation of Plastic Strain and Strain Hardening (Part OA of Fig. 7.15).*
With an increased rate of the applied stress ($\dot{\sigma} > 0$), plastic strain is accumulated and the effect due to thermal softening may be neglected. In this case, the plastic strain rate $\dot{\varepsilon}^p$ may be identified with the hardening rate of the stress-strain curve. That is

$$\dot{\varepsilon}^p = \left(\frac{\partial\bar{\sigma}}{\partial t}\right) / \left(\frac{\partial\bar{\sigma}}{\partial\varepsilon}\right)_t \qquad (7.73)$$

(ii) *Strain Hardening Is in Equilibrium with Thermal Softening.*
In this case, the applied stress is kept constant ($\dot{\sigma} = 0$) and the flow stress may be assumed to be zero ($\bar{\sigma} = 0$). Thus, the stationary state creep flow ε_c may be considered to occur according to the relation

$$\dot{\varepsilon}_c = -\left(\frac{\partial\bar{\sigma}}{\partial t}\right)_\varepsilon / \left(\frac{\partial\bar{\sigma}}{\partial\varepsilon}\right)_t \qquad (7.74)$$

(iii) *Thermal Softening (No Creep Strain; Part BC in Fig. 7.15).*
If the stress is reduced rapidly from a stationary state condition such that $\bar{\sigma} < \sigma$, no creep strain will be observed. In this situation, however, strain will increase with time reflecting thermal softening. With increased thermal softening, the flow stress $\bar{\sigma}$ is expected to decrease until it becomes equal to σ. At this time, creep strain may resume again, but at a reduced rate. The time corresponding to this stage (where no creep is observed) is often referred to as the "*incubation period*"; e.g., Leckie (1981).

A large number of constitutive models have been proposed by different researchers to describe the creep response of metals. The majority of these models are phenomenological, e.g., the response models by Odqvist (1966), Malinen and Khadjinsky (1972), Leckie and Ponter (1974), Hart (1976), Miller (1976, 1987), Swearengen and Rhode (1977), Larsson and Storakes (1978), and Mroz and Trampczynski (1984), among others.

Ashby (1972) and Frost and Ashby (1973,1982) constructed deformation mechanism maps for various metals and alloys. The maps are based on available experimental data and show the fields of stress and temperature in which a given deformation mechanism may be dominant and the strain yield it may yield. Thus, the maps may be considered convenient for a normalized comparison of the behaviour of the materials being dealt with. Furthermore, they illustrate the effects of changes in the controlling parameters in a manner that could be useful in the development and use of engineering solids. Figures 7.16 to 7.18 (after Frost and Ashby, 1973) show examples of the deformation maps for various metals.

As pointed out by Frost and Ashby (1973), deformation maps, such as those introduced by the above figures, suffer the limitations of steady-state flow formulations in presenting a view of plastic deformation behaviour. Although steady-state flow may be considered to be an accurate representation of "*diffusion*" creep (*e.g.*,Mukherjee, 1974), it does neglect primary and tertiary creep.

Figure 7.19 compares the creep rupture properties of a range of engineering alloys. It demonstrates the gain in creep strength that may be obtained by alloying additions to steel. As shown in the Figure, nickel base alloys may be the best candidates for design applications involving very high temperatures, e.g., gas turbines combustor components.

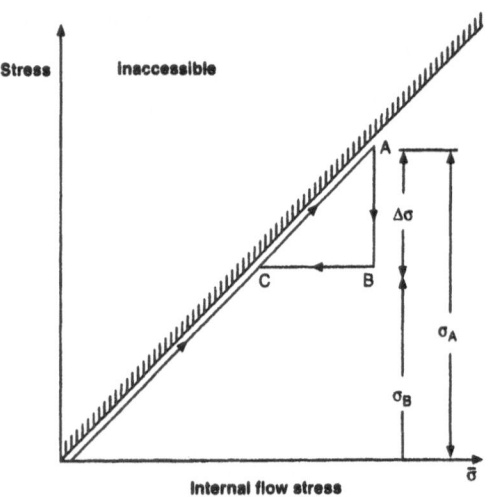

Figure 7.15. State space. From: Leckie, FA. (1981) Advances in Creep Mechanics, 3rd Symposium, IUTAM, Eds. A.R.S. Ponter and D.R. Hayhurst, Leicester, UK, Sept. 8-12, 1980, Springer-Verlag, Berlin, pp. 13-47. Reprinted with kind permission of Springer-Verlag.

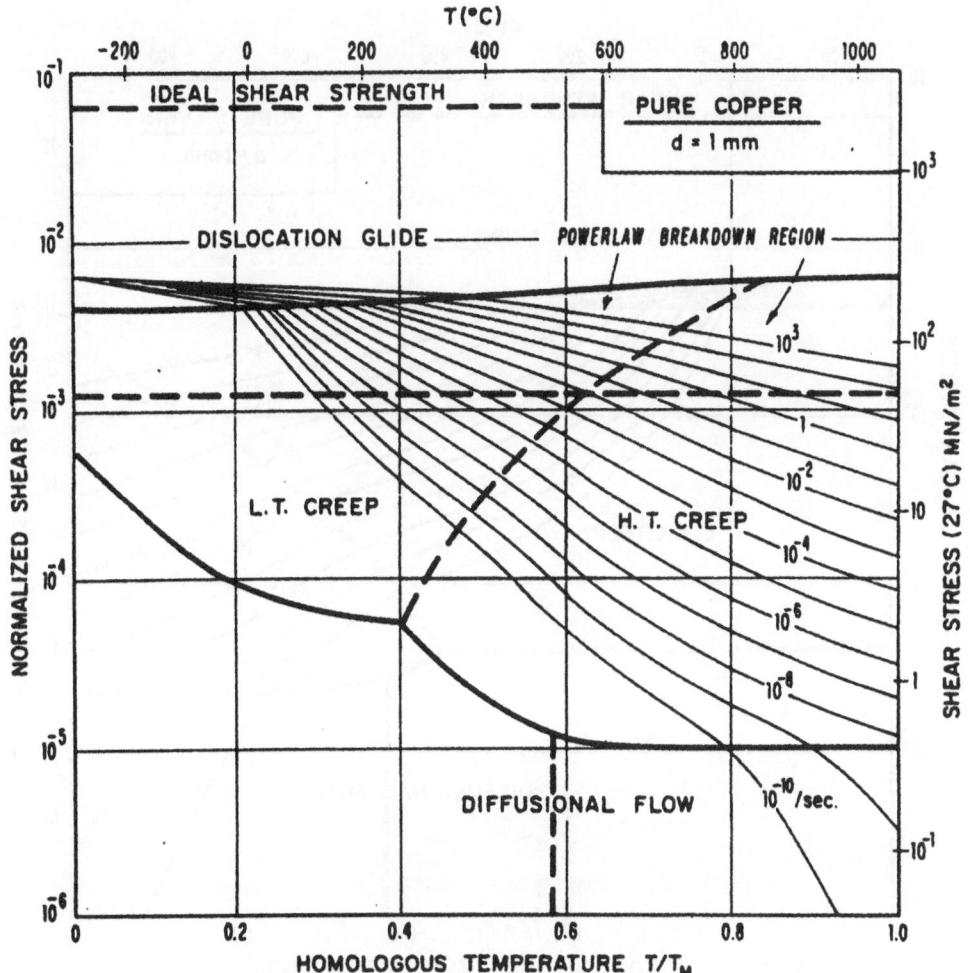

Figure 7.16. Deformation map for pure copper of grain size 1 mm. Reprinted with kind permission from: Frost, N.J. and Ashby, M.F. (1973) A Second Report on Deformation Mechanism Maps, Division of Engineering and Applied Physics, Harvard University, Cambridge, Massachusetts.

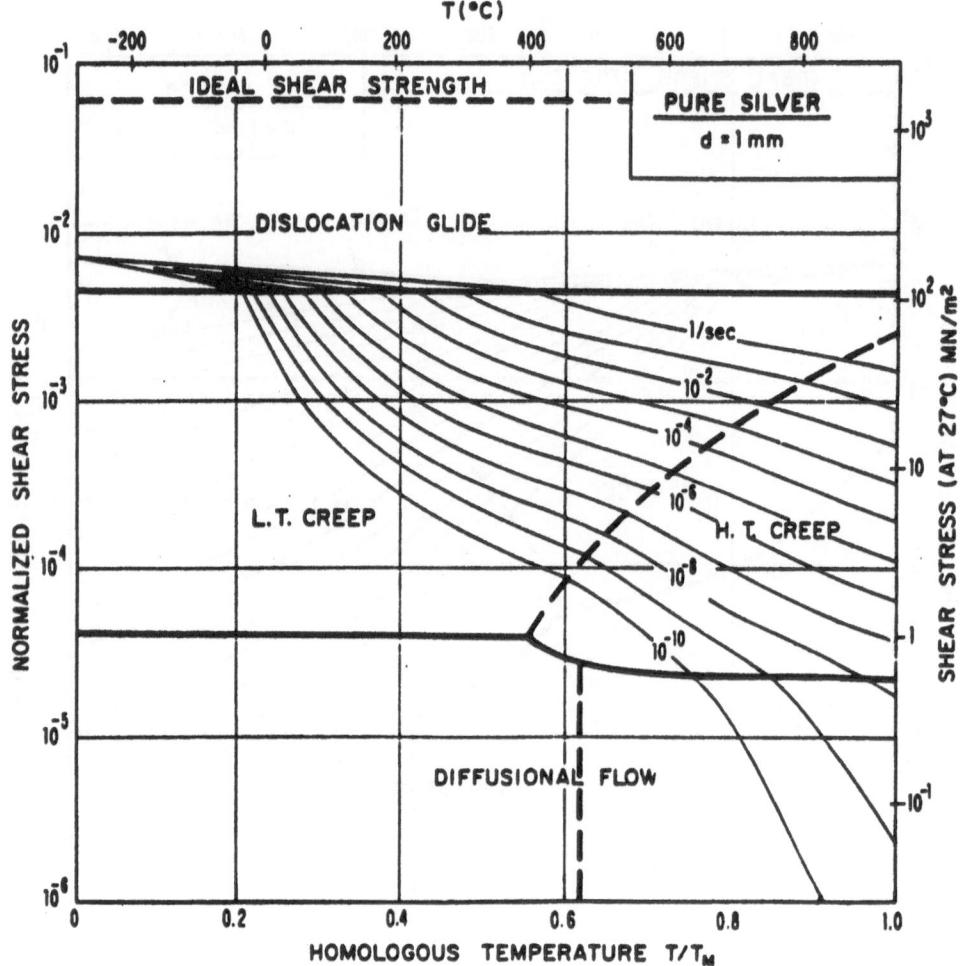

Figure 7.17. Deformation map for pure silver of grain size 1 mm. Reprinted with kind permission from: Frost, N.J. and Ashby, M.F. (1973) A Second Report on Deformation Mechanism Maps, Division of Engineering and Applied Physics, Harvard University, Cambridge, Massachusetts

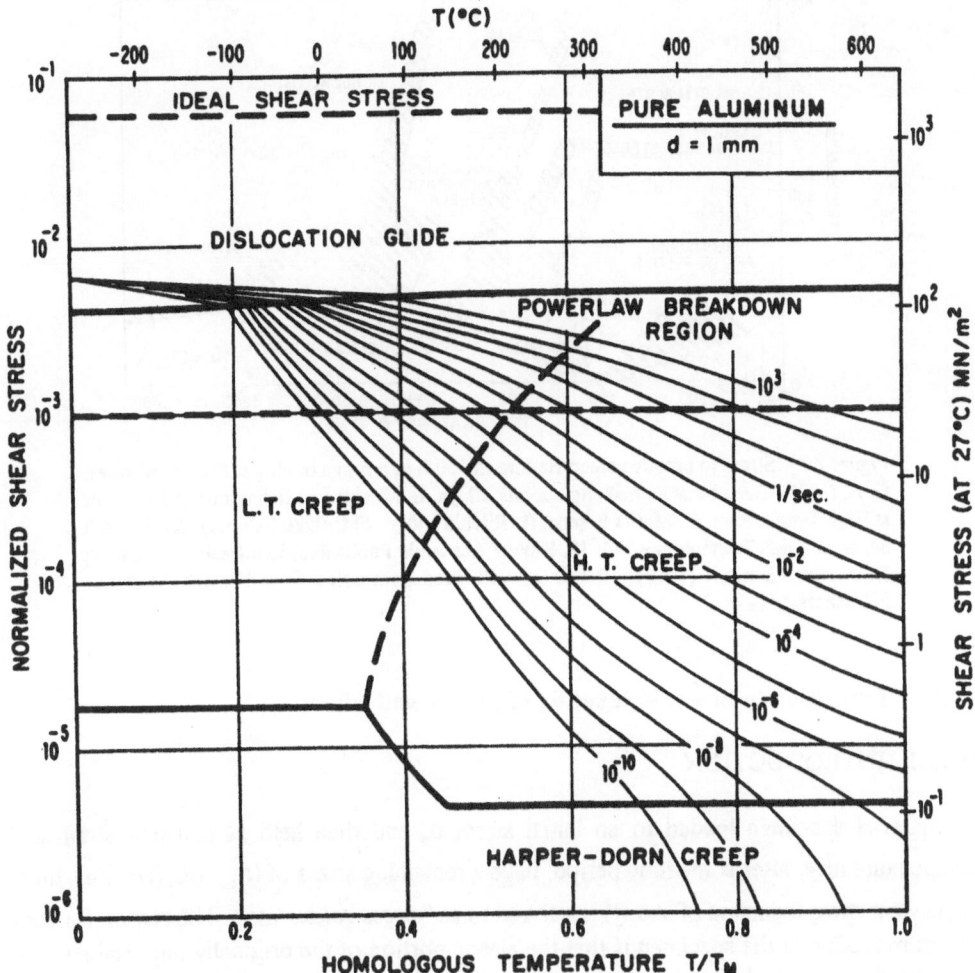

Figure 7.18. Deformation map for pure aluminum of grain size 1 mm. Reprinted with kind permission from: Frost, N.J. and Ashby, M.F. (1973) A Second Report on Deformation Mechanism Maps, Division of Engineering and Applied Physics, Harvard University, Cambridge, Massachusetts.

256

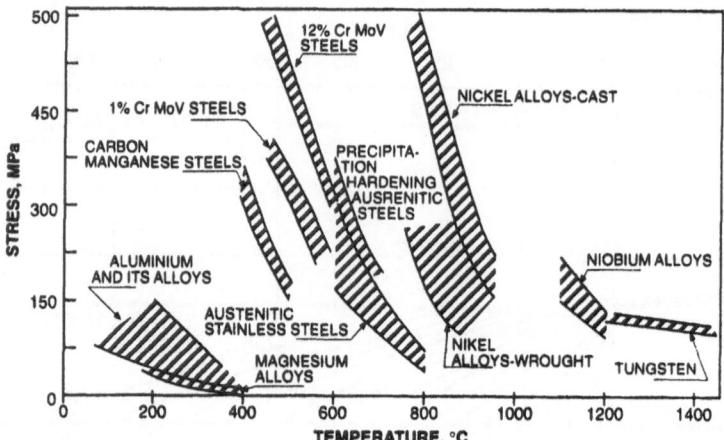

Figure 7.19. Stress to produce creep rupture in 100 h in a range of alloys. From: Webster, G. A. (1996) Creep behaviour of engineering alloys, in Mechanical Behaviour of Materials at High Temperature, C. Moura Branco, R. Ritchie and V. Skleníčka (editors), NATO ASI Series, 3. High Technology - Vol. 15, Kluwer Academic Publishers, Dordrecht, pp. 23-42; Data after Gemmill (1966). Reprinted with kind permission from Kluwer Academic Publishers.

7.13. Transition to Stress-Relaxation of Metals and Alloys

7.13.1. INTRODUCTION

A material specimen loaded to an initial stress σ_0 and then held at constant strain and temperature may, after some time period, have a remaining stress of $(\sigma_0 - \Delta\sigma_0(t))$. This time-dependent stress reduction of $\Delta\sigma_0(t)$ is referred to as *"stress-relaxation"*. The reason for such stress reduction in the specimen is that the elastic portion of the originally imposed strain is converted, partly or wholly, to inelastic strain, although the total strain remains constant. The initial stress σ_0 can be induced in, for instance, mechanical members and assemblies, by fabrication loads or service thermal gradients (e.g., ASTM DS 60, 1982).

Examples of stress relaxation in metallic components may include:

- Loss of pre-load of a bolt in a rigid flange connection and similar attachments.
- Relief of part or all of residual stresses and/or stress redistribution in mechanical or structural members.

Figure 7.20 (afterASTM DS 60, 1982) shows a comparison of the 1000-hour relaxation strength for several classes of alloys.

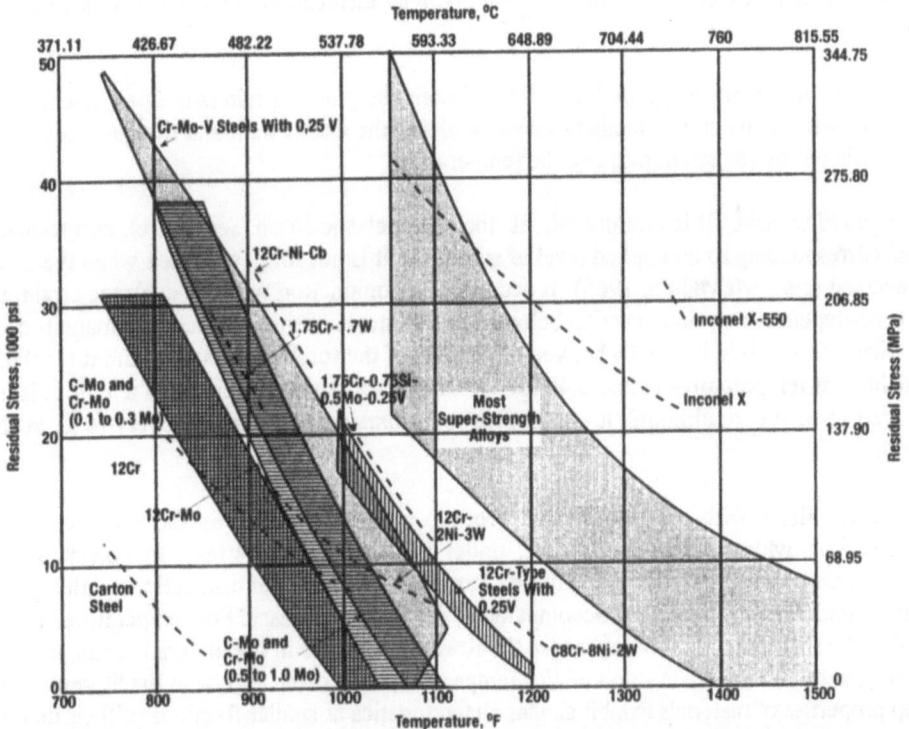

Figure 7.20. Comparative 1000-hour relaxation strengths for several classes of alloys. From: ASTM Data Series Publication DS 60 (1982), *Compilation of Stress-Relaxation Data for Engineering Alloys,* by Manjoine, M. J. and Voorhees, H. R., Copyright ASTM. Reprinted with permission.

7.13.2. MECHANICS OF STRESS-RELAXATION

Deformation Mechanisms

As mentioned in the foregoing, the total strain, in the stress relaxation experiment, is constant and the stress reduction, at a fixed temperature, occurs as elastic strain is converted partly or wholly to inelastic strain. As advanced by Manjoine and Voorhees (1982), the resulting inelastic strain could be due to *"anelasticity"*, *"plasticity"*, *"microplasticity"* and *"creep"*. One of these mechanisms might act solely or in combination with the others, pending on the level of stress, stress-rate, temperature, the nature of the microstructure, among others..

Anelastic strain. It is interpreted as the transient strain due to stress variation in the material. It is recovered when the incurred stress change is reversed (e.g., Zener, 1948). The anelastic strain would not usually exceed 5% of the total elastic strain (e.g., Zener, 1948 and Hill et al., 1961).

As anelasticity results from internal friction, anelastic strain is out of phase with the stress and it contributes to internal heating during deformation, particularly if high loadfrequencies are involved (e.g., Manjoine and Landerman, 1982). Anelastic strain is, in general, a function of stress, stress rate and temperature. It can be affected by, among others, the composition, magnetic properties, degree of order, and elastic fields in the material. The anelasticity of the material is often expressed in view of the involved dislocation mechanisms (e.g., LeMay, 1981).

Plastic strain. As presented earlier in this chapter, the plastic strain is the permanent strain measured when a material is loaded to a stress above the elastic limit and then unloaded. It is often affected by the strain rate and the temperature.

Microplastic strain. It is interpreted, as the the anelastic strain (see above), as a transient strain corresponding to an applied level of stress, but it is not fully recovered when the stress is reversed (e.g., McMahon, 1968). It accounts, in conjunction with the anelastic strain, for the time-dependent strain occurring below the yield stress of the material. The magnitude of the microplastic strain is usually between 5 to 15% of the total elastic strain, and it reaches a finite limit at temperatures below $0.40T_m$. As the microplastic strain reaches a limit, it is not referred to as *creep*, although it contributes to *"primary creep"* and recovery (e.g., ASM, 1957).

Creep. As dealt with in the previous section, creep is defined as the time-dependent deformation, which persists with time, under applied constant stress. In polycrystalline materials, creep is often interpreted to be a result of the motion of dislocations within grain, grain boundary sliding and any accompanying diffusion processes. For temperatures above $0.4T_m$, the thermal activation enhances the flow mechanisms in the material. Thus, creep is accelerated by an increase in stress and/or temperature. It is often argued in the literature that creep properties of materials exhibit similar characteristics at similar fractions of their melting temperature T_m in degrees absolute (e.g., Webster, 1996). On unloading, a small portion of creep strain is recoverable, although, for the most part, creep strain is permanent.

Traditionally, stress-relaxation has been investigated to interpret the mechanisms of flow occurring in the material. Thus, other loading histories or states of stress may be analyzed. The major effort has been to determine a correlation between creep and relaxation data, so that the contributions of the different types of strain, mentioned in the foregoing, can be evaluated (*see Chapter 8*).

Analysis of the relaxation data, and, hence, the prediction of the stress-relaxation behaviour of the material represent an important part of the design process of mechanical assemblies and structures, particularly those involved in high temperature applications (ASTM DS 60, 1982). For instance,

- Comparison of the relaxation and creep data allows the assessment of the inter-related roles of the two response phenomena in the analysis and design of structural members (e.g., Manjoine and Mudge, 1954, and Henderson and Snedden, 1971).

- As mentioned in the foregoing, analysis of relaxation data is particularly useful in the design of bolted joints of pressure vessels, and similar pressure retaining structures, whereby leakage at such joints may be prevented by applying the appropriate pre-load on the bolts, and maintaining the appropriate levels of loading in them afterwards. The required pre-load and the allowance for relaxation can be calculated from the relaxation data of the involved materials (ASTM DS 60, 1982). The same would apply to applications involving, for instance, press-fitted joints, springs, and clamps (e.g., Davis, 1952 and Manjoine, 1975).

7.13.3. MECHANICAL STRESS-RELAXATION TESTING

The stress-time history of a mechanical component for elevated temperature service can be very complex during start up, operation and shut-down cycles. Representative examples of such components are combustor components in turbo-engines, and blades of turbines and compressors operating in a high temperature service environment. In an aeroengine, for instance, HP turbine blades are situated in the hot gas stream emerging from the combustor. Turbine entry temperatures in modern aeroengines can reach 1400°C, and, thus , such components may be required to survive in a temperature environment somewhat above their own melting point. In addition to the essential internal cooling design requirements, which ensures that the peak metal temperatures are maintained at less than 1100°C, prime physical and mechanical requirements of blade materials include, among others, good creep strength and predictable stress-relaxation behaviour over a wide range of temperature (e.g., Harrison and Winstone, 1996, and Harrison and Tranter, 1996)). The damages from stress and strain histories, in such mechanical components, are a function of the strain rate, stress state, temperature and the nature of the microstructure among other factors (e.g., Manjoine, 1970, 1974, 1975, and Henderson and Snedden, 1971).

The relaxation test is often carried out by adopting one of the following loading procedures (e.g., ASTM DS 60, 1982):

a)   *Initial stress, at fixed test temperature, with the resulting total strain to be maintained constant during the entire experiment.* This loading procedure is considered to be the most common in mechanical stress-relaxation testing. In this, the tension test, of a specimen with uniform cross section, instrumented with a sensitive extensometer, is employed.

b)   *Initial total strain, at a fixed test temperature, with the imposed total strain is maintained constant during the entire experiment.* This loading procedure is used to approximate the service constraint, when an initial strain results from service displacements.

c)   *Initial total strain at room temperature, followed by heating to a peak test temperature for a time period.* This procedure simulates the loading of a bolt at room temperature and the subsequent relaxation after an elevated service temperature is reached.

d)   *Repeated loadings to selected stress levels, with the resulting total strain is maintained constant during the entire experiment.* This loading procedure is particularly used to obtain design data pertaining to the tightening frequency of bolted joints in pressure vessels and similar pressure bearing structures.

e)   *Maintaining a given strain in a cyclic stress-strain loop.* This loading procedure is used to determine design data pertaining to reverse straining that may occur in the material due to thermal transients, whereby the relaxation characteristics depend on the inelastic strain history (e.g., Krempl, 1979).

Stress-relaxation at temperatures below $0.4T_m$

The stress relaxation at temperatures below about $0.4T_m$ is a result of inelastic strain which after a time period reaches a limit, referred-to as the *"relaxation limit"*. The latter, for a particular microstructure, is a function of the initial stress and temperature. This inelastic strain, as discussed earlier in this section, is often due to micro- and macro-plasticity, and anelasticity.

An example of the *relaxation limit* is given in Figure 7.21 for an annealed Type 304 stainless steel for which T_m is about 800°F. The remaining stress reaches a limiting value within 100 hours for temperatures up to 600°F (315°C) and 90% of this relaxation takes place within 24 hours. The remaining stress is given in Fig. 7.21 as a function of the initial stress.

With reference to Figure 7.21 (ASTM DS 60, 1982), the following is observed:

- The material has significant relaxation at room temperature that decreases with the initial stress.
- Little relaxation is observed for stresses below about one-half of the yield strength for virgin monotonic loading.
- At 600°F (315°C), the amount of relaxation is greater than that mentioned above for a given initial stress, and no relaxation at stresses below about one-half of the yield strength. In this low temperature range, sufficient stress must be applied to initiate micro-plasticity. However, after a reversed stress above the proportional limit *(lowest curve of Figure 7.21)*, relaxation is observed for stress levels below one-half of the

yield strength and negative relaxation (*increase in stress*) may occur at low forward stresses where the anelastic strain is dominant (e.g., Krempl, 1979 and Swindeman, 1979).

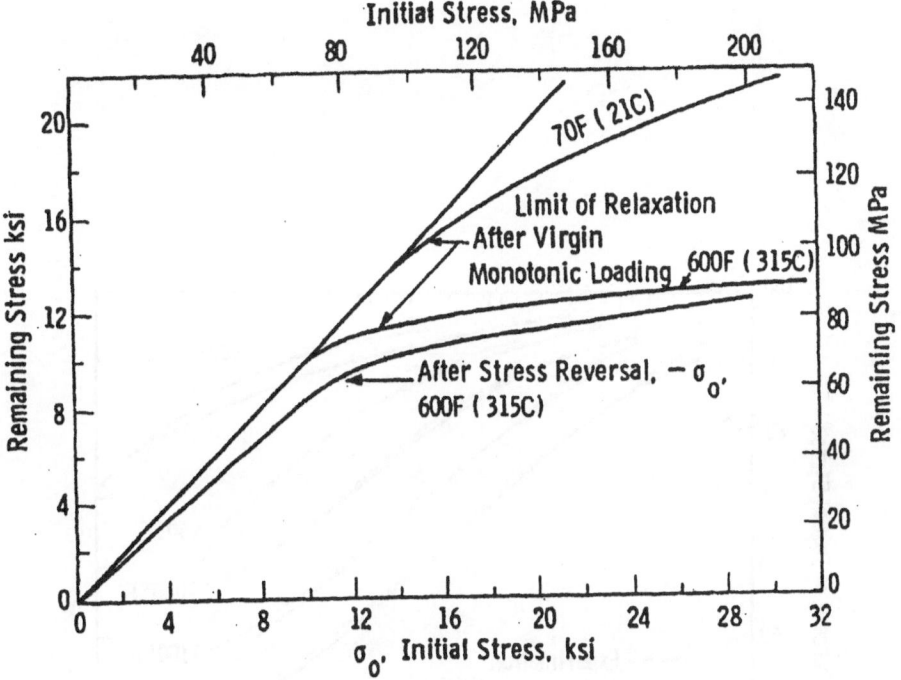

Figure 7.21. Relaxation of solution annealed Type 304 stainless steel. From: ASTM Data Series Publication DS 60 (1982), *Compilation of Stress-Relaxation Data for Engineering Alloys,* by Manjoine, M. J. and Voorhees, H. R., Copyright ASTM. Reprinted with permission.

Stress-relaxation at temperatures above 0.4T$_m$

When creep strain is the dominate inelastic strain, stress-relaxation occurs continually with time and as a function of the stress and temperature (see the preceding section concerning creep of metals and alloys).

For a structure which is given an initial strain, the percent relaxation can be measured as a function of time at a given temperature. As an illustration, the percent relaxation for specimens, of a 20% cold-worked Type 304 stainless steel, loaded to an initial 0.07 per cent strain and tested at several temperatures is plotted as the solid curves in Figure 7.22 (ASTM DS 60, 1982).

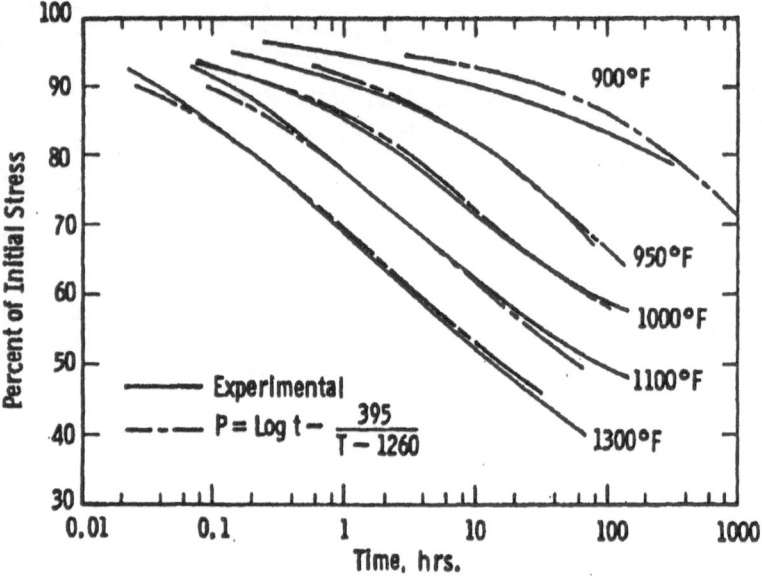

Figure 7.22. Relaxation of a 20% cold-worked Type 304 stainless steel from 900 to 1300 °F (482 to 704 °C) for an initial inelastic strain of 0.07%.   From: ASTM Data Series Publication DS 60 (1982), *Compilation of Stress-Relaxation Data for Engineering Alloys,* by Manjoine, M. J. and Voorhees, H. R., Copyright ASTM. Reprinted with permission.

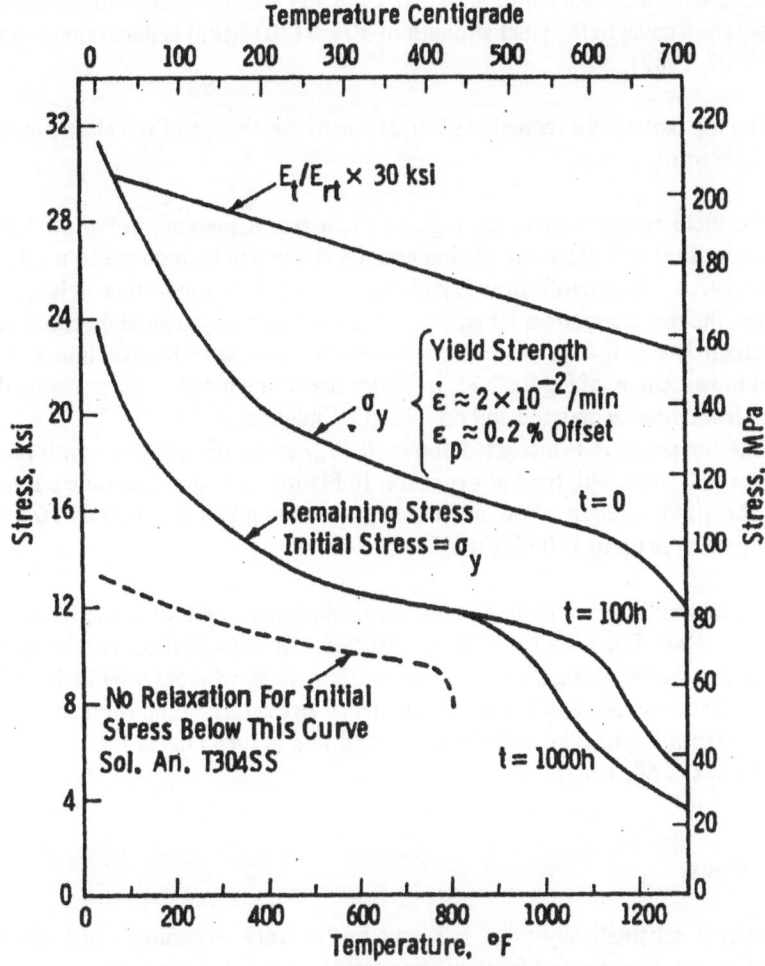

Figure 7.23. Remaining stress for an annealed Type 304 stainless steel bar at constant strain as a function of temperature and time. From: ASTM Data Series Publication DS 60 (1982), *Compilation of Stress-Relaxation Data for Engineering Alloys,* by Manjoine, M. J. and Voorhees, H. R., Copyright ASTM. Reprinted with permission.

Illustrative Example of Stress-Relaxation under Fixed Total Strain

The relaxation of the stress in a bar under constant total strain can result from a thermal expansion, inelastic flow, and metallurgical changes. The initial stress may be a result of an external load or displacement, or from a residual stress due to differential plastic strains. Since the initial strain in the bar is constant, the stress is a product of the elastic strain and the elastic modulus. The latter decreases with rising temperature; therefore, the stress is reduced on heating. The initial stress under simple tension can be as high as the flow stress for the initial total strain (ASTM DS 60, 1982).

The case for heating a bar of annealed Type 304 stainless steel with an initial stress at room temperature equal to the yield strength of 30 ksi (207 MPa) is illustrated in Figure 7.23 (ASTM DS 60, 1982):

- The top curve is the reduction of stress due to the change of the elastic modulus with temperature.

-The yield strength curve for a given strain rate is marked, in Figure 7.23, by σ_y. Plastic flow will occur on heating and the stress will be reduced to a value near the yield curve. Since plastic flow is initiated, stress-relaxation further reduces the stress. Thus, the remaining stress for an initial stress becomes approximately equal to the yield strength. Below $0.4T_m$, this stress reaches a limit value within 100 hours. The dashed and lowest curve, of Figure 7.23, indicates that if the initial stress is below this curve, no plastic flow is initiated and no relaxation may occur.
If the temperature is increased above $0.4T_m$, creep will continue with time and the stress will relax with time of exposure. In Figure 7.23, the zero-time curve is shown as the yield strength curve; additional curves are given for 100 and 1000 hours for temperatures up to 1300°F (704°C).

- For materials which undergo metallurgical changes with time and temperature, the stress relaxation response will be affected. If precipitation results in a volume decrease, the remaining stress will increase for a time, whereas volume increases (such as irradiation swelling) will cause an additional stress relaxation. Metallurgical changes may also influence the creep strength and this in turn affects the rate of relaxation (ASTM DS 60, 1982).

7.14. Problems

1. Illustrate schematically what is meant by (i) creep experiment and (ii) relaxation experiment, in the case of a crystalline solid.
2. Comment, using appropriate illustrations, on the microstructural mechanisms involved in both low and high temperature creep of metals.
3. Explain briefly the following terms. Support your comments by particular examples and schematic illustrations:
    - Low *vs.* high temperature barriers.
    - Classical "idealized" creep curve (*for a crystalline solid*).
    - *"Primary"* vs. *"instantaneous"* stage of creep behaviour.
    - *"Secondary"* vs. *"tertiary"* stage of creep.
4. Comment, using appropriate illustrations, on the possible microstructural mechanisms that may be involved in both *"secondary"* and *"tertiary"* stages of creep.
5. Comment, with appropriate reasons, on the possible variations which one may encounter concerning the "primary" creep stage of a crystalline solid.
6. What is meant by the "Recovery Model" as pertaining to the creep of crystalline solids? Describe the various deformation processes that might be involved.
7. What is meant by the so-called "incubation period"? Describe, with appropriate

illustrations the role of this term in determining to the mechanics of creep of crystalline solids.

8. What is meant by *"Diffusion Creep"*? Illustrate such behaviour for the case of three different crystalline solids of your choice.

9. Why nickel base alloys are often regarded as the best candidates for applications that may involve very high temperatures?

10. Explain briefly, with appropriate examples, the probable sources of anelasticity in a crystalline solid. Make the comparison between two different constitutive models *(from the literature)* which link the anelastic strain to the occurring dislocation mechanisms.

11. Comment on the differences between *"anelasticity"* and *"microplasticity"* as possible mechanisms of stress-relaxation in a crystalline solid.

12. Give at least five examples of the use of stress-relaxation data in engineering applications. Illustrate how one uses such data in one the examples you have already provided.

13. Comment, using at least two illustrative examples, on the inter-related roles of creep and stress-relaxation data in the design and analysis and design of structural members.

14. Describe three different loading procedures for stress-relaxation testing of a crystalline solid. Comment on the applicability domain of each method.

7.15. References

ASM (1957) *Creep and Recovery*, American Society for Metals, Cleveland, OH.

ASTM Data Series Publication DS 60 (1982) *Compilation of Stress-Relaxation Data for Engineering Alloys*, by Manjoine, M. J. and Voorhees, H. R., ASTM Publication Code Number (PCN) 05-060000-30, ASTM, Philadelphia.

Ashby, M.F. (1972) A first report on deformation-mechanism maps, *Acta Met.* 20, 887-97.

Bailey, R.W. and Roberts, AM. (1933) Testing of materials for service in high-temperature steam-plant, *Proc. of the Inst. of Mech. Engg.* 122, pp.209-84.

Bird, J.E., Mukherjee, A.K. and Dorn, J.E. (1969) Quantitative relation between properties and microstructure, *Israel Univ. Press*, Jerusalem, pp. 255-342.

Costa Andrade, E.N. (1910) On the viscous flow in metals and allied phenomena, *Proc. Roy. Soc. A* 84, 1-12.

Davies, R.G. (1963) Steady-state creep in FE-2 to 11 at. pct Si alloys, *Trans. AIME* 227, 665-68.

Davis, E. A. (1952) Relaxation of stress in a heat-exchanger tube of ideal material, *Trans. ASME* 74, 381-5, April 1952.

De St. Venant, B. (1870) Mémoire sur l'établissement des équations différentielles des mouvements intérieurs opérés dans les corps solides ductiles au déjà des limites où l'élasticité pourrait les rammener à leurs premiers états, *Compt. rend.* 70, 473-80.

De St. Venant, B. (1871) Mémoire sur l'établissement des équations différentielles des mouvements intérieurs opérés dans les corps solides ductiles au déjà des limites où l'élasticité pourrait les rammener à leurs premiers états, *J. math. pures appl.* II, 16.

Dorn, J.E. and Mote, J. (1963) *High Temperature Structures and Materials*, Pergamon, Oxford, pp. 95-168.

Dushman, S., Dunbar, L. and Huthsteiner, H. (1944) Creep of metals, *J. Appl. Phys.* 15(2), 108-24.

Finnie, I. (1959) *Creep of Engineering Materials*, McGraw-Hill, New York.

Frost, N.J. and Ashby, M.F. (1973) *A Second Report on Deformation Mechanism Maps*, Division of Engineering and Applied Physics, Harvard University, Cambridge, Massachusetts.

Frost, H.J. and Ashby, M.F. (1982) *Deformation-Mechanism Maps, The Plasticity and Creep of Metals and Ceramics*, Pergamon Press, New York.

Garafalo, F. (1965) *Fundamentals of Creep and Creep Rupture of Metals*, McMillan, New York.

266

Geiringer, H. (1953) Some recent results in the theory of an ideal plastic body, in *Advances in Applied Mechanics* III, von Mises, R. and von Karman, T. (editors), Academic Press, New York, pp.197-294.

Gemmill, M. G. (1966) *The Technology and Properties of Ferrous Alloys for High Temperature Use*, Newnes, London.

Gooch, D.J. and How, I.M., Eds. (1986) *Techniques for Multiaxial Creep Testing*, Proc. of a Symp., General Electricity Research Laboratories, Letherhead, Great Britain, Sept. 25-26, 1985, Elsevier, London.

Grant, N.J. (1971) Fracture under conditions of hot creep rupture, in *Fracture*, Ed. H. Liebowitz, Vol.111, Chapter 8, Academic Press, New York, pp. 483-533.

Grant, N.J. and Mullendore, A.W. (1965) *Deformation and Fracture at Elevated Temperatures*, M.I.T. Press, Cambridge, Massachusetts.

Greenfield, P. (1972) *Creep of Metals at High Temperature*, Mills, New York.

Harrison, G. F. and Tranter, P.H.. (1996) Stressing and lifing techniques for high temperature aeroengine components, in: *Mechanical Behaviour of Materials at High Temperature*, NATO ASI Series (3. High Technology - Vol.15), Branco, C.M., Ritchie, R. and Sklenička, V. (eds.) (1996), Kluwer, Dordrecht, pp. 327-45.

Harrison, G. F. and Winstone, M. R. (1996) Aeroengine applications of advanced high temperature materials, in: *Mechanical Behaviour of Materials at High Temperature*, NATO ASI Series (3. High Technology Vol.15), Branco, C.M., Ritchie, R. and Sklenička, V. (eds.) (1996), Kluwer, Dordrecht, pp. 309-25.

Hart, E.W. (1976) Constitutive relations for the nonelastic deformation of metals, *ASME J. Engg. Materials and Technology* 98, 193-202.

Hazlett, T. and Hansen, R.D. (1954) Influence of substructure on the shape of the creep curve, *Trans. Amer. Soc. Metals* 47, 508-19.

Henderson, J. and Snedden, J.O. (1971) In *Advances in Creep Design, The A. E. Johnson Memorial Volume*, A. I. Smith and A. M. Nicolson (editors), Applied Science Publishers, Londo, pp. 163.

Hill, R. (1950) *The Mathematical Theory of Plasticity*, Clarendon Press, Oxford.

Hill, W. H., Shimmin, D. L. and Wilcox, B.A. (1961) Elevated temperature dynamic moduli of metallic materials, Proc. ASTM 61.

Hodge, G.P., Jr. (1950) An Introduction to the Mathematical Theory of Perfectly Plastic Solids, ONR, NR-041-032, Providence.

Hult, J.A.H. (1966) *Creep in Engineering Structures*, Blaisdell, Waitham, Mass.

Johnson, A.E. (1960) Complex-stress creep of metals, *Metallurgical Reviews* 5(20), 447-506.

Kachanov, L.M. (1967) *The Theory of Creep*, National Lending Library of Science and Technology, Boston.

Kanter, J.J. (1936) Interpretation and use of creep results, *Transactions of ASM* 24, 870-918.

Kanter, J.J. (1938) Problems of temperature-coefficient of tensile creep rate, *J. of Metals, Trans. AIME* 131, 385-418.

Kauzmann, W. (1941) Flow of solid metals from the standpoint of the chemical-rate theory, *Metals Technology* 6, 1301-17.

Krempl, E. (1979) An experimental study of room-temperature rate-sensitivity, creep and relaxation of AISI type 304 stainless steel, *J. Mech. Phys. Solids* 27, 363-75.

Krieg, R.D., Swearengen, J.C. and Rhode, R.W. (1976) *Special Issue ASME Pressure Vessel Piping Section*, PVPPB 028.

Lagenborg, R. (1972) Dislocation mechanisms in creep, *Met. Rev.* 17, 130-46.

Larsson, B. and Storakes, B. (1978) A state variable interpretation of some rate-dependent inelastic properties of steel, *ASME J. Engg. Materials and Technology* 100, 395-401.

Lawley, A., Coll, J. and Cahn, R.W. (1960) Influence of crystallographic order on creep of iron-aluminum solid solutions, *Trans. AIME* 218, 166-79.

Leckie, FA. (1981) Advances in creep mechanics, in *Creep in Structures*, 3rd Symposium, IUTAM, Eds. A.R.S. Ponter and D.R. Hayhurst, Leicester, UK, Sept. 8-12, 1980, Springer-Verlag, Berlin, pp. 13-47.

Leckie, F.A. and Ponter, A.R.S. (1974) On the state variable description of creeping Materials, *Ing. Archiv.* 43, 158-67

LeMay, I. (1981) *Principles of Mechanical Metallurgy*, Elsevier North Holland Inc., New York.

Lévy, M. (1870) Extrait du mémoire sur les équations générales des mouvements intérieurs des corps solides où l'élasticité pourrait les rammener à leur premier état, *Compt. rend.*70, 1323-5.

Lévy, M. (1871) Extrait du mémoire sur les équations générales des mouvements intérieurs, *J. math.pures appl.* **II**, 16.

Malinin, N.N. and Khadjinsky, G.M. (1972) Theory of creep with anisotropic hardening, *Int. J. Mech. Sci.* 14, 235-46.

Manjoine, M. J. (1970) Multiaxial stress and fracture, Fracture, Vol. 3: Engineering Fundamentals and Environmental Effects, Academic Press, New York.

Manjoine, M. J. (1974) Elevated temperature mechanics of metals, 1974 Symposium on Mechanical Behaviour of Materials, Society of Computer Science, Kyoto, Japan.

Manjoine, M. J. (1975) Ductility indices at elevated temperature, *Trans. ASME, J. Engineering Materials and Technology* 97(14)(2), April 1975, 156-61.

Manjoine, M. J. and Landerman, E. I. (1982) Techniques for fatigue testing and extrapolation of fatigue life for austentic stainless steels, *ASTM J. of Testing and Evaluation*, May 1982.

Manjoine, M. J. and Mudge, W. L. (1954) Creep properties of annealed unalloyed zirconium, *Proc. ASTM* 54, 1050-67.

Manjoine, M. J. and Voorhees, H. R . (1982) ASTM Data Series Publication DS 60, *Compilation of Stress-Relaxation Data for Engineering Alloys*, ASTM Publication Code Number (PCN) 05-060000-30, ASTM, Philadelphia.

Mclean, D. (1966) The physics of high-temperature creep in metals, *Rep. Prog. Phys.* 29(1),1-33.

McMahon, C. J. (Ed.) (1968) *Microplasticity, Advances in Materials Research*, Vol. 2, John Wiley & Sons, New York.

Miller, A. (1976) An inelastic constitutive model for monotonic, cyclic and creep deformation, *ASME J. Engg. Materials and Technology* 98, 97-113.

Miller, A.K. (1987) *Unified Constitutive Equations for Creep and Plasticity*, Elsevier, London.

Mitra, S.K. and Mclean, D. (1966) Work hardening and recovery in creep, *Proc. Royal Soc. London A* 295, 288-99.

Mroz, Z. and Trampczynski, W.A. (1984) On the creep-hardening rule for metals with a memory of maximal prestress, *Int. J. Solids and Structures* 20, 467-86.

Mukherjee, A.K. (1974) High-temperature creep, in: *Plastic Deformation of Materials, Treatise on Materials Science and Technology*, Ed. H. Herman,Vol.6, Academic Press, New York, pp.163-224.

Mukherjee, A.K., Bird, J.E. and Dorn, J.E. (1969) Experimental correlations for high-temperature creep, *ASM, Trans. Quart.* 62, 155-79.

Mukherjee, S. and Ohno, N. (1982) A constitutive equation of creep based on the concept of a creep-hardening surface, *Int. J. Solids and Structures* 18, 597-609.

Nadai, A. (1931) Plasticity, *Mechanics of the Plastic State of Matter*, London.

Nadai, A. (1950) *Theory of Flow and Fracture of Solids*, Vol.1, McGraw Hill, New York.

Norton, F.H. (1929) *Creep of Steel at High Temperatures*, McGraw-Hill, New York.

Nowick, A.S. and Machlin, E.S. (1946) Quantitative treatment of creep of metals by dislocation and rate-process theories, National Advisory Committee Aeronautics, Washington, Technical Note No.1039, Apr. 1946, pp.1-21.

Oding, I.A., Ed. (1965) *Creep and Stress Relaxation in Metals*, translated from Russian by E.Bishop, Oliver and Boyd, London.

Odqvist, F.K.G. (1966) *Mathematical Theory of Creep and Creep Rupture*, Clarendon, Oxford.

Odqvist, F.K.G. (1981) Historical survey of the development of creep mechanics from its beginnings in the last century to 1970, in: *Creep in Structures*, 3rd Symposium, IUTAM, Eds. A.R.S. Ponter and D.R. Hayhurst, Leicester, U.K., Sept. 8-12, 1980, Springer-Verlag, Berlin, pp. 1-12.

Onat, E.T.(1976) Oak Ridge National Lab.Report, SUB-3863-7.

Ponter, A.R.S. and Leckie, F.A. (1976) Constitutive relationships for the time-dependent deformation of metals, *Trans. ASME, J. of Eng. Materials and Tech.* 98, 47-51.

Prager, N. and Hodege, G.P. Jr. (1951) *Theory of Perfectly Plastic Solids*, Wiley, New York.

Rabotnov, Y.N. (1968) *Creep Problems in Structural Member*, translated from Russian, English translation edited by F.A. Leckie, North Holland, Amsterdam.

Raymond, L. and Dorn, J.E. (1964) Recovery of creep resistant substructures, *Trans.AIME* 230, 560-67.

Sellers, C.M. and Quarrell, A.G. (1961) The high temperature creep of gold-nickel alloys, *J. Inst. Metals* 90, 329-36.

Sherby, O.D. (1962) Factors affecting the high temperature strength of polycrystalline solids, *Acta Met.* 10, 135-47.

Sherby, O.D. and Burke, P.M. (1967) Mechanical behaviour of crystalline solids at elevated temperature, *Prog. Mater. Sci.* **13**, 325-90.

Sherby, O.D., Orr, R.L. and Dorn, J.E. (1954) Creep correlations of metals at elevated temperatures, *J. of Metals, Transactions of AIME* **6**(1), pp.71-80.

Sokolovsky, V.V. (1947) *Theory of Plasticity*, Acad. Sci. U.S.S.R., Moscow.

Swearengen, J.C. and Rhode, R.W. (1977) Application of mechanical state relations at low and high homologous temperatures, *Met. Trans.* **8A**, 577-82.

Swindeman, R. W. (1979) Isocronous relaxation curves for Type 304 stainless steel after monotonic and cyclic strain, *ASTM J. of Testing and Evaluation* **2**, 192.

Tresca, H. (1872) Mémoire sur l'écoulement des corps solides, *Mém. pres. par div. Sav.* **18**, 733-99.

von Mises, R. (1913) *Mechanik der fester Koerper in plastisch deformablem Zustand*, Goettinger Nachrichten, pp. 582-92.

von Mises, R. (1949) *Three remarks on the theory of the ideal plastic body*, Reissner Anniversary Volume, Edwards Bros., Ann Arbor, Mich., pp. 215-29.

Webster, G. A. (1996) Creep behaviour of engineering alloys, in Mechanical Behaviour of Materials at High Temperature, C, Moura Branco, R. Ritchie and V. Sklenička (editors), NATO ASI Series, 3. High Technology - Vol. 15, Kluwer, Dordrecht, pp. 23-42.

Webster, G. A. and Ainsworth, R.A. (1994) *High Temperature Component Life Assessment*, Chapman and Hall, London.

Webster, G. A. and Piearcey, B.J. (1967) An interpretation of the effects of stress and temperature on the creep properties of nickel-base superalloys, *Metal Sci. J.* **1**, 97-104.

Weertman, J. (1968) Dislocation climb theory of steady state creep, *ASM (Amer. Soc. Metals), Trans. Quart.* **61**, 681-94.

Zener, C. (1948) *Elasticity and Anelasticity*, University of Chicago Press, Chicago.

7.16. Further Reading

ASME Boiler and Pressure Vessel Code (1991) *Case N-47 (29, Class 1 Components in Elevated Temperature Service*, Section III, Division I, ASME, New York.

ASTM (1982) *Standard Recommended Practices for Stress-Relaxation Tests for Materials and Structures*, Annual Book of ASTM Standards, E328-72, American Society for Testing and Materials, Philadelphia.

Aifantis, E.C. (1987) The physics of plastic deformation, *Int. J. Plasticity* **3**, 211-47.

Agatonovic, P. (1996) New developments in engineering applications of materials at high temperatures, in: *Mechanical Behaviour of Materials at High Temperature*, NATO ASI Series (3. High Technology - Vol.15), Branco, C.M., Ritchie, R. and Sklenička, V. (eds.) (1996), Kluwer, Dordrecht, pp. 683-702.

Anand, L., Dillon, O., Place, T.A. and Von Turkovich, B.F. (1990) Report of the NSF workshop on localized plastic instabilities and failure criteria, *Int. J. Plasticity* **6**, I-IX.

Arutiunian, N.K. (1966) *Some Problems in the Theory of Creep*, Pergamon, New York

Bardet, P. (1990). Finite element analysis of plane strain bifurcation within compressible Solids, *Computers & Structures* **36**, 993-1007.

Batra, R.C. and Liu, De-Shin (1989) Adiabatic shear banding in plane strain problems, *J. Appl. Mech.* **56**, 527-34.

Batra, R.C. and Liu, De-Shin (1990) Adiabatic shear banding in dynamic plane strain compression of a viscoplastic material, *Int. J. Plasticity* **6**, 231-46.

Batra, R.C. and Zhang, X.T. (1990) Shear band development in dynamic loading of a viscoplastic cylinder containing two voids, *Acta Mech.* **85**, 221-34.

Batra, R.C. and Zhu, Z.G. (1991) Dynamic shear band development in a thermally softening bimetallic body containing two voids, *Acta Mech.* **86**, 31-52.

Bazant, Z.P., Belytschko, T. and Cheng, T.P. (1984) Continuum theory for strain softening, *Trans. ASCE J. Engng. Mech. Div.* **110**, 1666-92.

Bodner, S.R. and Partom, Y. (1975) Constitutive equations for elastic-viscoplastic strain-hardening materials, *J. Appl. Mech.*, June 1975, 385-9.

Borst, R. de (1987) Integration of plasticity equation for singular yield conditions, *Computers & Structures*

26, 823-9.

Borst, R. de (1989) Numerical methods for bifurcation analysis in geomechanics, *Ing. Arch.* **59**, 160-74.

Borst, R. de and Mühlhause, H.B. (1992) Gradient dependent plasticity: Formulation and algorithmic aspects, *Int. J. Num. Meth. Engng* **35**, 521-39.

Boyce, M.C., and Arruda, M. (1990) An experimental and analytical investigation of the large strain compressive and tensile response of glassy polymers, *Pol. Engng Sci.* **30**, 1288-98.

Boyce, M.C., Parks, D.M. and Argon, A.S. (1988) Large inelastic deformation of glassy polymers, Part I: Rate dependent constitutive model, *Mech. Mater.* **7**, 15-33.

Boyle, J.T. (1983) *Stress Analysis for Creep*, Butterworths, Toronto

Boyle, J. and Spence, J. (1981) Generalized structural models in creep mechanics, in *Creep Structures*, 3rd Symposium, IUTAM, Eds. A.R.S. Ponter and D.R. Hayhurst, Leicester, UK, Sept. 8-12, 1980, pp. 233-48.

Branco, C.M., Ritchie, R. and Sklenička, V. (eds.) (1996) *Mechanical Behaviour of Materials at High Temperature*, NATO ASI Series (3. High Technology - Vol.15), Kluwer, Dordrecht.

Budiansky, B. (1959) Assessment of deformation theory, *J. Appl. Mech.* **26**, 259-64.

Cabot, G.P., Bazant, Z.P. and Tabbarn, M. (1988) Comparison of various models for strain softening, *Eng. Comput.* **5**, 141-50.

Campbell, J.D., Eleiche, A.M. and Tsao, M.C.C. (1977) *Fundamental Aspect of Structural Alloy Design*, Plenum Publ., New York.

Chateau, X. and Ngugen, Q.S. (1991) Buckling of elastic structures in unilateral contact with or without friction, *Eur. J. Mech. A/Solids* **10**, 71-89.

Chater, E. and Hutchinson, J.W. (1984) On the propagation of bulges and buckles, *J. Appl. Mech.* **51**, 269-77.

Christoffersen, J. and Hutchinson, J.W. (1979) A Class of phenomenological corner theories of plasticity, *J. Mech. Phys. Solids* **27**, 465-87.

Chu, C.C. (1979) Bifurcation of elastic-plastic cylindrical shells under internal pressure, *J. Appl.Mech.* **46**, 889-94.

Chung, K. and Wagoner, R. (1986) Invariance of neck formation to material strength and strain rate for power-law materials, *Metall Trans.* **A17**, 1632-3.

Conway, J.B. (1960) *Numerical Methods for Creep and Rupture Analyses*, Gordon, New York.

Conway, J.B. (1969). Stress Rupture Parameters: Origin, Calculation and Use, Gordon, New York.

Cottrell, A.H. (1964) *The Mechanical Properties of Matter*, John Wiley & Sons, New York.

Dafalias, J.F. (1983) Corotational rates for kinematic hardening at large plastic deformations, *J. Appl. Mech.* **50**, 561-5.

Dafalias, J.F. and Aifantis, E.C. (1990) On the macroscopic origin of the plastic spin, *Acta Mech.* **82**, 31-48.

Dubey, R.N. (1970) Variational method for nonconservative problems, *J. Appl. Mech.* **37**, 133-6.

Evans, H.E. (1984) *Mechanisms of Creep Fracture*, Elsevier Applied Science, London.

Freeman, J.W. and Voorhees, H.R. (1956) *Relaxation Properties of Steels and Super-Strength Alloys at Elevated Temperatures*, ASTM-American Society for Testing and Materials, Special Technical Publication No. 187, August 1956. Also, ASTM publication DS-114 (August 1961).

Gilman, J.J. (1969) *Micromechanics of Flow in Solids*, McGraw-Hill, New York.

Gittus, J. (1975) *Creep, Viscoelasticity and Creep Fracture in Solids*, John Wiley and Sons, New York.

Goodall, I. W. (Ed.) (1990) *Assessment Procedure for the High Temperature Response of Structures*, Nuclear Electric Procedure R5, Issue 1.

Gotoh, M. (1985) A Class of plastic constitutive equation with vertex effect-I, *Int. J. Solids & Structures* **21**, 1101-16.

G'Sell, G. and Jonas, C. (1979) Determination of the plastic behaviour of solid polymers at constant true strain rate, *J. Mater Sci.* **14**, 583-91.

Gurson, A.L. (1977) Continuum theory of ductile rupture by void nucleation and growth, Part I. Yield criteria and flow rules for porous ductile media, *J. Eng. Mater Tech.* **99**, 2-15.

Hill, R. (1958) A General theory of uniquences and stability in elastic-plastic solids, *J. Mech. Phys. Solids* **8**, 236-49.

Hill, R. (1962a) Uniqueness criteria and extremum principles in self-adjoint problems of continuum mechanics, *J. Mech. Phys. Solids* **10**, 185-94.

Hill, R. (1962b) Acceleration waves in solids, *J. Mech. Phys. Solids* **10**, 1-16.

270

Hill, R. and Hutchinson, J.W. (1975) Bifurcation phenomena in the plane tension test, *J. Mech. Phys. Solids* **23**, 236-64.

Hughes, T.J.R. and Winget, J. (1980) Finite rotation effects in numerical integration of rate constitutive equations arising in large-deformation analysis, *Int. J. Num. Meth. Engng.* **15**, 1862-7.

Hunter, S.C. (1961) Tentative equations for the propagation of stress, strain and temperature fields in visco-elastic solids, *J. Mech. and Phys. Solids* **4**(1), 219-34.

Hutchinson, J.W. (1973a) Post bifurcation behaviour in the plastic range, *J. Mech. Phys. Solids* **21**, 163-90.

Hutchinson, J.W. (1973b) Finite strain analysis of elasto-plastic solids and structures, in *Numerical Solution of Nonlinear Structural Problems*, Hartung, R.F. (ed.), *ASME*, New York, 17-29.

Hutchinson, J.W. and Neale, K.W. (1973) Sheet necking-II. Time independent behaviour, in *Mechanics of Sheet Metal Forming*, Koistinen, D.P. and Wang, N.M.(eds.), Plenum Press, New York, 127-53.

Hutchinson, J.W. and Neale, K.W. (1983) Neck propagation, *J. Mech. Phys. Solids*, **31**, 405-26.

Hutchinson, J.W. and Tvergaard, V. (1980) Surface instabilities on statically strained solids, *Int. J. Mech. Sci.* **26**, 339-54.

Kanter, J. J. (1936) Interpretation and use of creep results, *Trans. ASM* **24**, 900.

Kennedy, A.J. (1962) *Processes of Creep and Fatigue in Metals*, Wiley, New York.

Kim, K.H. and Anand, L. (1987) A note on adiabatic flow localization in visco-plastic solids, *Computational Methods for Predicting Material Processing Defects*, Predeanu, M. (ed.), Elsevier, London, pp. 181-92.

Kitagawa, H., Seguchi, Y. and Tomita, Y. (1972) An incremental theory of large strain and large displacement problems and its finite element application, *Ing. Arch.* **41**, 213-24.

Kraus, H. (1980) *Creep Analysis*, Wiley, Toronto.

Krieg, R.D. and Krieg, D.B. (1977) Accuracies of numerical solution methods for the elastic-perfectly plastic model, *Trans. ASME, J. Press Vessel Tech.* **99**, 510-15.

Lagneborg, R. (1972) A modified recovery-creep model and its evaluation, *Institute of Metals*, London, Paper No. M5270, pp. 127-33.

Larrson, M., Needleman, A., Tvergaard, V. and Storakers, B. (1982) Instability and failure of internally pressurized ductile metal cylinder, *J. Mech. Phys. Solids* **30**, 121-54.

Lee, D. and Hart, E. W. (1971) Stress relaxation and mechanical behavior of metals, *Met. Trans.* **2**, 1245.

Lemonds, J. and Needleman, A. (1986a) Finite element analyses of shear localization in rate and temperature dependent solids, *Mech. Materials* **5**, 339-61.

Lemonds, J. and Needleman, A. (1986b) An analysis of shear band development incorporating heat conduction, *Mech. Materials* **5**, 363-73.

Leroy, Y.M. (1991) Linear stability analysis of rate-dependent discrete systems, *Int. J. Solids Structures* **27**, 783-808.

Leroy, Y.M. and Chapuis, O. (1991) Localization in strain-rate-dependent solids, *Comp. Meth. Appl. Mech. Engng.* **90**, 969-86.

Leroy, Y.M. and Ortiz, M. (1989) Finite element analysis of strain localization in frictional materials, *Int. J. Num. Anal Meth. Geomech.* **13**, 53-74.

Leroy, Y.M. and Ortiz, M. (1990) Finite element analysis of transient strain localization phenomena in frictional solids, *Int. J. Num. Anal. Meth. Geomech.* **14**, 93-124.

Loret, B. (1983) On the effects of plastic rotation in finite deformation of anisotropic elastoplastic materials, *Mech. Materials* **2**, 287-304.

Lubahn, J.D. (1961) *Plasticity and Creep of Metals*, Wiley, New York.

Manjoine, M.J. and Voorhees, H.R. (1982) *Compilation of Stress-Relaxation Data for Engineering Alloys*, ASTM Data Series Publication DS60.

Mear, M.E. and Hutchinson, J.W. (1985) Influence of yield surface curvature on flow localization in dilatant plasticity, *Mech. Mater.* **4**, 395-407.

Meitzner, C. F. (1972) Cause and prevention of stress-relief cracking in quenched and tempered steel weldments, Trans. ASME, J. of Engrg. For Industry, Feb. 1972, 336-42.

Mimura, K. and Tomita, Y. (1991) Constitutive relations of mild steel and alpha-titanium at high rates under multiaxial loading condition, *J. de Physique* IV, *DYMAT* 91, *Les éditions de Physique*, Paris, 813-20.

Mühlhaus, H. and Aifantis, E.C. (1991) A variational principal for gradient plasticity, *Int. J. Solids Structures* **28**, 845-57.

Naghdi, P.M. (1972) The theory of shells and plates, *Handbuch der Physik*, VIa/2, Springer, Verlag, Berlin.

Nagtegaari, J.C., Parks, D.M. and Rice, J.R. (1974) On numerically accurate finite element solutions in the fully plastic range, *Comp. Meth. Appl. Mech. Engng.* **4**, 153-77.

Neale, K.W. (1980) Phenomenological constitutive laws in finite plasticity, *SM Arch.* **6**, 79-128.

Needleman, A. (1989) Dynamic shear band development in plane strain, *J. Appl. Mech.* **56**, 1-9.

Needleman, A. and Ortiz, M. (1991) Effect of boundaries and interfaces on shear band localization, *Int. J. Solids Structure* **28**, 859-77.

Needleman, A. and Rice, J.R. (1978) Limits to ductility set by plastic flow localization, *Mechanics of Sheet Metal Forming*, Koistinen, D.P. and Wang, N.M. (eds.), Plenum Press, New York, pp. 237-67.

Needleman, A. and Tvergaard, V. (1984) Finite element analysis of localization in plasticity, in *Finite Strains Special Problems in Solid Mechanics*, Oden, J.T. and Carey, G.F. (eds.), Prentice-Hall, New York, pp. 94-157.

Nemat-Nasser, S. (1988) Micromechanics of failure at high strain rate: Theory, experiments and computations, *Computers & Structures* **30**, 95-104.

Nemat-Nasser, S., Chung, D.T. and Taylor, L.M. (1989) Phenomenological modelling of rate-dependent plasticity for high strain rate problems, *Mech. Materials* **7**, 319-44.

Nemat-Nasser, S. and Li, Y.F. (1992) A New explicit algorithm for finite-deformation elastoplasticity and elastoviscoplasticity: Performance evaluation, *Computers & Sturctures* **44**, 937-63.

Neville, A.M. (1970) *Creep of Concrete:Plain, Reinforced and Prestressed*, North Holland, Amsterdam.

Ohtani, R., Ohnami, M. and Inoue, T., Eds. (1988) *High Temperature Creep-Fatigue*, Current Japanese Materials Research, Vol.3, The Society of Materials Science, Japan, Elsevier Applied Science, London.

Orava, R.N., Stone, G. and Conrad, H. (1966) The Effects of temperature and strain rate on the yield and flow stresses of A-titanium, *Transactions of the American Society of Metals* **59**, pp. 171-84.

Ortiz, M. and Popov, E.P. (1985) Accuracy and stability of integration algorithms for elastoplastic constitutive relations, *Int. J. Num. Meth. Engng.* **21**, 1561-76.

Ortiz, M. and Simo, J.C. (1986) An analysis of a new class of integration algorithms for elastoplastic constitutive relations, *Int. J. Num. Meth. Engng.* **23**, 353-66.

Paslay, P.R. and Wells, C.H. (1976) Uniaxial creep behaviour of metals under cyclic temperature and stress and strain variation, *ASME J. Applied Mechanics* **98**, 445-9.

Peirce, D., Asaro, R.J. and Needleman, A. (1983) Material rate dependence and localized deformation in crystalline solids, *Acta Metall.* **31**, 1951-76.

Penny, R.K. (1971) *Design for Creep*, McGraw-Hill, London.

Perzina, P. (1966) Fundamental problems in viscoplasticity, *Advan. Appl. Mech.* **9**, 243-377.

Poirier, J.P. (1985) *Creep of Crystals:High-Temperature Deformation Processes in Metals, Ceramics and Minerals*, Cambridge University Press, Cambridge.

Raniecki, B. (1979) Uniqueness criteria in solids with nonassociated plastic flow laws at finite deformations, *Bull Acad Pol Ser Sci Tech.* **27**, 721-9.

Raniecki, B. and Brunhs, O. (1981) Bounds to bifurcation stresses in solids with non-associated plastic flow law at finite strain, *J. Mech. Phys. Solids* **29**, 153-72.

Rashid, M.M. and Nemat-Nasser, S. (1992) A Constitutive algorithm for rate-dependent crystal plasticity, *Comp. Meth. Appl. Mech. Engng.* **94**, 201-28.

Reddy, B.D. (1982) A deformation theory analysis of the bifurcation of pressurized thick-walled cylinders, *Q. J. Mech. Appl. Math.* **35**, 183-96.

Rice, J.R. (1976) The localization of plastic deformation, *Proc. 14th IUTAM Congress*, Koiter, W. (ed.), North Holland, Amsterdam, 207-20.

Rice, J.R. and Tracey, D.M. (1973) *Numerical and Computational Methods in Structural Mechanics*, Academic Press, New York.

Robinson, E. L. (1939) The resistance to relaxation of materials at high temperatures, *Trans. ASME* **61**, 543.

Rudnicki, J.W. and Rice, J.R. (1975) Conditions for the localization of deformation in pressure-sensitive dilatant materials, *J. Mech. Phys. Solids* **23**, 271-94.

Runesson, K., Samuelsson, A. and Bernspang, L. (1986) Numerical techinque in plasticity including solution advancement control, *Int. J. Num. Meth. Engng.* **22**, 769-88.

Runesson, K., Sture, S. and William, K. (1988) Integration in computational plasticity, *Computers & Structures* **30**, 119-30.

Sehitoglu, H. (1996) Thermo-mechanical deformation of engineering alloys and components - Experimental and modeling, in: *Mechanical Behaviour of Materials at High Temperature*, NATO ASI Series (3.

High Technology - Vol.15), Branco, C.M., Ritchie, R. and Sklenička, V. (eds.) (1996), Kluwer, Dordrecht, pp. 349-77.

Sewell, M.J. (1967) On configuration dependent loading, *Arch. Mech. Anal.* **23**, 327-51.

Simo, J.C., Kennedy, J.G. and Govindjee, S. (1988) Non-smooth multisurface plasticity and viscoplasticity. Loading/unloading conditions and numerical algorithms, *Int. J. Num. Meth. Engng.* **26**, 2161-85.

Simo, J.C. and Taylor, R.L. (1985) Consistent tangent operators for rate-independent elastoplasticity, *Comp. Meth. Appl. Mech. Engng.* **48**, 101-18.

Smith, D. J. (1996) Aspects on the assessment of the mechanical behaviour of metallic materials at high temperature, in: *Mechanical Behaviour of Materials at High Temperature*, NATO ASI Series (3. High Technology - Vol.15), Branco, C.M., Ritchie, R. and Sklenička, V. (eds.) (1996), Kluwer, Dordrecht, pp. 239-58.

Smith, J.O. (1965) *Inelastic Behavior of Load Carrying Members*, Wiley, New York.

Spencer, A.J. (1982) Deformation of ideal granular materials, in *Mechanics of Solids*, The Rodney Hill 60th Anniversary Volume, Hopkins, H. and Sewell, M.J. (eds.), pp. 607-52, Pergamon Press, Oxford.

Strang, A., Ed. (1991) Rupture ductility of creep resistant steels, Proc.of a Conf., York Univ., Dec. 4-5, 1990, *Inst. of Metals*, London.

Sukue, L. (1969) *Rheological Aspects of Soil Mechanics*, Wiley-Interscience, London.

Tomita, Y. and Hayashi, K. (1991) Deformation behaviour in elasto-viscoplastic polymeric Bars under Tension, *Proc. Plasticity '91*, Boehler, J.P. and Khan, A.S. (eds), Elsevier, London, pp. 524-27.

Townley, C.H.S. (1979) *Creep of Engineering Materials and Structures*, G. Bernasconi and G. Piatti, Eds., Chapter 10, Applied Science Publishers, London.

Triantafyllidis, N. (1985) The Localization of deformation in finitely strained shells, *Proc. Considere Memorial Symposium*, Salencon, J. (ed.), Ecole Nationale des Pont et Chausses, Paris, pp.115-24

Triantafyllidis, N. and Aifantis, E.C. (1986) A gradient approach to localization of deformation, *J. Elasticity* **16**, 225-37.

Tugcu, P. and Neale, K.W. (1987) Analysis of plane-strain neck propagation in viscoplastic polymeric films, *Int. J. Mech. Sci.* **29**, 793-805.

Tugcu, P. and Neale, K.W. (1988) Analysis of neck propagation in polymeric fibers including the effects of viscoplasticity, *J. Engng Mat. Tech.***110**, 395-400.

Tvergaard, V. (1978) Effect of kinematic hardening on localization necking in biaxially stretched sheets, *Int. J. Mech. Sci.* **20**, 651-58.

Tvergaard, V. (1981) Influence of voids on shear band instabilities under plane strain localization, *Int. J. Fracture* **17**, 389-437.

Tvergaard, V. (1987) Effect of yield surface curvature and void nucleation on plane flow localization, *J. Mech. Phys. Solids* **35**, 43-60.

Tvergaard, V., Needleman, A. and Lo, K.K. (1981) Flow localization in the plane strain tensile test, *J. Mech. Phys. Solids* **29**, 115-42.

Wang, X. and Lee, L.H.N. (1989) Wrinkling of an unevenly stretched sheet metal, *J. Engng Materials and Technology* **111**, 235-42.

Wilshire, B. and Owen, D.R.J., Eds. (1981) *Creep and Fracture of Engineering Materials and Structures*, Proc. of Int. Conf., Univ. College, Swansea, Mar. 24-27, 1981, Pineridge, Swansea, UK.

Wilshire, B. and Owen, D.R.J.,Eds. (1983) *Engineering Approaches to High Temperature Design*, Vol.2 in the Series: Recent Advances in Creep and Fracture of Engineering Materials and Structures, Pineridge Press, Swansea, UK.

Wright, T.W. and Walter, J.W. (1987) On stress collapse in adiabatic shear bands, *J. Mech. Phys. Solids* **35**, 701-20.

Zbib, H.M. and Aifantis, E.C. (1988a) On the localization and post-localization behaviour of plastic deformation, I. On the initiation of shear bands, *Res. Mechanica* **23**, 261-77.

Zbib, H.M. and Aifantis, E.C. (1988b) On the localization and post-localization behaviour of plastic deformation, II. On the evolution and thickness of shear bands, *Res. Mechanica* **23**, 279-91.

Zbib, H.M. and Jubran, J.S (1992) Dynamic shear banding: A Three-dimensional analysis, *Int. J. Plasticity* **8**, 619-41.

CHAPTER 8

VISCOELASTIC RESPONSE BEHAVIOUR

8.1. Introduction

With the recent advances in material science and the parallel extensive industrial demands on advanced industrial materials such as high polymers and polymeric base composite systems, the subject of *Viscoelasticity* has gained recently a strong momentum in the realms of engineering techniques and applications.

High polymeric materials are organic substances of high molecular weight, the technical importance of which depends on their particular microstructure (e.g., Leaderman, 1943 and Bernal, 1958). This class of materials may include, for example, rubber in its various forms, synthetic rubber-like materials, commercial plastics, and natural and synthetic textile fibres. Other few examples of a viscoelastic material would include a wide range of inorganic polymeric systems such as silicones and glass resins, constituents of polymeric base systems, natural fibres such as wood and the by-products of such fibres as, for instance, paper and board, building materials such as concrete, and a large class of biomaterials, among others. As will be demonstrated in this Chapter, these materials are *"time-dependent"* in response and possess a *"time-memory"* (e.g., Haddad, 1995).

In the mechanics of deformable media, the response behaviour of an elastic solid (*Chapter 6*) is dealt with within the realm of the classical theory of elasticity. The most direct description of such response, in the case of a linear elastic solid, is in accordance with the well-known Hooke's law[1]. This law forms the basis of the mathematical theory of linear elasticity[2]. That is, provided that the occurring deformations are small, the stress is considered to be directly proportional to the strain and it is independent of the strain rate. Such response is termed consequently as *"perfectly elastic"* or *"Hookean"* solid.

In a simple uniaxial test, the load-deformation curve of the perfectly elastic solid will follow the same path for both loading and unloading. Thus, the material test specimen will regain its original dimensions instantaneously upon the removal of the load. Under constant level of loading, the occurring deformation is constant, i.e., *time-independent*. Further, when such

---

[1]   Robert Hooke, De Potentia Restitutiva, London, 1678.

[2]   The reader is referred to Love (1944) for an introductory review of the history of the mathematical theory of elasticity.

a solid is subjected to a sinusoidally oscillating loading, the deformation will be found also to be sinusoidal and practically in phase with the load. All the energy is stored and recovered in each cycle.

On the other hand, the mechanical response of a viscous fluid is dealt with within the realm of the classical theory of fluid dynamics. In this case, the most direct description of the response is in accordance with Newton's law whereby the stress is considered to be proportional to the occurring rate of strain, but, independent of the strain itself. This is provided that the rate of strain is small. When a "Newtonian viscous fluid" is subjected to a sinusoidally oscillating load, the deformation will be found to be 90° out of phase with the load.

The classical theories of linear elasticity and Newtonian fluids, though impressively well structured, do not adequately describe the response behaviour and flow of a large class of real materials; particularly polymeric systems and alike as introduced above. That is, between the above two described responses of the elastic solid and the viscous fluid, a polymeric type of material may exhibit, even if both strain and strain rate are infinitesimal, the combined response characteristics of the elastic and viscous media. Attempts to characterize the behaviour of such real materials under the action of external loading, consequently, gave rise to the science of rheology within which the phenomenon labelled "*Viscoelasticity*" is well defined and intended to convey mechanical behaviour combining response characteristics of both an elastic solid and a viscous fluid. A viscoelastic material is, thus, characterized by a certain level of rigidity of an elastic solid body, but, at the same time, it flows and dissipates energy by frictional losses as a viscous fluid. A few characteristics of a viscoelastic material may be cited as follows:

–   When a viscoelastic material specimen is subjected to a constant stress, it does not hold a constant deformation (as it would be the case for an elastic solid), but it continues to flow with time, i.e., it "*creeps*". Immediately, upon the removal of the load, the specimen is found to have taken an amount of "residual" strain the magnitude of which depends on the level of loading and the length of time for which the load is applied, among other factors. Following the removal of the load, a noticeable reduction in the amount of residual strain gradually takes place with the passage of time. This residual strain may even disappear entirely in the course of time. The latter phenomenon which occurs following the removal of the load is referred to as "*creep recovery*". A specimen of viscoelastic material, tested as mentioned above, would eventually regain its original dimensions. Consequently, the creep of such material under load cannot be regarded as a phenomenon of plasticity, as in the case of polycrystalline solids (*Chapter 7*), but rather as a "*delayed elasticity*"", (e.g., Leaderman, 1943). In a simple uniaxial test of a viscoelastic material, a load-deformation loop (*hysteresis*) is obtained, i.e., the descending load curve corresponds to a larger amount of strain than the ascending load curve. Neither of the two curves is completely linear. *The shape of the resulting hysteresis loop is dependent on the magnitude of load, rate of application and removal of the load, and temperature, among other variables.*

– Further, when a viscoelastic material is subjected to sinusoidally oscillating stress, the resulting strain, though sinusoidal, is neither exactly in phase with the stress (as it would be the case for a perfectly elastic solid) nor 90° out of phase (as it would be for a perfectly viscous fluid); it is somewhere between. The magnitude of the strain and the phase angle between the stress and strain are generally frequency- and temperature-dependent. On loading and unloading a viscoelastic material specimen, some of the energy input is stored and recovered in each cycle and some of it is dissipated as heat.

The particular nature of viscoelastic response, as considered in the foregoing, proves the existence of a property of *"passive resistance"* in such materials. This is in contrast to the instantaneous response and reversibility that usually characterize pure elastic behaviour (*Chapter 6*). This passive resistance is of viscous nature and reflects what is usually called the property of *"hereditary response"* of the material. *That is, the present state of response depends not only upon the present state of loading input, but also upon previous states* (Boltzmann 1874,1877,1878). This property is revealed experimentally, in different time-dependent phenomena pertaining to the viscoelastic response behaviour such as creep, stress-relaxation and intrinsic attenuation and dispersion of waves propagating in a viscoelastic medium (*Chapter 14*).

The phenomenological theory of Viscoelasticity dates from the nineteenth century[3] (e.g. Leaderman, 1943a&b, and Markovitz, 1977), but unfortunately, the application of the theory to actual engineering applications is only a development of the last fifty years. This contrasts with the situation of the related field of linear elasticity whereby technological requirements have traditionally simulated significant research over the last two centuries. Such technological stimulus was lacking for a formal theory of Viscoelasticity to develop as engineering design has traditionally made use of materials whose mechanical response behaviour would be adequately described by the laws of classical elasticity. Research in the theory of Viscoelasticity has been, however, recently enhanced by the introduction of engineering components that are fabricated from advanced industrial materials such as those mentioned at the beginning of this introduction. A large class of these newly developed materials exhibit mechanical response behaviour outside the scopes of the more conventional theories of elasticity, viscosity and plasticity. Consequently, a development of the theory of Viscoelasticity has become of parallel necessity with the gradual introduction of such new materials.

It is often considered that the response behaviour of a viscoelastic material is a fundamental

---

[3]   Leaderman (1943) and Markovitz (1977) present interesting reviews of the history of Viscoelasticity.

property of its molecular structure. Hence, the prediction of the viscoelastic response of polymeric systems[4], for instance, has been often considered from the point of view of reaction-rate theory (Tobolosky and Eyring, 1943 and Mark and Tobolosky, 1950), that is, by treating flow as a bond breakage/bond formation process (e.g., Peters, 1955). Most of the formulations, in this context, however, have referred solely to the deformation of the critical weak bonds in the microstructure by the reasoning that the deformation of the much stronger bonds is likely to be negligible at low levels of stress. This is, however, an oversimplification to the actual flow process occurring in the real, complex microstructure of viscoelastic material systems such as those mentioned earlier (e.g., Takehiro and San-Ichiro, 1955 and Bernal, 1958). It is well recognized that such materials behave in a manner which depends primarily on the material source, microstructure and previous history, in addition to the current state of loading and environmental conditions. In a large class of viscoelastic materials, such as natural amorphous or semi-amorphous types of materials, for example, environmental effects, e.g., moisture content, could enhance the deterioration of the internal microstructure and, hence, the amount of occurring deformation. Furthermore, the energy required to produce a certain deformation may change abruptly at particular temperatures due to internal transitions in the material (e.g., Kauman, 1966). In the case of polymeric materials, for instance, the level of order (crystallinity), the extent of alignment of the morphological units (orientation) and the degree of polymerization are among the many factors that could influence the viscoelastic response characteristics of such materials. Thus, while, in this chapter, the continuum mechanics approach is maintained primarily for the characterization of the response behaviour of viscoelastic materials, it is emphasized that such response is essentially dependent on the effects of a large number of significant microscopic and macroscopic parameters such as those introduced above.

In this chapter, the basic formalism of the mechanical response of the linear viscoelastic material is presented. Consequently, the formulations are considered entirely within the context of the infinitesimal linearized deformation theory. Here, the ideas set down by L. Boltzmann (1844-1906) and V. Volterra (1860-1940) are taken as fundamental within the context of linear superposition of input histories (Boltzmann, 1874, 1877, 1878; Volterra, 1913, Volterra and Peres, 1936 and Leaderman, 1943 a&b, 1958). The transition to "*dynamic viscoelasticity*" is then discussed whereby the formulation is extended to include the possibility of characterizing the linear viscoelastic behaviour of materials in frequency domain. In this context, the relationships between the material functions characterizing the viscoelastic response in both the time and frequency domains are considered. In this context, the author confines his attention to the one-dimensional linear theory of isothermal viscoelasticity under variable levels of stress and strain inputs. For further studies on the subject of linear viscoelasticity, the reader is referred to the books by Gross (1953), Eirich (1956), Bland

---

[4]   References which focus on polymeric materials include, for instance, Eirich (1956), Staverman and Schwarzl (1956), Ferry (1970) and Doi and Edwards (1986), among others.

(1960), Ferry (1970), Christensen (1971), Flügge (1975), Gittus (1975), Ward (1983), Tschoegl (1989), and Haddad (1995), among others.

Figure 8.1 illustrates the differences in strain response of elastic, viscous and viscoelastic specimens when the three specimens are subjected to a constant stress of unit magnitude. The stress is applied, at time t=0, to originally undisturbed specimens and maintained constant for a time duration t_1, Fig. 8.1a. As shown in Fig. 8.1b, the strain-time response of the elastic specimen has the same form as the applied stress. Upon the application of the load, the strain reaches instantaneously a certain level ϵ_0 and then remains constant. For the viscous fluid, Fig. 8.1c, the material flows at a constant rate and the strain response is proportional to the time. For the viscoelastic specimen, Fig. 8.1d, there is a relatively rapid increase in the strain response for small values of t immediately after the application of the load. As t increases, the slope of the curve decreases and as $t \to \infty$, the slope may approach zero or finite value provided that the applied stress is maintained constant. Upon the removal of the load at time t_1, the strains in the three specimens will recover in the manners shown in Fig. 8.1. The perfectly elastic solid will recover instantaneously upon the removal of the load (Fig. 8.1b), but the viscous fluid will not recover, i.e. it maintains the already acquired level of strain (Fig. 8.1c). Meantime, upon the removal of the load, the viscoelastic specimen will recover immediately its elastic deformation, however, the retarded part of the response will require time for partial or complete recovery (Fig. 8.1d).

Under constant stress, the creep strain in a viscoelastic material may be divided, with reference to Fig. 8.2, into the following three components (e.g., Lethersich, 1950):

(i)   Instantaneous (immediate) elastic strain $\epsilon_e(0^+)$. In a polymeric material, for instance, this part of strain is attributable to bond stretching and bending including the deformation of weak Van der Waals bonds between the molecular chains. This strain is *reversible* and disappears upon the removal of stress.

(ii)   Delayed elastic strain $\epsilon_d(t)$. The rate of increase of this part of strain decreases steadily with time. It is also elastic, but, after the removal of the load, it requires time for complete recovery. It is often termed *"primary creep"* or *"elastic after-effect"*, among other terms. In a polymeric material, the delayed elastic strain is attributable to, for instance, chain uncoiling.

(iii)   Viscous flow $\epsilon_v(t)$. It is an *irreversible* component of strain which may or may not increase linearly with time of stress application. In a polymeric substance, it is characteristic of inter-chain slipping. It is often referred to as *"secondary creep"* or *"non-recoverable strain"*.

On unloading the viscoelastic specimen at t_1, the instantaneous elastic response recovers immediately, the delayed elastic response recovers gradually, but the viscous flow remains (e.g., Ward, 1983).

278

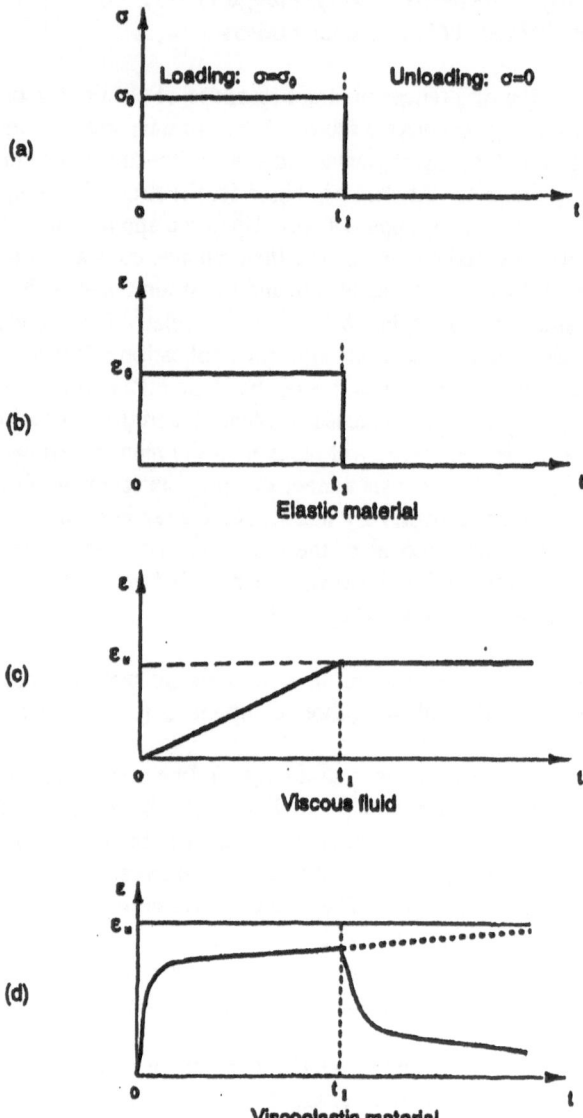

Figure 8.1. Comparison of strain response for elastic, viscous and viscoelastic material specimens under constant stress of unit magnitude until time t_1.

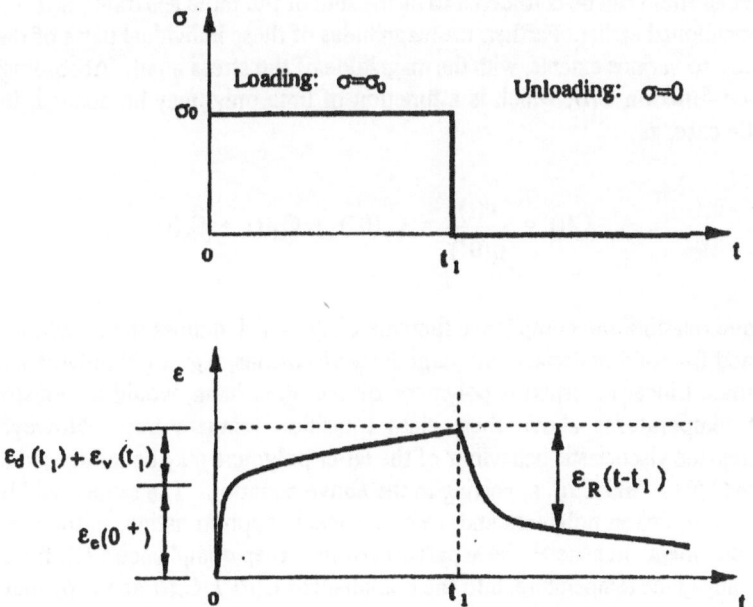

Figure 8.2. Creep and recovery of a viscoelastic material specimen subjected to a constant stress of unit magnitude until time t_1.

From a phenomenological point of view, two aspects of viscoelastic behaviour are dealt with, i.e., "**creep response under constant stress**" and "**stress-relaxation response under constant strain**". As will be dealt with later in this chapter, the correlation between these two aspects of response constitutes an important characteristic of the development of the linear viscoelasticity theory.

8.1.1. CREEP RESPONSE

In a simple creep experiment, the undisturbed material specimen is subjected initially, at time $t=0$, to a stress $\sigma_0 = \sigma(0^+)$ which is maintained constant during the experiment, meanwhile the time-dependent strain $\epsilon(t)$ is observed. In the linear viscoelastic case, the creep response follows, in general, the pattern discussed above in conjunction with Figure 8.2. In this case, the total creep strain can be considered to be the sum of the three separate parts $\epsilon_e(0^+)$, $\epsilon_d(t)$, and $\epsilon_v(t)$ mentioned earlier. Further, the magnitudes of these individual parts of the strain are proportional, to various extents, with the magnitude of the stress input. Accordingly, a creep compliance function $C(t)$, which is a function of time only, may be defined, in the linear viscoelastic case, as

$$C(t) = \frac{\epsilon(t)}{\sigma(0^+)} = C_e(0^+) + C_d(t) + C_v(t)$$

In the above relation, the compliance function $C_v(t)$ which defines the Newtonian flow can be neglected for solid materials with large flow viscosities, e.g., rigid polymers at ordinary temperatures. Linear amorphous polymers, on the other hand, would demonstrate a finite $C_v(t)$ at temperatures above their glass transition temperatures. However, at low temperatures, the viscoelastic behaviour of the latter polymers may be influenced only by the compliance $C_e(0+)$, and $C_d(t)$, appearing in the above equation. The same could be valid for the case of high linked polymers and, to a reasonable approximation, in the case of highly crystalline polymers. In general, the separation of the creep compliance $C(t)$, for a particular material at any given temperature, into the compliances $C_e(0^+)$, $C_d(t)$ and $C_v(t)$ may not be an easy task and could involve an arbitrary division.

8.1.2. CREEP AND RECOVERY

With reference to Fig. 8.2, consider the case where the stress σ_0 is applied to, an originally undisturbed specimen, at time t=0 and removed at time $t=t_1$. Thus, on the assumption of linear viscoelastic behaviour, the total creep strain $\epsilon(t)$ at any instant of time $t>t_1$ is given by the superposition of the two individual strains, i.e., $\epsilon_e = \sigma_0 C(t)$ corresponding to loading the specimen at t=0 and $\epsilon_R(t-t_1) = -\sigma_0 C(t-t_1)$ corresponding to unloading at $t=t_1$. That is,

$$\epsilon(t) = \sigma_0 C(t) - \sigma_0 C(t - t_1)$$

The recovery strain, $\epsilon_R(t-t_1)$, is defined, in view of the relation above, as the difference between the anticipated creep under the initial stress and the actual measured creep strain. This is shown in Fig. 8.2. Examples of the creep response and creep recovery of a number of engineering materials are shown in Figures 8.3 and 8.4.

8.1.3. STRESS-RELAXATION

In a simple stress-relaxation experiment, the material specimen is subjected initially to a constant strain $\epsilon(t) = \epsilon(0^+)$ and the time-dependent stress response is observed, Fig. 8.5. As shown in the latter figure, the stress-relaxation response, of a linear viscoelastic response, is monotonously decreasing with time. On the assumption of a linear viscoelastic behaviour, a stress-relaxation modulus, which is a function of time only, is defined as

$$R(t) = \frac{\sigma(t)}{\varepsilon(0^+)}$$

In a stress-relaxation experiment, as described here, viscous flow affects the limiting value of stress. In the presence of viscous flow, the stress may decay to zero at sufficient long times. On the other hand, if there is no viscous flow, the stress decays to a finite value. This would result in a so-called *"equilibrium"* or *"relaxed"* modulus $R_\infty = R(\infty)$ at infinite time (e.g., Lockett, 1972 and Gittus, 1975). An example of the stress-relaxation response of a class of engineering material is given in Figure 8.6.

The particular nature of the class of viscoelastic materials considered in the above examples proves, as mentioned earlier, the existence of a property of *'passive resistance'* in such materials. This is in contrast to the instantaneous response and reversibility that usually characterize pure elastic behaviour. This passive resistance is of viscous nature and reflects what is usually called the property of *'hereditary response'* of the material. *That is, the present state of viscoelastic response depends not only upon the present state of loading input, but also upon previous states.*

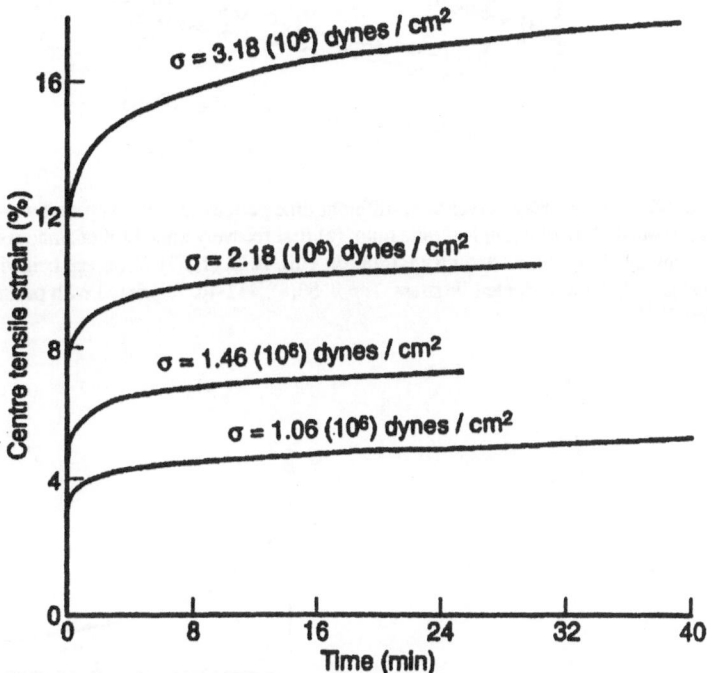

Figure 8.3. Effect of load on creep of a composite solid propellant at 75° F. (Reprinted with permission from: Blatz, P. J. (1956) Rheology of composite solid propellants, *J. Ind. Eng. Chem.* **48**(4), 737-9. Copyright (1956) American Chemical Society).

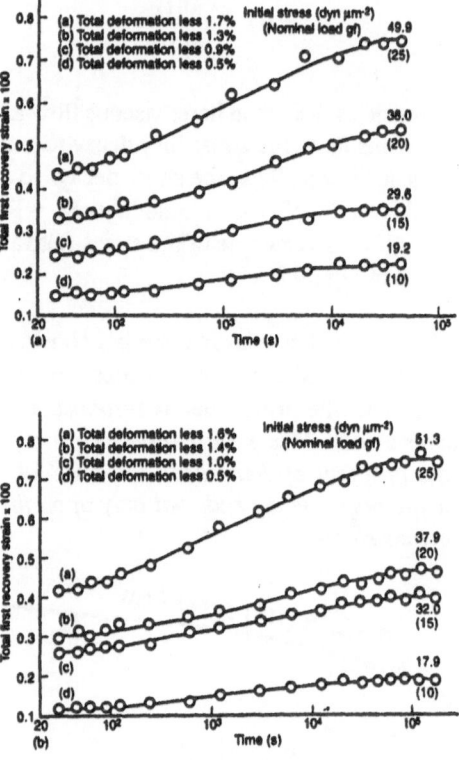

Figure 8.4. First recovery curves after different time periods of first creep. Individual dry suummerwood fibres of a longleaf pine pulp: (a) first recovery after 12 h of first creep; (b) first recovery after 48 h of first creep. (Source: Hill, R. L. (1967) The creep behaviour of individual pulp fibres under tensile stress. *Tappi* 50(8), 432-40. Reprinted with permission of Tappi).

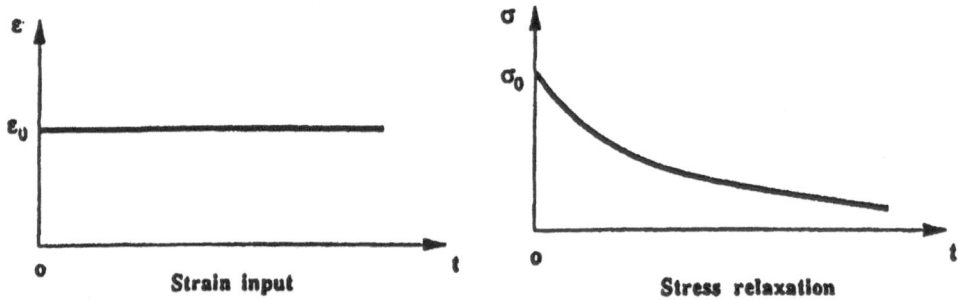

Figure 8.5. Stress relaxation of a viscoelastic material specimen subjected to constant strain of unit magnitude.

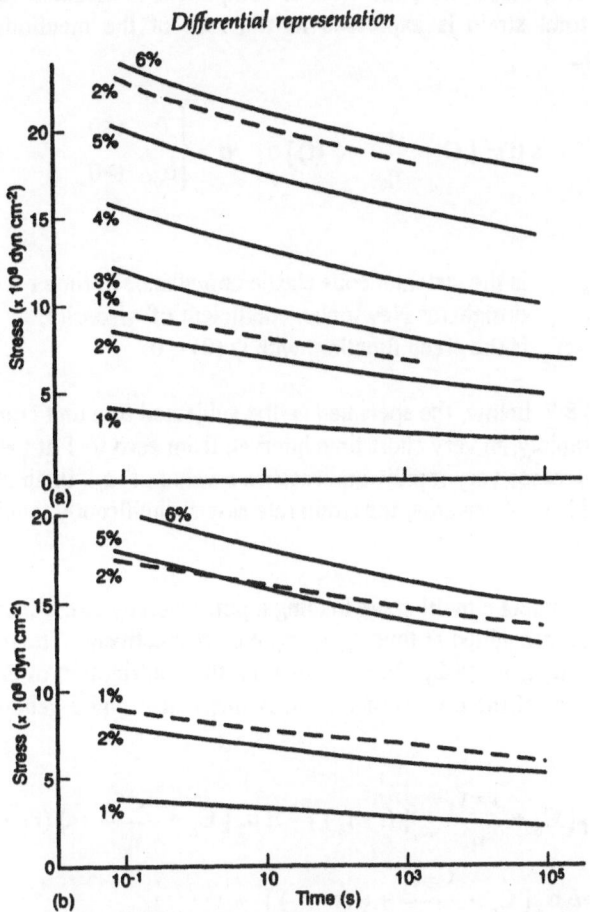

Figure 8.6. Relaxation of stress in cotton (——) and flax (----) at 65 % RH and 25°C: (a) first relaxation; (b) fourth relaxation. (Source: Meredith, R. (1954) Relaxation of stress in stretched cellulose fibres. *J. Textile Inst.*, **45**, T436-T460. Reprinted by permission of the British Textile Technology Group).

8.2. Linear Viscoelastic Behaviour

8.2.1. DESCRIPTION IN THE TIME DOMAIN

Creep Response

For a viscoelastic material, an application of a constant stress σ_0 produces a strain which is visualized, as mentioned earlier, to compose, in general, of three components; namely, *instantaneous*, *viscous* and *delayed* response components. For a linear viscoelastic response, the instantaneous response component is postulated to be instantaneously reversible upon the removal of the load, meanwhile, the viscous component is assumed to be Newtonian in character. The total strain is expressed as the sum of the mentioned three response components, i. e.,

$$\varepsilon(t) = [\, C_o + \frac{t}{\eta_0} + \psi(t)\,]\,\sigma, \quad \sigma = \begin{cases} 0, & t<0 \\ \sigma_0, & t>0 \end{cases} \tag{8.1}$$

where, C_o is the instantaneous elastic compliance at time $t = 0$,
η_0 designates Newtonian coefficient of viscosity,
$\psi(t)$ is the creep function, with $\psi(0) = 0$.

As shown in Fig. 8.7, below, the specimen is first subjected to a unit constant stress $\sigma_0 = 1$ which is applied rapidly, in very short time interval, from zero to 1 at $t = 0$. This results in a strain which increases very rapidly and reaches a value of ε_o in an almost straight line fashion at $t = 0$. Then, afterwards, the strain rate slows significantly tending asymptotically to a constant value.

In Fig. 8.8, one considers a multi-stage loading input, whereby stress-increments $\Delta\sigma_1, \Delta\sigma_2, \Delta\sigma_3$, etc. are added at time τ_1, τ_2, τ_3, etc., respectively. The total creep strain at time t is expressed by Eqn. (8.2). In this equation, the contribution of each loading step is given as the product of the corresponding stress-increment and a general time-dependent material function.

$$\varepsilon(t) = \Delta\sigma_1\,[\, C_o + \frac{t-\tau_1}{\eta_0} + \psi(t-\tau_1)\,] + \Delta\sigma_2\,[\, C_o + \frac{t-\tau_2}{\eta_0} + \psi(t-\tau_2)\,]$$
$$+ \Delta\sigma_3\,[\, C_o + \frac{t-\tau_3}{\eta_0} + \psi(t-\tau_3)\,] + \cdots\cdots \tag{8.2}$$

Allowing for the continuity of the loading, by letting the time-interval to go to zero, the response equation (8.2) can be written in terms of the integral over the involved time-history as

$$\varepsilon(t) = \int_{-\infty}^{t} \frac{d\sigma(\tau)}{d\tau}\,[\, C_o + \frac{t-\tau}{\eta_0} + \psi(t-\tau)\,]\,d\tau \tag{8.3}$$

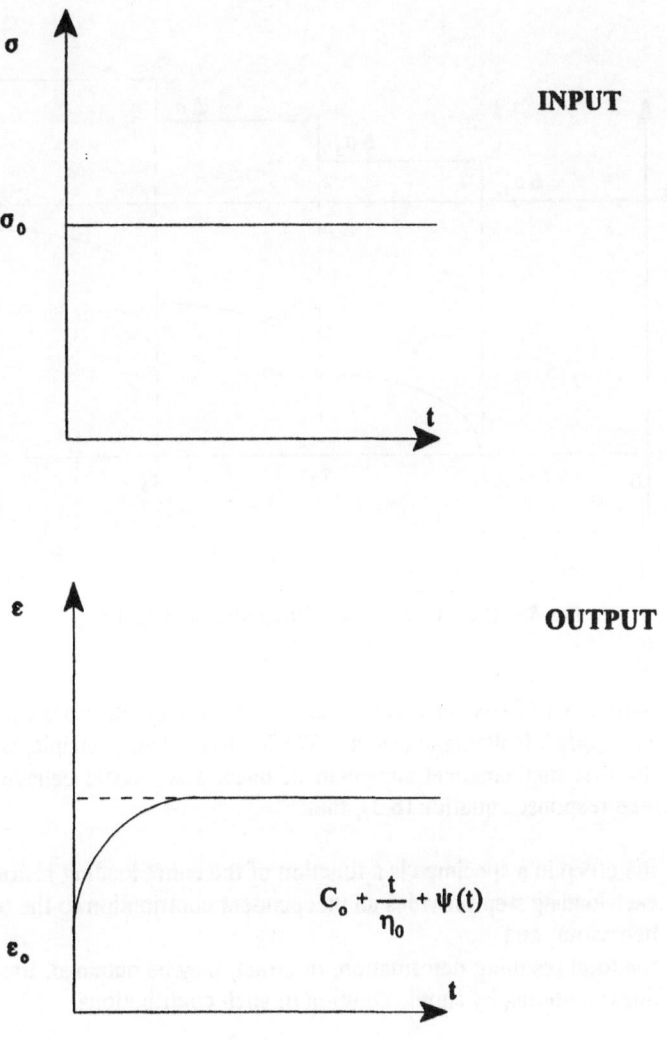

Figure 8.7. Creep response behaviour.

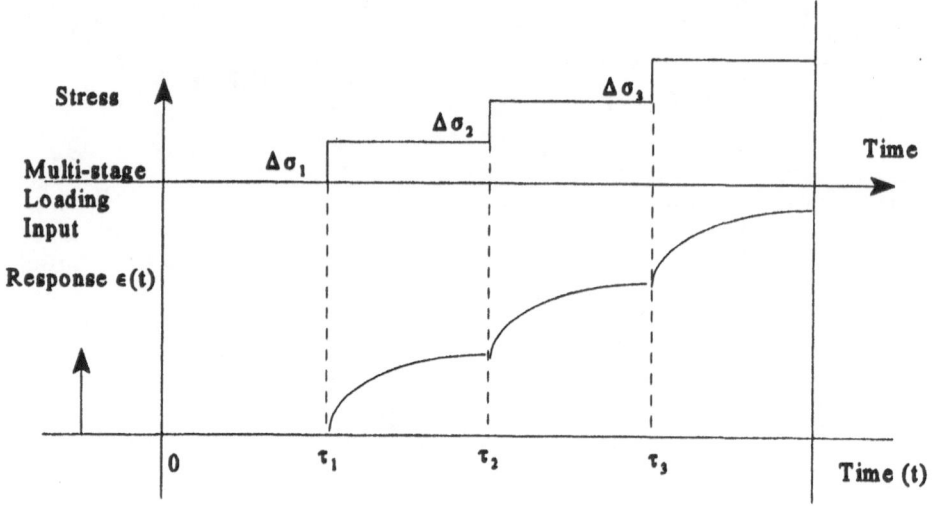

Fig. 8.8. Creep behaviour of a linear viscoelastic solid.

The response equation (8.3) is, in essence, an expression of the classical "*Boltzmann Superposition Principle*"; Boltzmann (1874, 1877 & 1878). This principle, which is often considered as the first mathematical statement of linear viscoelastic behaviour, implies, inview of the creep response equation (8.3), that:

- the creep in a specimen is a function of the entire loading history,
- each loading step provides an independent contribution to the total response behaviour, and
- the total resulting deformation, or strain, may be obtained, for the case of a linear material, by simple addition of such contributions.

Fig. 8.9 illustrates, for instance, a unit constant stress which is applied at $t=0$ and unloaded at $t=t_1$.

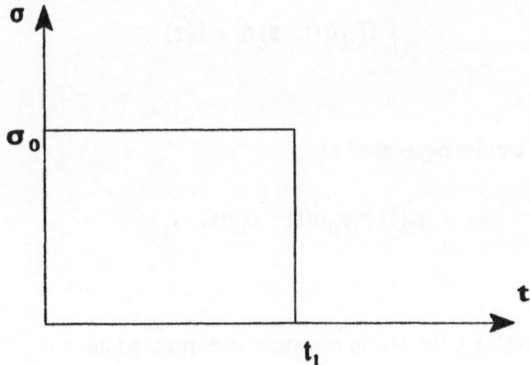

Figure 8.9. Unit constant stress.

Introducing a unit step function u(t), and its time-derivative, the delta function δ(t), i.e.,

$$u(t) = \begin{cases} 0, & t < 0 \\ 1, & t \geq 0 \end{cases} \tag{8.4}$$

$$\frac{du(t)}{dt} = \delta(t) \tag{8.5}$$

$$\delta(t) = \begin{cases} 0, & t \neq 0 \\ \infty, & t = 0 \end{cases} \tag{8.6}$$

$$\int_{-\infty}^{\infty} \delta(t)\, dt = 1 \tag{8.7}$$

$$\int_{-\infty}^{\infty} f(t)\delta(t-\tau)\, dt = f(\tau) \tag{8.8}$$

then, the load input can be expressed as:

$$\sigma(t) = \sigma_0 u(t) - \sigma_0 u(t-t_1) \tag{8.9}$$

Combining (8.3) and (8.9), the strain response is expressed as

For $t < 0$,

$$\varepsilon(t) = 0 \qquad ; t<0 \tag{8.10}$$

For $0 < t < t_1$,

$$\begin{aligned}
\varepsilon(t) &= \int_{-\infty}^{t} [\, C_o + \frac{t-\tau}{\eta_0} + \psi(t-\tau)\,] \frac{d\sigma(\tau)}{d\tau}\, d\tau \\
&= \int_{-\infty}^{t} [\, C_o + \frac{t-\tau}{\eta_0} + \psi(t-\tau)\,] \sigma_0\, \delta(\tau)\, d\tau \\
&= [\, C_o + \frac{t-\tau}{\eta_0} + \psi(t)\,] \sigma_0
\end{aligned} \tag{8.11}$$

For $t > t_1$,

$$\varepsilon(t) = \int_{-\infty}^{t} [\, C_o + \frac{t-\tau}{\eta_0} + \psi(t-\tau)\,] \frac{d\sigma(\tau)}{d\tau}\, d\tau$$

$$= \int_{-\infty}^{t} [\, C_o + \frac{t-\tau}{\eta_0} + \psi(t-\tau)\,]\, \sigma_0\, [\, \delta(\tau) - \delta(\tau - t_1)\,]\, d\tau \qquad \text{(8.12)}$$

$$= \sigma_0 \frac{t_1}{\eta_0} + \sigma_0\, [\, \psi(t) - \psi(t-t_1)\,]$$

The dealt-with above responses are illustrated in Fig. 8.10 below.

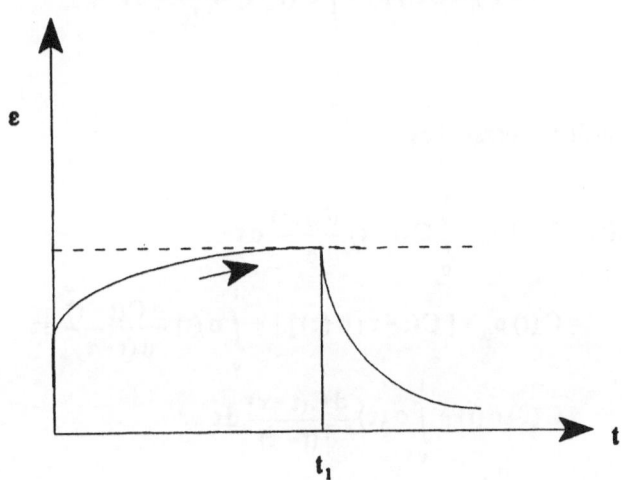

Figure 8.10. Creep and recovery.

Denoting,

$$C(t) = [\, C_o + \frac{t}{\eta_0} + \psi(t) \,] \tag{8.13}$$

then, equation (8.1) becomes

$$\varepsilon(t) = C(t)\,\sigma, \qquad \sigma = \begin{cases} 0, & t<0 \\ \sigma_0, & t>0 \end{cases} \tag{8.14}$$

Meantime, equation (8.3) can be expressed as

$$\varepsilon(t) = C(t)\,\sigma_0 + \int_0^t C(t-\tau)\,\frac{d\sigma(\tau)}{d\tau}\,d\tau \tag{8.15}$$

which can be further expressed as

$$
\begin{aligned}
\varepsilon(t) &= C(t)\,\sigma_0 + \int_0^t C(t-\tau)\,\frac{d\sigma(\tau)}{d\tau}\,d\tau \\
&= C(t)\,\sigma_0 + [\,C(t-\tau)\,\sigma(\tau)\,]\Big|_0^t + \int_0^t \sigma(t)\,\frac{dC(t-\tau)}{d(t-\tau)}\,d\tau \\
&= C(0)\,\sigma(t) + \int_0^t \sigma(\tau)\,\frac{dC(t-\tau)}{d(t-\tau)}\,d\tau
\end{aligned}
\tag{8.16}
$$

Equations (8.3) and (8.16) show that the strain at any instant of time t is a function of the history of stress input and not just its current value.

Relaxation Response
Application of constant strain ε_0, on the material specimen, produces a time-dependent stress given by

$$\sigma(t) = [\, E_0 + \phi(t)\,]\,\varepsilon \qquad\qquad \varepsilon = \begin{cases} 0 & t < 0 \\ \varepsilon_0 & t > 0 \end{cases} \qquad\qquad (8.17)$$

in which,

    E_0 is the instantaneous elastic modulus; at time $t = 0$, and

    $\phi(t)$ designates the relaxation function, with $\phi(\infty) = 0$

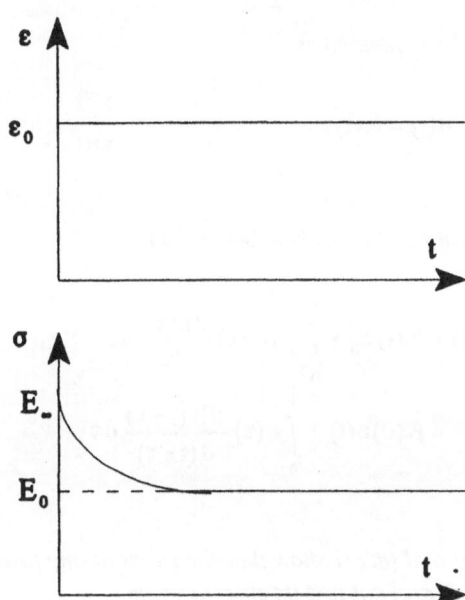

Figure 8.11. Relaxation response behaviour under idealized constant strain input.

A similar treatment to that carried out above for determining the creep response may be performed, as illustrated below, for the case of relaxation.

As shown in Fig. 8.11, a constant strain is applied rapidly, in a very short time interval, from zero to ε_0 at t=0. This results in a stress which increases very rapidly and reaches a value of σ_0 in an almost straight line fashion at t=0. Afterwards, the stress decreases asymptotically to reach ultimately a constant value σ_∞. The general expression of the relation between strain, stress and time is expressed as

$$\sigma(t) = \int_{-\infty}^{t} \frac{d\varepsilon(\tau)}{d\tau} [E_0 + \phi(t-\tau)] \tag{8.18}$$

Denoting

$$R(t) = [E_0 + \phi(t-\tau)] \tag{8.19}$$

then the Eq. (8.17) can be expressed as:

$$\sigma(t) = R(t)\varepsilon \qquad \varepsilon = \begin{cases} 0 & t<0 \\ \varepsilon_0 & t>0 \end{cases} \tag{8.20}$$

which may be further written as (e.g., Haddad, 1995)

$$\sigma(t) = R(t)\,\varepsilon_0 + \int_0^t R(t-\tau)\frac{d\varepsilon(\tau)}{d\tau}\, d\tau$$
$$= R(0)\varepsilon(t) + \int_0^t \varepsilon(\tau)\frac{dR(t-\tau)}{d(t-\tau)}\, d\tau \tag{8.21}$$

Again, Equations (8.18) and (8.21) show that the stress at any instant of time t is a function of the history of strain input and not just its current value.

Equations (8.3) and (8.18) represent "*a pair of integral equations*". They are not independent of each other. The inversion of Eq. (8.3) is given by Eq. (8.18) and vise versa. So, if we substitute one equation into the other one, an expression for the inter-relationship between the creep and relaxation functions may be written as follows

$$E_0 \varepsilon_\infty - 1 + C_o \phi(t) + E_0 [\frac{t}{\eta_0} + \psi(t)] + \int_0^t \phi(\tau) [\frac{1}{\eta_0} + \frac{d \psi(t-\tau)}{d(t-\tau)}] d\tau = 0 \qquad (8.22)$$

8.2.2. DESCRIPTION IN THE "FOURIER SPECTRUM" DOMAIN

Creep Response
Recalling the creep response equation (8.1), i.e.,

$$\varepsilon(t) = [C_o + \frac{t}{\eta_0} + \psi(t)] \sigma, \qquad \sigma = \begin{cases} 0, & t < 0 \\ \sigma_0, & t > 0 \end{cases} \qquad (8.23)$$

Using the concept of unit step function (*Appendix B*), it follows that

$$\varepsilon(t) = [C_o + \frac{t}{\eta_0} + \psi(t)] u(t) \sigma_0$$
$$= C(t) \sigma_0 \qquad\qquad -\infty < t < \infty \qquad (8.24)$$

where,

$$C(t) = [C_o + \frac{t}{\eta_0} + \psi(t)] u(t) \qquad (8.25)$$

with,

$$\dot{C}(t) = [C_o \delta(t) + \frac{u(t)}{\eta_0} + \frac{t\delta(t)}{\eta_0} + \psi(t)\delta(t) + \dot{\psi}(t) u(t)] \qquad (8.26)$$

Further, if we denote the Fourier-spectrum of $\dot{C}(t)$ by $J^*(i\omega)$, then

$$J^*(i\omega) = \int_{-\infty}^{\infty} e^{-i\omega t} \, \dot{C}(t) \, dt$$

$$= \int_{-\infty}^{\infty} e^{-i\omega t} \left[\varepsilon_\infty \, \delta(t) + \frac{u(t)}{\eta_0} + \dot{\psi}(t) \, u(t) \right] dt \qquad (8.27)$$

$$= C_0 + \frac{1}{i\omega \eta_0} + \int_{-\infty}^{\infty} e^{-i\omega t} \, \dot{\psi}(t) \, u(t) \, dt$$

$$= C_0 + \frac{1}{i\omega \eta_0} + \int_{-\infty}^{\infty} e^{-i\omega t} \, \dot{\psi}(t) \, dt$$

However, it is often convenient to write

$$J_1(\omega) + i J_2(\omega) = \int_0^{\infty} e^{-i\omega t} \, \dot{\psi}(t) \, dt \qquad (8.28)$$

with,

$$J_1(\omega) = \int_0^{\infty} \dot{\psi}(t) \, \cos(\omega t) \, dt \qquad (8.29)$$

and,

$$J_2(\omega) = \int_0^{\infty} \dot{\psi}(t) \, \sin(\omega t) \, dt \qquad (8.30)$$

It is apparent from (8.27) that $J^*(i\omega)$ is an integral transform of the creep function $C(t)$..

In order to illustrate the physical significance of $J^*(i\omega)$, we consider the following case:

$$J^*(i\omega) = \left[C_0 + J_1(\omega) \right] + i \left[J_2(\omega) - \frac{1}{\omega \eta_0} \right] \tag{8.31}$$

If we apply an alternating stress with frequency ω:

$$\sigma = \sigma_0 e^{i\omega t} \tag{8.32}$$

then, by combining equations (8.3) and (8.32), it follows that

$$
\begin{aligned}
\sigma(t) &= \int_{-\infty}^{t} \frac{d \sigma_0 e^{i\omega \tau}}{d\tau} \left[C_0 = \frac{t-\tau}{\eta_0} + \psi(t-\tau) \right] d\tau \\
&= \sigma_0 e^{i\omega t} i\omega \int_0^\infty e^{-i\omega \tau} \left[C_0 = \frac{\tau}{\eta_0} + \psi(t-\tau) \right] d\tau \\
&= \sigma_0 e^{i\omega t} i\omega \left[\frac{C_0}{i\omega} + \frac{1}{(i\omega)^2 \eta_0} + \int_0^\infty \psi(\tau) e^{-i\omega \tau} d\tau \right] \\
&= \sigma_0 e^{i\omega t} \left[C_0 + \frac{1}{i\omega \eta_0} + i\omega \int_0^\infty \psi(\tau) e^{-i\omega \tau} d\tau \right] \\
&= J^*(i\omega) \varepsilon
\end{aligned}
\tag{8.33}
$$

It is, then, evident from (8.33) that $J^*(i\omega)$ expresses the relation between the stress and strain when an alternating stress is applied on the specimen. Equation (8.33) represents the creep response in Fourier-spectrum domain. It also shows that the application of an alternating stress produces an alternating strain response of the same frequency. The amplitude of the response is

$$\varepsilon_0 = J^*(i\omega) \sigma_0 \tag{8.34}$$

$J^*(i\omega)$ is referred-to as the "*complex compliance*". Meanwhile, $J_1(\omega)$ is referred-to as the "*storage compliance*" or the "*dynamic compliance*", and $J_2(\omega)$ is known by the "*loss* *compliance*" or the "*friction compliance*".

Relaxation Response
Recalling the relaxation response equation (8.17), one has

$$\sigma(t) = [E_0 + \phi(t)]\varepsilon \qquad \varepsilon = \begin{cases} 0 & t<0 \\ \varepsilon_0 & t>0 \end{cases} \tag{8.35}$$

Using, further, the concept of unit step function (*Appendix B*), one can write

$$\begin{aligned} \sigma(t) &= [E_0 + \phi(t)]u(t)\varepsilon_0 \qquad -\infty<t<\infty \\ &= R(t)\varepsilon_0 \end{aligned} \tag{8.36}$$

in which,

$$R(t) = [E_0 + \phi(t)]u(t) \tag{8.37}$$

with,

$$\dot{R}(t) = E_0\,\delta(t) + \dot{\psi}(t)u(t) + \bar{\psi}(t)\delta(t) \tag{8.38}$$

Denoting the Fourier-spectrum of R(t) by $E^*(i\omega)$, it follows that

$$E^*(i\omega) = \int_{-\infty}^{\infty} \dot{R}(t)\,e^{-i\omega t}\,dt \tag{8.39}$$

It is, however, convenient to write:

$$E^*(i\omega) = \int\limits_{-\infty}^{\infty} [E_0\,\delta(t) + \dot\phi(t)\,u(t) + \phi(t)\,\delta(t)]\,dt$$

$$= E_0 + \int\limits_{-\infty}^{\infty} \dot\phi(t)\,e^{-i\omega t}\,u(t)\,dt + \phi(0)$$

$$= E_0 + \phi(0) + \int\limits_{-\infty}^{\infty} \dot\phi(t)\,e^{-i\omega t}\,dt \tag{8.40}$$

$$= E_0 + \phi(0) + \phi(t)\,e^{-i\omega t}\,\Big|_0^{\infty} + i\omega \int\limits_0^{\infty} \phi(t)\,e^{-i\omega t}\,dt$$

$$= E_0 + i\omega \int\limits_0^{\infty} \phi(t)\,e^{-i\omega t}\,dt$$

where,

$$E_1 + iE_2 = i\omega \int\limits_0^{\infty} \phi(t)\,e^{-i\omega t}\,dt \tag{8.41}$$

and,

$$E_1 = \omega \int\limits_0^{\infty} \phi(t)\,\sin(\omega t)\,dt \tag{8.42}$$

$$E_2 = \omega \int\limits_0^{\infty} \phi(t)\,\cos(\omega t)\,dt \tag{8.43}$$

Thus,

$$E^*(i\omega) = [E_0 + E_1(\omega)] + iE_2(\omega) \tag{8.44}$$

In order to determine the physical significance of E*(iω) in the Fourier-spectrum domain, w apply, on the material specimen, an alternating strain with frequency ω, i.e.,

$$\varepsilon = \varepsilon_0 e^{i\omega t} \tag{8.45}$$

Thus, by substituting (8.45) into equation (8.18), it follows that

$$\sigma(t) = \int_{-\infty}^{t} \varepsilon_0 i\omega e^{i\omega\tau} [E_0 + \phi(t-\tau)] d\tau \tag{8.46}$$

$$t - \tau = \tau'$$

$$\sigma(t) = \int_0^{\infty} \varepsilon_0 i\omega e^{i\omega(t-\tau)} [E_0 + \phi(\tau)] d\tau$$

$$= \varepsilon_0 e^{i\omega t} i\omega \int_0^{\infty} [E_0 + \phi(\tau)] e^{i\omega\tau} d\tau$$

$$= \varepsilon_0 e^{i\omega t} i\omega [E_0 \frac{1}{i\omega} + \int_0^{\infty} \phi(\tau) e^{i\omega\tau} d\tau] \tag{8.47}$$

$$= \varepsilon(t) [E_0 + i\omega \int_0^{\infty} \phi(\tau) e^{-i\omega\tau} d\tau]$$

$$= \varepsilon(t) E^*(i\omega)$$

It is clear from (8.47) that the Fourier-spectrum of E* (iω) expresses the relation between th strain and stress when we apply an alternating strain on the specimen. Equation (8.47) represent the relaxation behaviour in frequency domain, and shows that the application of an alternatin; strain produces an alternating stress response of the same frequency. The amplitude of th response is:

$$\sigma_0 = E^*(i\omega)\,\varepsilon_0 \tag{8.48}$$

$E^*(i\omega)$ is referred-to as the "*complex modulus*". Meanwhile $E_1(\omega)$ is referred-to as the '*storage modulus*" or the "*dynamic Modulus*", and $E_2(\omega)$ is called the "*loss modulus*" or the '*friction modulus*".

From equations (8.34) and (8.48), it follows that one can write the following relation

$$E^*(i\omega)\,J^*(i\omega) = 1 \tag{8.49}$$

which represents the well known relation between the complex compliance $J^*(i\omega)$ and the complex modulus $E^*(i\omega)$ in linear viscoelasticity.

8.3. Inverse-relations between "Fourier Spectrum" and "Creep and Relaxation Functions"

In equation (8.41), we denote

$$E(i\omega) = E_1 + iE_2 \tag{8.50}$$

then,

$$\frac{E(i\omega)}{i\omega} = \int_{-\infty}^{\infty} \phi(t)\,u(t)\,e^{-i\omega t}\,dt \tag{8.51}$$

Meantime, by using the inverse Fourier transform, it follows that

$$\phi(t)\,u(t) = \frac{1}{2\pi} \int_{-\infty}^{\infty} \frac{E(i\omega)}{i\omega}\,e^{i\omega t}\,d\omega \tag{8.52}$$

The determination of the above integration may be simplified by expressing $E(i\omega)$ in terms of $E_1(\omega)$ (*dynamic modulus*) and $E_2(\omega)$ (*dynamic friction*). Thus,

$$\frac{E(i\omega)}{i\omega} e^{i\omega t} = \frac{E_1(\omega) + i E_2(\omega)}{i\omega} [\cos(\omega t) + i \sin(\omega t)]$$

$$= \frac{E_1(\omega) + i E_2(\omega)}{\omega} [\sin(\omega t) - i\cos(\omega t)] \tag{8.53}$$

$$= \frac{E_1}{\omega} \sin(\omega t) + \frac{E_2}{\omega} \cos(\omega t) - i [\frac{E_2}{\omega} \sin(\omega t) - \frac{E_1}{\omega} \cos(\omega t)]$$

From our previous discussion, equations (8.42) and (8.43), we know that:

$E_1(\omega)/\omega$ is an odd function of ω

$E_2(\omega)/\omega$ is an even function of ω

Hence,

$$\phi(t) u(t) = \frac{1}{2\pi} \int_{-\infty}^{\infty} [\frac{E_1(\omega)}{\omega} \sin(\omega t) + \frac{E_2(\omega)}{\omega} \cos(\omega t)] d\omega$$

$$= \frac{1}{\pi} \int_{0}^{\infty} [\frac{E_1(\omega)}{\omega} \sin(\omega t) + \frac{E_2(\omega)}{\omega} \cos(\omega t)] d\omega \tag{8.54}$$

and,

$$\phi(t) u(t) + \phi(-t) u(-t) = \frac{1}{\pi} \int_{0}^{\infty} [\frac{E_1(\omega)}{\omega} \sin(\omega t) + \frac{E_2(\omega)}{\omega} \cos(\omega t)] d\omega$$

$$+ \frac{1}{\pi} \int_{0}^{\infty} [-\frac{E_1(\omega)}{\omega} \sin(\omega t) + \frac{E_2(\omega)}{\omega} \cos(\omega t)] d\omega \tag{8.55}$$

$$= \frac{2}{\pi} \int_{0}^{\infty} \frac{E_1(\omega)}{\omega} \sin(\omega t) d\omega$$

ıntime, if we confine our attention to t>0, then,

$$\phi(t) = \frac{2}{\pi} \int_0^\infty \frac{E_2(\omega)}{\omega} \cos(\omega t) \, d\omega \qquad\qquad t>0 \qquad\qquad (8.56)$$

ilarly,

$$\phi(t) u(t) - \phi(-t) u(-t) = \frac{1}{\pi} \int_0^\infty [\, \frac{E_1(\omega)}{\omega} \sin(\omega t) + \frac{E_2(\omega)}{\omega} \cos(\omega t)\,] \, d\omega$$

$$- \frac{1}{\pi} \int_0^\infty [\, -\frac{E_1(\omega)}{\omega} \sin(\omega t) + \frac{E_2(\omega)}{\omega} \cos(\omega t)\,] \, d\omega \qquad (8.57)$$

$$= \frac{2}{\pi} \int_0^\infty \frac{E_1(\omega)}{\omega} \sin(\omega t) \, d\omega$$

t≥ 0, it follows that,

$$\phi(t) = \frac{2}{\pi} \int_0^\infty \frac{E_1(\omega)}{\omega} \sin(\omega t) \, d\omega \qquad\qquad t>0 \qquad\qquad (8.58)$$

similar manner, for the creep case, we denote

$$J(i\omega) = J_1(\omega) + i J_2(\omega) = \int_0^\infty e^{-i\omega t} \, \dot{\psi}(t) \, dt \qquad\qquad (8.59)$$

.s,

302

$$\dot{\psi}(t)\,u(t) = \frac{1}{2\pi}\int_{-\infty}^{\infty} J(i\omega)\,e^{i\omega t}\,dt \qquad (8.60)$$

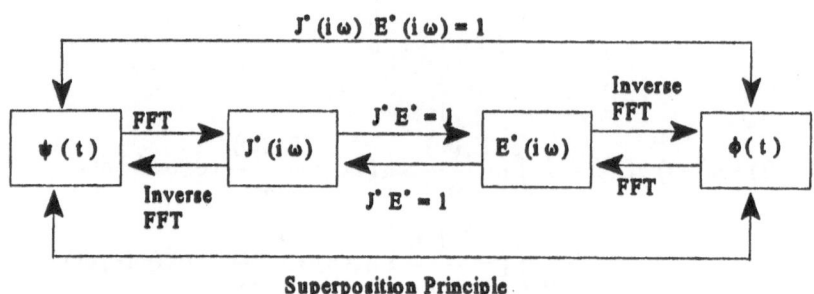

Figure 8.12. Inter-relations between material functions in time- and frequency-domains.

Up to this point, the properties of the creep and relaxation functions in both the time and Fourier spectrum domains have been dealt with. It is obvious that the Fourier transform is the bridge between these two domains. For example, if the creep function is obtained by experiment, the corresponding complex compliance can be easily obtained by using Fourier transform (FT), or alternatively, the Fast Fourier Transform (FFT), then by simple algebraic calculation, the complex elastic modulus can be obtained. Also, from the complex elastic modulus, the relaxation function can be determined by using the inverse Fourier transform, or, alternatively, the inverse FFT. In this context, Fig.8.12, below, shows the inter-relations between the functions in Fourier spectrum domain and the corresponding interrelations between the time domain and Fourier spectrum domain.

8.4. Inter-relations between "Retardation-time" and "Relaxation-time" *Spectra*

In addition to the Fourier-spectrum description, the so-called *"retardation-time"* and *"relaxation-time"* spectra play an important role in linear viscoelasticity. The pertaining analysis is introduced in the following subsections.

8.4.1. CREEP BEHAVIOUR

As previously introduced, the creep function is a continuous, monotonously increasing function. It can be represented by the integral:

$$\psi(t) = \int_0^\infty \beta F(\tau) \left[1 - e^{-\frac{t}{\tau}} \right] d\tau \tag{8.61}$$

in which $F(\tau)$ is the so-called "*retardation-time spectrum*", and β is a normalization factor. The latter is often taken as

$$\beta = \psi(\infty) \tag{8.62}$$

thus, one has

$$\int_0^\infty F(\tau) \, d\tau = 1 \tag{8.63}$$

Equation (8.61) may be also expressed as

$$\psi(t) = \int_0^\infty \beta F(\tau) \, d\tau - \int_0^\infty \beta F(\tau) \, e^{-\frac{t}{\tau}} \, d\tau$$

$$= \psi(\infty) - \int_0^\infty \psi(\infty) F(\tau) \, e^{-\frac{t}{\tau}} \, d\tau \tag{8.64}$$

i. e.,

$$1 - \frac{\psi(t)}{\psi(\infty)} = \int_0^\infty F(\tau) \, e^{-\frac{t}{\tau}} \, d\tau \tag{8.65}$$

or, alternatively,

$$\psi(\infty) - \psi(t) = \psi(\infty) \int_0^\infty F(\tau) e^{-\frac{t}{\tau}} d\tau \qquad (8.66)$$

Therefore, $F(\tau)$ represents the distribution of retardation-time spectrum of a linear viscoelasti material in an indirect manner.

Differentiating (8.66) with respect to the time t, one has

$$-\dot\psi(t) = -\psi(\infty) \int_0^\infty F(\tau) e^{-\frac{t}{\tau}} \frac{1}{\tau} dt \qquad (8.67)$$

and by introducing another function N (s) of the form

$$N(s) = \beta F(\frac{1}{s}) \frac{1}{s^2} \qquad (8.68)$$

it follows that

$$\psi(t) = \int_0^\infty sN(s) e^{-ts} ds \qquad (8.69)$$

It is clear that (8.69) is a representation of Laplace transform, where the parameter s may b interpreted as the pertaining "*retardation frequency*". Accordingly, the function N(s) may b interpreted as a "*frequency-spectrum*" characterizing the creep phenomenon.

8.4.2. RELAXATION BEHAVIOUR

Unlike the creep function which is a continuous, monotonously increasing function, the relaxatior function is a continuous, monotonously decreasing function. The latter may tend to 0 as the tim t goes to ∞. For the relaxation response phenomenon, one may define the relaxation- tim spectrum as:

$$\phi(t) = \int_0^\infty \gamma G(\tau) e^{-\frac{t}{\tau}} d\tau \qquad (8.70)$$

Again, γ is just a normalization factor, and it is equal to

$$\gamma = \phi(0) \qquad (8.71)$$

Therefore,

$$\int_0^\infty G(\tau) d\tau = 0 \qquad (8.72)$$

Similarly, as in the case of creep, by introducing another function of the form

$$Q(s) = \gamma G\left(\frac{1}{s}\right) \frac{1}{s^2} \qquad (8.73)$$

We can transform the definition (8.70) into a representation of Laplace integral or Laplace transform, i.e,

$$\phi(t) = \int_0^\infty Q(s) e^{-ts} ds \qquad (8.74)$$

In which s may be interpreted as the *"relaxation frequency"*. Meantime, $Q(s)$ is a *"frequency spectrum"* characterizing the relaxation phenomenon.

Referring to our earlier presentation, both equations (8.69) and (8.74) are representations of Laplace integral. Thus, when a creep or relaxation equation is given as an analytical expression, one can determine the corresponding spectrum N(s) or Q(s) by applying the inverse Laplace transform, and, consequently, the corresponding retardation-time and relaxation-time spectra can be obtained.

The direct relation between *"retardation-time"* and *"relaxation-time"* spectra is, in fact, rarel: used in practical applications. Here, we give the latter referred-to relation, as was obtained b: Gross (1953), with no derivation.

$$\gamma G(\tau) = \frac{1}{\pi \tau^2} \frac{\pi \beta F(\tau)}{[K(\tau)]^2 + [\pi \beta F(\tau)]^2} \tag{8.75}$$

Where the functions K(t) and κ(t) appearing in the above relations are identified as follows:

$$\beta F(\tau) = \frac{1}{\pi \tau^2} \frac{\pi \gamma G(\tau)}{[\kappa(\tau)]^2 + [\pi \beta F(\tau)]^2} \tag{8.76}$$

$$K(\tau) = \frac{\varepsilon_\infty}{\tau} - \frac{1}{\eta} + \int_0^\infty \beta F(u) \frac{1}{\tau - u} du \tag{8.77}$$

$$\kappa(\tau) = \frac{E_0}{\tau} - \int_0^\infty \gamma G(u) \frac{u}{\tau(\tau - u)} du \tag{8.78}$$

Fig. 8.13 shows the relations among the functions in retardation-time spectrum and relaxation time spectrum domains.

8.5. Inter-relations between "Fourier", "Retardation-time" and "Relaxation-time" Spectr:

8.5.1. CREEP BEHAVIOUR

Because the complex compliance modulus $J^*(i\omega)$ and the retardation time spectrum $F(\tau)$, or th: frequency spectrum $N(s)$, are related to the creep function, one can establish the relation betwee: $J^*(i\omega)$ and $N(s)$.

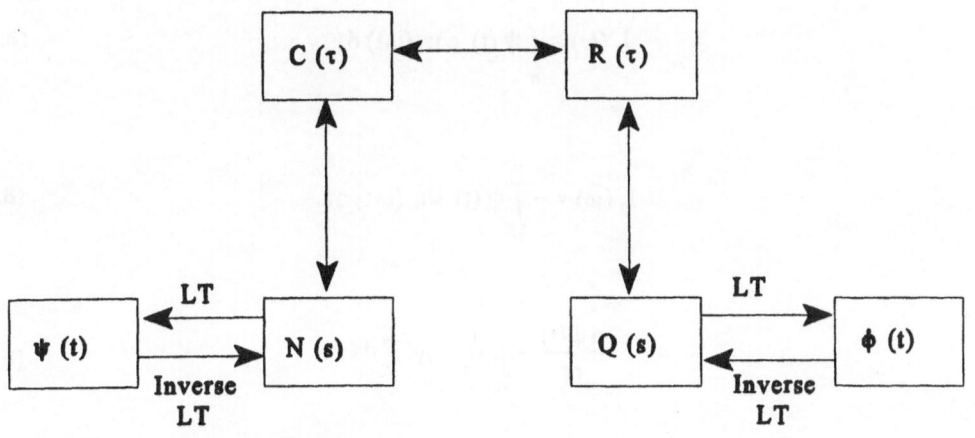

Figure 8.13. Material functions pertaining to "Retardation-time" and "Relaxation-time" spectra.

From the above analysis, the complex compliance modulus $J^*(i\omega)$ can be expressed in terms of the creep function $\psi(t)$ as

$$J^*(i\omega) = C_o + \frac{1}{i\omega\eta_0} + \int_0^\infty e^{-i\omega\tau} \frac{d\psi(\tau)}{d\tau} d\tau \qquad (8.79)$$

and, also,

$$J^*(i\omega) = J_1(\omega) + iJ_2(\omega) = \int_0^\infty e^{-i\omega\tau} \frac{d\psi(\tau)}{d\tau} d\tau \qquad (8.80)$$

But, on the other hand, by definition,

$$J_1(\omega) = \int_0^\infty \dot\psi(t) \cos(\omega t) dt \tag{8.81}$$

$$J_2(\omega) = -\int_0^\infty \dot\psi(t) \sin(\omega t) dt \tag{8.82}$$

$$\frac{d\psi(t)}{dt} = \int_0^\infty s N(s) e^{-ts} ds \tag{8.83}$$

Thus, the relation between $J^*(i\omega)$ and $N(s)$ may be determined as

$$
\begin{aligned}
J^*(i\omega) &= \int_0^\infty e^{-i\omega t} \int_0^\infty s N(s) e^{-ts} ds d\tau \\
&= \int_0^\infty s N(s) \left[\int_0^\infty e^{-i\omega \tau} e^{-ts} d\tau \right] ds \\
&= \int_0^\infty \frac{s N(s)}{s + i\omega} ds
\end{aligned}
\tag{8.84}
$$

Separating the real and imaginary parts, one has

$$J_1(\omega) = \int_0^\infty \frac{s^2 N(s)}{s^2 + \omega^2} ds \tag{8.85}$$

$$J_2(\omega) = -\int_0^\infty \frac{\omega s N(s)}{s^2 + \omega^2}\, ds \tag{8.86}$$

8.5.2. RELAXATION BEHAVIOUR

The complex elastic modulus is expressed as:

$$E^*(i\omega) = E_0 + i\omega \int_0^\infty \phi(t)\, e^{-i\omega t}\, dt \tag{8.87}$$

which may be further written as

$$E^*(i\omega) = E_1 + iE_2(\omega) = i\omega \int_0^\infty \phi(t)\, e^{-i\omega t}\, dt \tag{8.88}$$

On the other hand, by definition,

$$\phi(t) = \int_0^\infty Q(s)\, e^{-ts}\, ds$$

$$\tag{8.89}$$

therefore, it follows that

$$E^*(i\omega) = i\omega \int_0^\infty [\int_0^\infty Q(s) e^{-ts} ds] e^{-i\omega t} dt$$

$$= i\omega \int_0^\infty Q(s) [\int_0^\infty e^{-ts} e^{-i\omega t} dt] ds$$

$$= i\omega \int_0^\infty Q(s) \frac{1}{s + i\omega} ds \qquad (8.90)$$

$$= \int_0^\infty Q(s) \frac{i\omega}{s + i\omega} ds$$

Separating the real and imaginary part, then,

$$E_1(\omega) = \int_0^\infty Q(s) \frac{\omega^2}{\omega^2 + s^2} ds \qquad (8.91)$$

8.6. Inverse-relations between "Fourier", "Retardation-time"and "*Relaxation-time*" Spectr

In Section 8.2.4 above, we have expressed the Fourier spectrum in terms of the retardation-time or relaxation-time spectra. If the retardation-time spectrum or relaxation-time spectrum i given, then the corresponding Fourier spectrum can be determined by integration.

Now, if the situation is different, that is, when Fourier spectrum J^* ($i\omega$) or E^* ($i\omega$) is given, th retardation-time spectrum or relaxation-time spectrum may be obtained, under the followin considerations.

1)   If J^* ($i\omega$) or E^* ($i\omega$) is given numerically, then the integral equation (8.84) or (8.90) woul be solved numerically to obtain N(s) or Q(s).

2)   If J^* ($i\omega$) or E^* ($i\omega$) is given analytically, i.e., then, N(s), or Q(s), may be obtained a follows:

SOLVED PROBLEM 8.1

$$E_2(\omega) = \int_0^\infty Q(s) \frac{\omega^2}{\omega^2 + s^2}\, ds \qquad (8.92)$$

Determine the energy dissipated per cycle, for a linear viscoelastic material, in terms of the loss compliance $J_2(\omega)$.

Solution:

$$\Delta W = \int \sigma\, d\varepsilon$$

$$= W \varepsilon_0^2 \int_0^{\frac{2\pi}{W}} [\, E_1 \sin(\omega t) + E_2 \cos^2(\omega t)\,]\, dt$$

$$= \pi E_2 \varepsilon_0^2$$

$$\because E_2 = \frac{J_2}{J_1^2 + J_2^2}$$

$$\therefore \Delta W = \pi \left(-\frac{J_2}{J_1^2 + J_2^2} \right) \varepsilon_0^2$$

$$\Delta W = -\frac{\pi J_2 \varepsilon_0^2}{J_1^2 + J-2^2}$$

From Eq. (8.84), one has

$$J^*(i\omega) = \int_0^\infty \frac{sN(s)}{s + i\omega} ds \qquad (8.93)$$

and substituting in the above expression

$$i\omega \rightarrow \omega \pm i\varepsilon \qquad (8.94)$$

Accordingly,

$$
\begin{aligned}
J^*(-\omega - i\varepsilon) &= \int_0^\infty sN(s) \frac{1}{s - \omega \pm i\varepsilon} ds \\
&= \int_0^\infty sN(s) \frac{s - \omega \mp i\varepsilon}{(s-\omega)^2 + \varepsilon^2} ds \\
&= \int_0^\infty sN(s) [\frac{(s-\omega)}{(s-\omega)^2 + \varepsilon^2} \mp \frac{i\varepsilon}{(s-\omega)^2 + \varepsilon^2}] ds
\end{aligned}
\qquad (8.95)
$$

then, where the integral is a principle value.

$$\lim_{\varepsilon \to 0} \int_0^\infty sN(s) \frac{s - \omega}{(s-\omega)^2 + \varepsilon^2} ds = \int_0^\infty \frac{sN(s)}{s - \omega} ds \qquad (8.96)$$

and,

$$\underset{\varepsilon \to 0}{\text{Lim}} \int_0^\infty s N(s) \frac{i\varepsilon}{(s-\omega)^2 + \varepsilon^2} \, ds = \int_0^\infty s N(s) \, (i\pi) \, [\, \underset{\varepsilon \to 0}{\text{Lim}} \frac{1}{\pi} \frac{\varepsilon}{(s-\omega)^2 + \varepsilon^2} \,] \, ds$$

$$= \int_0^\infty s N(s) \, (i\pi) \, \delta \, (s - \omega) \, ds \qquad (8.97)$$

$$= i \, \pi \, \omega \, N \, (\omega)$$

Therefore,

$$J(-\omega) = \underset{\varepsilon \to 0}{\text{Lim}} J(-\omega \pm i \, \varepsilon)$$

$$= \int_0^\infty \frac{s N(s)}{s - \omega} \, ds \mp i \, \pi \, \omega \, N(\omega) \qquad (8.98)$$

and,

$$N(\omega) = \pm \frac{1}{\pi \, \omega} \, \text{Im} \, J(-\omega) \qquad (8.99)$$

In the above equation, the choice of "+" or "-" sign is to be decided so that $N(\omega)$ be non negative. Similarly, from Eqn. (8.90), one has

$$E^*(i\omega) = i\omega \int_0^\infty \frac{Q(s)}{s + i\omega} \, ds \qquad (8.100)$$

By substituting

$$\omega = i\omega + \varepsilon, \qquad \text{i.e.,} \qquad i\omega = -\omega \pm i \, \varepsilon \qquad (8.101)$$

and noting that

$$E^*(-\omega+i\varepsilon) = (-\omega+i\varepsilon) \int_0^\infty \frac{Q(s)}{s-\omega\pm i\varepsilon} \, ds$$

$$= (-\omega\pm i\varepsilon) \int_0^\infty Q(s) \left[\frac{s-\omega}{(s-\omega)^2+\varepsilon^2} \mp \frac{i\varepsilon}{(s-\omega)^2+\varepsilon^2} \right] ds \qquad (8.102)$$

$$\lim_{\varepsilon\to 0} \frac{1}{\pi} \frac{\varepsilon}{(s-\omega)^2+\varepsilon^2} = \delta(s-\omega) \qquad (8.103)$$

we have

$$E(-\omega) = \lim_{\varepsilon\to 0} E(-\omega\pm i\omega)$$

$$= (-\omega) \int_0^\infty \frac{Q(s)}{s-\omega} \, ds \pm i\pi\,\omega Q(\omega) \qquad (8.104)$$

Thus,

$$Q(\omega) = \pm \frac{1}{\pi\,\omega} \operatorname{Im} E(-\omega) \qquad (8.105)$$

If we have the expression for $J(i\omega)$ or $E(i\omega)$, a simple substitution of variable $i\omega$ by $(-\omega)$ woul make it possible to determine $N(\omega)$ or $N(\omega)$. But there are some situations when the simpl substitution would cause the imaginary component to disappear. In this case, one has indeed t take the rigorous way and calculate the limit precisely. Fig. 8.14 gives the interrelations amon the above considered three domains; between "*Fourier*", "*Retardation-time*" and "*Relaxation time*" spectra

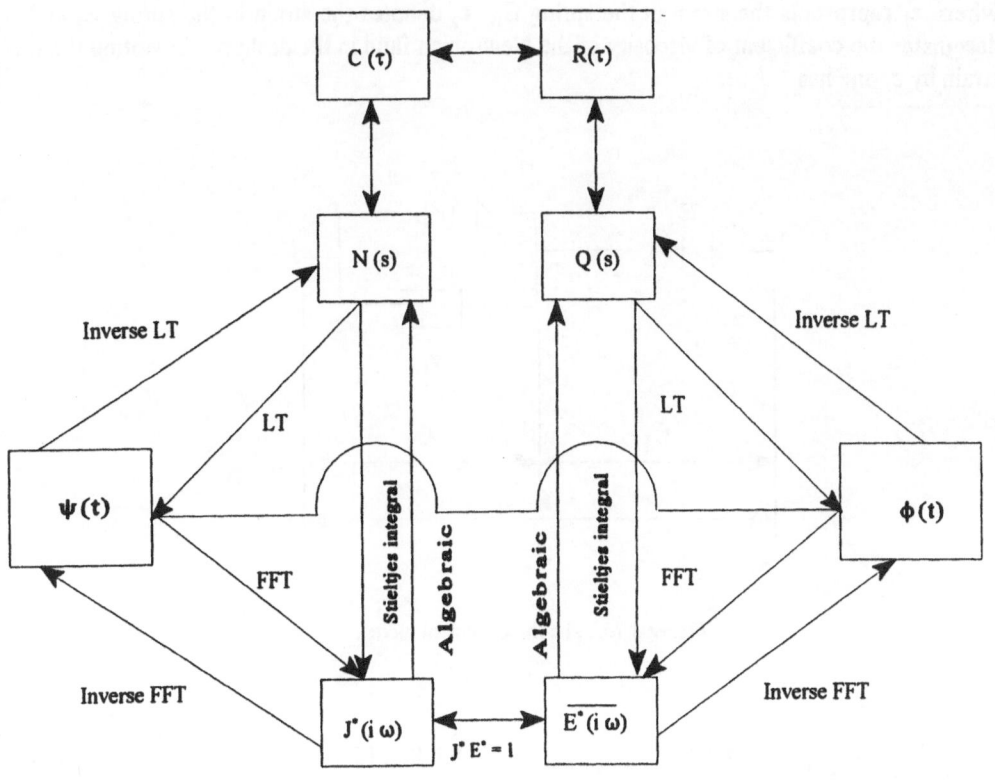

Figure 8.14. Intr-relations between "*Fourier*", "*Retardation-time*" and "*Relaxation-time*" spectra

8.7. Applications

In this section, a three-element model, Fig. 8.15 below, is chosen as an example, in order to illustrate how one can employ the analytical approach above. We designate the uniaxial stress and the strain by σ and ε, respectively.

From the constitutive laws for the spring and dashpot, the relation between stress and strain for the tree element model can be expressed as

$$\sigma = E_1 \, \varepsilon_1 \qquad\qquad \sigma = E_2 \, \varepsilon_2 + \eta \, \dot{\varepsilon}_2 \qquad\qquad (8.106)$$

where ε_1 represents the strain in the spring E_1, ε_2 denotes the strain in the spring E_2 and η designates the coefficient of viscosity of the Newtonian fluid in the dashpot. Denoting the total strain by ε, one has

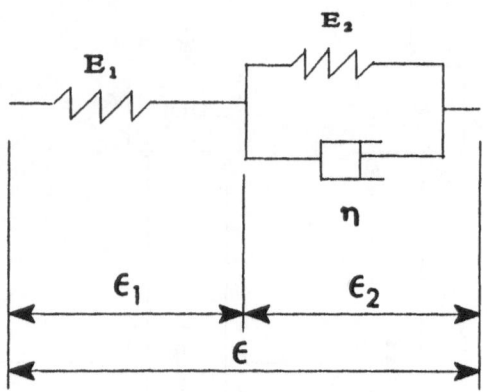

Figure 8.15. The three-element model.

$$\varepsilon = \varepsilon_1 + \varepsilon_2 \qquad\qquad \dot{\varepsilon} = \dot{\varepsilon}_1 + \dot{\varepsilon}_2 \qquad\qquad (8.107)$$

and, for the rate of the total strain,

$$\dot{\varepsilon}_1 = \frac{\dot{\sigma}}{E_1}$$

$$\dot{\varepsilon}_2 = \frac{1}{\eta}(\sigma - E_2\,\varepsilon_2)$$

$$= \frac{1}{\eta}\left[\sigma - E_2\,\varepsilon + \frac{E_2}{E_1}\,\sigma\right] \qquad\qquad (8.108)$$

$$\dot{\varepsilon} = \frac{\dot{\sigma}}{E_1} + \frac{1}{\eta}\left(\sigma - E_2\,\varepsilon + \frac{E_2}{E_1}\,\sigma\right)$$

$$= \frac{\dot{\sigma}}{E_1} + \frac{1}{\eta}\left(\frac{E_1 + E_2}{E_1}\,\sigma\right) - \frac{E_2}{\eta}\,\varepsilon \tag{8.109}$$

Meantime, the differential governing equation for the stress-strain relation of the three-element model can be obtained as

$$\eta\,E_1\,\dot{\varepsilon} + E_1\,E_2\,\varepsilon = \eta\,\dot{\sigma} + (E_1 + E_2)\,\sigma \tag{8.110}$$

8.7.1. CREEP RESPONSE

For a constant stress input σ, we have

$$\eta\,E_1\,\dot{\varepsilon} + E_1\,E_2\,\varepsilon = (E_1 + E_2)\,\sigma \tag{8.111}$$

Meantime, by applying Laplace transform to (8.111), it follows that

$$\varepsilon(s) = \frac{\sigma_0\,\eta}{\eta\,E_1\,s + E_2\,E_1} + \sigma_0\,\frac{E_1 + E_2}{s}\,\frac{1}{E_1\,s + E_2\,E_1}$$

$$= \frac{\sigma_0\,\eta}{\eta\,E_1\,s + E_1\,E_2} + \sigma_0\,\frac{1}{s}\,\frac{1}{\dfrac{\eta\,E_1}{E_1 + E_2}\,s + \dfrac{E_1\,E_2}{E_1 + E_2}} \tag{8.112}$$

Letting

$$\lambda = \frac{\eta}{E_2} \qquad \xi_1 = \frac{E_1\,\eta}{E_1 + E - 2} \qquad \xi_2 = \frac{\eta}{E_1 + E_2} \tag{8.113}$$

then, the Laplace transform of the strain response can be expressed as

$$\varepsilon(s) = \frac{\sigma_0}{E_1}\left(s + \frac{1}{\lambda}\right) + \frac{\sigma_0 \lambda}{\xi_1}\left[\frac{1}{s} - \frac{1}{s + \frac{1}{\lambda}}\right] \tag{8.114}$$

Applying the inverse Laplace transform of (8.114), the strain response of the three-element model can be further written as

$$\varepsilon(t) = \frac{\sigma_0}{E_1} e^{-\frac{t}{\lambda}} + \sigma_0 \frac{\lambda}{\xi_1}(1 - e^{-\frac{t}{\lambda}})$$

$$= \sigma_0 \left[\frac{e^{-\frac{t}{\lambda}}}{E_1} + \frac{\lambda}{\xi_1} - \frac{\lambda e^{-\frac{t}{\lambda}}}{\xi_1}\right]$$

$$= \frac{\lambda}{\xi_1} \sigma_0 \left[\frac{\xi_1 e^{-\frac{t}{\lambda}}}{\lambda E_1} + 1 - e^{-\frac{t}{\lambda}}\right] \tag{8.115}$$

$$= \frac{\lambda \sigma_0}{\xi_1}\left[1 + \left(\frac{\xi_2}{\lambda} - 1\right)e^{-\frac{t}{\lambda}}\right]$$

At $= 0$,

$$\varepsilon(0) = \frac{\lambda}{\xi_1}\sigma\left[1 + \frac{\xi_2}{\lambda} - 1\right]$$

$$= \frac{\xi_2}{\xi_1}\sigma_0 \tag{8.116}$$

$$= Co\,\sigma_0$$

Meanwhile, if no Newtonian viscous flow is assumed to exist, then

$$\frac{\varepsilon(t)}{\sigma_0} = C_o + \psi(t) \tag{8.117}$$

By employing Equations (8.116) and (8.117), the creep function of this model is expressed as

$$\psi(t) = \frac{\varepsilon(t)}{\sigma_0} - \varepsilon_\infty$$

$$= \frac{\lambda}{\xi_1}[1 + (\frac{\xi_2}{\lambda} - 1)e^{-\frac{t}{\lambda}}] - \frac{\xi_2}{\xi_1}$$

$$= \frac{1}{\xi_1}[(\lambda - \xi_2) + (\xi_2 - \lambda)e^{-\frac{t}{\lambda}}] \qquad (8.118)$$

$$= \frac{\lambda - \xi_2}{\xi_1}[1 - e^{-\frac{t}{\lambda}}]$$

Thus, for the three-element model, we have

$$C_o = \frac{\xi_2}{\xi_1} \qquad \psi(t) = \frac{\lambda - \xi_2}{\xi_1}(1 - e^{-\frac{t}{\lambda}}) \qquad \frac{1}{\eta_0} = 0 \qquad (8.119)$$

"Fourier-spectrum" of the Creep-rate
The creep function may be expressed as

$$C(t) = [\varepsilon_\infty + \psi(t)]u(t)$$

$$= [\frac{\xi_2}{\xi_1} + \frac{\lambda - \xi_2}{\xi_1}(1 - e^{-\frac{t}{\lambda}})]u(t) \qquad (8.120)$$

with a time-rate,

$$\dot{C}(t) = \frac{\xi_2}{\xi_1}\delta(t) + \frac{\lambda - \xi_2}{\xi_1\lambda}e^{-\frac{t}{\lambda}}u(t) + \frac{\lambda\xi_2}{\xi_1}(1 - e^{-\frac{t}{\lambda}})\delta(t) \qquad (8.121)$$

Meanwhile, the creep compliance is expressed as

$$J^*(i\omega) = \int_{-\infty}^{\infty} e^{-i\omega t} \left[\frac{\xi_2}{\xi_1} \delta(t) + \frac{\lambda - \xi_2}{\xi_1 \lambda} e^{-\frac{t}{\lambda}} u(t) + \frac{\lambda - \xi_2}{\xi_1} (1 - e^{-\frac{t}{\lambda}}) \delta(t) \right] dt$$

$$= \frac{\xi_2}{\xi_1} + \int_{-\infty}^{\infty} \frac{\lambda - \xi_2}{\xi_1 \lambda} e^{-\frac{t}{\lambda}} e^{-i\omega t} dt \qquad (8.122)$$

$$= \frac{\xi_2}{\xi_1} + \frac{\lambda - \xi_2}{\xi_1 \lambda} \frac{1}{\frac{1}{\lambda} + i\omega}$$

Thus, $J(i\omega)$ is given by

$$J(i\omega) = \frac{\lambda - \xi_2}{\xi_1 \lambda} \frac{1}{\frac{1}{\lambda} + i\omega} \qquad (8.123)$$

"Retardation-time" Spectrum

Here, we attempt to determine N(s) from $J(i\omega)$. Thus, by substituting $i\omega$ by $-\omega - i\varepsilon$ in (8.123), i follows that

$$J(-i\omega - i\varepsilon) = \frac{\lambda \xi_2}{\xi_1 \lambda} \frac{1}{\frac{1}{\lambda} - \omega - i\varepsilon}$$

$$= \frac{\lambda - \xi_2}{\xi_1 \lambda} \frac{\frac{1}{\lambda} - \omega + i\varepsilon}{(\frac{1}{\lambda} - \omega)^2 + \varepsilon^2} \qquad (8.124)$$

$$= \frac{\lambda - \xi_2}{\xi_1 \lambda} \frac{\frac{1}{\lambda} - \omega}{(\frac{1}{\lambda} - \omega)^2 + \varepsilon^2} + \frac{\lambda \xi_2}{\xi_1 \lambda} \frac{i\varepsilon}{(\frac{1}{\lambda} - \omega)^2 + \varepsilon^2}$$

$$\lim_{\varepsilon \to 0} J(-\omega - i\varepsilon) = \frac{\lambda - \xi_2}{\xi_1 \lambda} \frac{1}{\frac{1}{\lambda} - \omega} + i \frac{\lambda - \xi_2}{\xi_1 \lambda} \pi \delta (\frac{1}{\lambda} - \omega) \tag{8.125}$$

$$\begin{aligned} N(\omega) &= \frac{1}{\pi \omega} \operatorname{Im} [\operatorname*{Lim}_{\varepsilon \to 0} J(-\omega - i\omega)] \\ &= \frac{\lambda - \xi_2}{\xi_1 \lambda} \frac{1}{\omega} \delta (\frac{1}{\lambda} - \omega) \\ &= \frac{\lambda - \xi_2}{\xi_1} \delta (\frac{1}{\lambda} - \omega) \end{aligned} \tag{8.126}$$

Thus, the retardation-time spectrum can be derived from N(s) as

$$\begin{aligned} C(\tau) &= \frac{1}{\beta} N(\frac{1}{\tau}) \frac{1}{\tau^2} \\ &= \frac{\lambda - \xi_2}{\beta \xi_1} \frac{1}{\tau^2} \delta (\frac{1}{\lambda} - \frac{1}{\tau}) \\ &= \frac{\lambda - \xi_2}{\beta \xi_1} \frac{1}{\tau^2} \delta [\frac{1}{\lambda \tau} (\tau - \lambda)] \\ &= \frac{1}{\beta} \frac{\lambda - \xi_2}{\xi_1} \frac{\lambda \tau}{\tau^2} \delta (\tau - \lambda) \\ &= \frac{1}{\beta} \frac{\lambda - \xi_2}{\xi_1} \delta (\tau - \lambda) \end{aligned} \tag{8.127}$$

$$\beta = \psi\,(\infty) = \frac{\lambda - \xi_2}{\xi_1} \tag{8.128}$$

Therefore, retardation-time spectrum of the three-element model is given as

$$C(\tau) = \delta\,(\tau - \lambda) \tag{8.129}$$

8.7.2. RELAXATION RESPONSE

"Fourier-spectrum" of the Rate of Relaxation
From (8.49),

$$J^*(i\omega)\,E^*(i\omega) = 1 \tag{8.130}$$

one has

$$E^*(i\omega) = \frac{1}{J^*(i\omega)} \tag{8.131}$$

Meantime, since

$$
\begin{aligned}
J^*(i\omega) &= \frac{\xi_2}{\xi_1} + \frac{\lambda - \xi_2}{\xi_1 \lambda}\,\frac{1}{\dfrac{1}{\lambda} + i\omega} \\
&= \frac{1}{E_1} + \frac{1}{E_2 + i\eta\omega}
\end{aligned}
\tag{8.132}
$$

it follows that

$$E^*(i\omega) = E_1 - \frac{E_1^2}{\eta}\,\frac{1}{\dfrac{E_1 + E_2}{\eta} + i\omega} \tag{8.133}$$

Relaxation Function

Through the inverse Fourier transform of E^*, the rate of relaxation is determined as

$$\dot{R}(t) = \frac{1}{2\pi} \int_{-\infty}^{\infty} (E_1 - \frac{E_1^2}{\eta} \frac{1}{\frac{E_1 + E_2}{\eta} + i\omega}) e^{i\omega t} d\omega$$

$$= \delta(t) E_1 \frac{E_1^2}{\eta} e^{-\frac{(E_1 + E_2)}{\eta}t} u(t)$$

(8.134)

or,

$$R(t) = \frac{E_1 E_2}{E_1 + E_2} + \frac{E_1^2}{E_1 + E_2} e^{-\frac{(E_1 + E_2)}{\eta}t} \qquad t \geq 0$$

(8.135)

Relaxation-time Spectrum

As in the case of creep, one may determine $Q(s)$ directly from $E(i\omega)$ by substituting $i\omega$ by $(-\omega + i\varepsilon)$, and calculating the limit. Thus, substituting $i\omega$ by $-\omega + i\varepsilon$ in the expression for $E(i\omega)$, one has:

$$E(-\omega + i\varepsilon) = -\frac{E_1^2}{\eta} \frac{1}{\frac{E_1 E_2}{\eta} - \omega + i\varepsilon}$$

$$= -\frac{E_1^2}{\eta} \frac{\frac{E_1 + E_2}{\eta} - \omega + i\varepsilon}{(\frac{E_1 + E_2}{\eta} - \omega)^2 + \varepsilon^2}$$

(8.136)

$$\lim_{\varepsilon \to \infty} (-\omega + i\varepsilon) = -\frac{E_1^2}{\eta} \frac{1}{\frac{E_1 + E_2}{\eta} - \omega} + i \frac{E_1^2}{\eta} \pi \delta (\frac{E_1 + E_2}{\eta} - \omega)$$

(8.137)

324

Therefore:

$$Q(\omega) = \frac{E_1^2}{\eta} \frac{1}{\omega} \delta \left(\frac{E_1 + E_2}{\eta} - \omega \right)$$

(8.138)

and,

$$G(\tau) = \frac{1}{\gamma} \frac{1}{\tau^2} N \left(\frac{1}{\tau} \right)$$

(8.139)

$$\gamma = \phi(0) = \frac{E_1^2}{E_1 + E_2}$$

(8.140)

Thus, the relaxation-time spectrum is determined as

$$G(\tau) = \frac{E_1 + E_2}{\eta} \frac{1}{\tau} \left[\frac{1}{\tau} \frac{E_1 + E_2}{\eta} \left(\tau - \frac{\eta}{E_1 + E_2} \right) \right]$$

$$= \delta \left(\tau - \frac{\eta}{E_1 + E_2} \right)$$

(8.141)

EXAMPLE PROBLEM 8.2

Prove that the relation between the dynamic complex compliance $J^*(i\omega)$ and the creep function $C(t)$ is given by

$$J^*(i\omega) = \int_0^\infty \exp(-i\omega\tau) \frac{dC(\tau)}{d\tau} d\tau$$

Solution:

Since $\varepsilon(t) = \int_0^\infty C(t-\tau) \frac{d\sigma(\tau)}{d\tau} d\tau$

Substituting $\sigma(t) = \sigma_0 \exp(i\omega t)$ into the above relation, it follows that

$$\varepsilon(t) = \int_0^\infty C(t-\tau) \frac{d[\sigma_0 \exp(i\omega\tau)]}{d\tau} d\tau$$

$$t - \tau = \tau$$

$$\varepsilon(t) = \int_0^\infty C(\tau) \frac{d[\sigma_0 \exp(i\omega(t-\tau))])}{d(t-\tau)}$$

$$= C(\tau)\sigma_0 \exp(i\omega(t-\tau)) \Big|_0^\infty + \int_0^\infty \sigma_0 \exp(i\omega(t-\tau)) \frac{dC(\tau)}{d\tau} d\tau$$

$$= C(0)\sigma_0 \exp(\tau i\omega) + \int_0^\infty \sigma_0 \exp(i\omega\tau) \exp(i\omega t) \frac{dC(\tau)}{d\tau} d\tau$$

$$= [C_0 + \int_0^\infty \exp(-i\omega\tau) \frac{dC(\tau)}{d\tau} d\tau] \sigma(t)$$

$$J^*(i\omega) = C_0 + \int_0^\infty \exp(-i\omega\tau) \frac{dC(\tau)}{d\tau} d\tau$$

EXAMPLE PROBLEM 8.3

For a creep frequency $s = 1/\lambda$ and a frequency distribution $N(s) = \alpha C(1/s)/s^2$ where α is a constant, prove that the time-rate of the creep function is given by

$$\frac{dC(t)}{dt} = \int_0^\infty s N(s) \exp(-ts) ds$$

Solution: We have

$$C(t) = \int_0^\infty \alpha C(\lambda)[1 - \exp(-t/\lambda)] d\lambda$$

$$\frac{dC(t)}{dt} = \int_0^\infty \frac{\alpha C(\lambda)}{\lambda} \exp(-t/\lambda d\lambda$$

(A)

$$s = \frac{1}{\lambda}$$

$$N(s) = \alpha\, C\left(\frac{1}{s}\right) / s^2 \qquad \text{(B)}$$

From (A) and (B) we obtain:

$$\frac{dC(t)}{dt} = \int_0^\infty s\, N(s) \exp(-ts)\, ds \qquad \text{(C}$$

8.8. Problems

1. Determine the creep constitutive equation in both the time and frequency domains of the following mechanical models (which represent the response of a linear viscoelastic material):

    (a) A linear spring in series with a Newtonian dashpot (*The Maxwell Model*).
    (b) A linear spring in parallel with a Newtonian dashpot (*The Kelvin-Voigt Model*).

2. Use the "*Superposition Principle*" to determine the creep recovery response of the model mentioned in Problem #1 above.

3. Determine the creep constitutive equation in both the time and frequency domains of the three-element model (Fig. 8.15).

4. Solve Problem #1 above in the relaxation phase of the indicated models.

5. Solve Problem #2 above in the relaxation phase of the three-element model.

6. For a material with a single relaxation time λ show that E_1 is proportional to $\omega^2\lambda^2/(1 + \omega^2\lambda^2)$ and E_2 is proportional to $\omega\lambda/(1 + \omega^2\lambda^2)$ where ω is the frequency.

7. If $\phi(\lambda)$ denotes a spectrum of relaxation times, show that the following relations are valid

$$E_1(\omega) = \int_0^\infty \phi(\lambda) \frac{\omega^2\lambda^2}{1 + \omega^2\lambda^2} \, d\lambda$$

and

$$E_2(\omega) = \int_0^\infty \phi(\lambda) \frac{\omega\lambda}{1 + \omega^2\lambda^2} \, d\lambda$$

where $\phi(\cdot)$ is a normalized distribution function with

$$\int_0^\infty \phi(\lambda) \, d\lambda = 1$$

8. Derive the expressions for the components of the complex modulus in terms of the relaxation spectrum.

9. Following Example Problem #8.3, show that the complex compliance function can be expressed in terms of the distribution function of retardation times $N(s)$ as

$$J^*(i\omega) = \int_0^\infty \frac{sN(s)}{s + i\omega} \, ds$$

Express the components of the complex compliance function, i.e., $J_1(\omega)$ and $J_2(\omega)$, for this case.

8.9. Transition to Thermoviscoelasticity

The viscoelastic response behaviour appropriate to the formulation of the constitutive equations as dealt with in the foregoing generally exhibits a very strong dependence upon the service temperature. The simplest and most direct situation of such dependence occurs when the constitutive equation is to be related or connected with different base temperatures within the context of the isothermal theory. In this, the material functions characterizing the viscoelastic behaviour of the material would be identified in an ambient temperature environment equal to the base temperature at which the constitutive equation is to be applied. On the other hand, when the constitutive equation is to be used in a non-isothermal situation, the identification problem is mor involved and must be dealt within the realm of an advanced topic of thermoviscoelasticity. Thus one generally deals with the following two classifications of the theory of thermoviscoelasticity

Classification I. A thermoviscoelastic treatment which examines the performance of the materia as related to a fixed reference temperature T_R. In this treatment, the effects due t infinitesimal temperature variation from T_R would be neglected.

Classification II. An advanced thermoviscoelastic theory within which the dependence of th material performance upon a transient temperature field is dealt with.

One is generally concerned with the thermomechanical process and its effect on the respons behaviour of the viscoelastic medium. For this purpose, the increase of the total energy of th medium is seen to be due to the work done by the external forces as well as the supply of energ (heat) from other sources. In such a study, however, one must differentiate between *reversibl* and *irreversible* effects. Reversible effects would include cases in which temporary changes ar produced in the material. These may include geometrical changes such as expansion an contraction as well as changes in the values of the material parameters or functions characterizin the constitutive equations of the material, but, the format of the constitutive equations woul essentially remain unchanged. On the other hand, irreversible effects include situations in whic permanent changes are occurring in the material. In case of a viscoelastic medium, such change may include, for instance, primary bond rupture and weight loss. Polymeric composite material and their constituents, for instance, are often subject to irreversible changes under a variety c environmental and chemical aging influences. In the latter context, temperature, moisture an water vapour are considered as important factors during the manufacturing process and in servic (e.g., Steel, 1965; Fried, 1970; Tsai, 1970 and Schapery, 1974). For the purpose of viscoelasti analysis, one simple case is where the response is considered over times that are short compare to the time scale over which changes due to environmental and/or aging effects would occu (Schapery, 1974) otherwise one must allow for the dependence of the material parameters an functions, characterizing the constitutive equations, on such changes.

8.9.1. RHEOLOGICAL EQUATIONS OF STATE

Significant research efforts have been undertaken in the last four decades towards the developmer of a rigorous thermomechanical theory concerning the response behaviour of materials wit memory as based on phenomenological considerations. In this context, theories have bee presented from different points of view by Biot (1958,1973), Coleman and Noll (1963), Colema (1964a&b), Schapery (1964, 1958 & 1969), Christensen and Naghdi (1967), Crochet and Naghe (1974),Crochet (1975) and Rivlin (1975), among others. Other work of interest includes that c

Coleman and Mizel (1963,1964), Breuer and Onat (1964), Breuer (1969) and Day (1970) concerning the free energy concept, recoverable work and related work bounds. The reader is also referred to Müller (1967), Meixner (1969) and others for developments in the subject of continuum thermodynamics.

8.9.2. THERMODYNAMICAL DERIVATION OF THE CONSTITUTIVE RELATIONS

In their derivation of the thermodynamic constitutive equation for a material with memory, Christensen and Naghdi (1967) and Christensen (1971) based their work on the balance of energy equation for the infinitesimal theory and the entropy production postulate (Truesdell and Toupin, 1960). The derivation parallels, in essence, the means of deriving the constitutive equation in the linear isothermal case. However, the situation here is more difficult since the free energy not only depends upon the strain history, but also depends upon the temperature history. At this point, a remark should be cited concerning the free energy relationships. In these relationships, the stress and deformation are conjugate variables. One, therefore, has to make a choice as to which will be the independent variable. If the stress is the independent variable, then the appropriate free energy function is the *Gibbs* free energy. On the other hand, if the strain is taken as the independent variable, then, the corresponding free energy function is the *Helmholz* free energy. Hence, considering the latter context, the local balance of energy equation for infinitesimal theory is given (Christensen, 1971) by

$$\rho r - \rho\left[\dot{A} + \dot{T}S + T\dot{S}\right] + \sigma_{ij}\dot{\varepsilon}_{ij} - Z_{i,i} = 0 \qquad (8.142)$$

In Eqn. (8.142), ρ is the mass density, r is the heat supply function per unit mass, A is the time-derivative of the Helmholtz free energy per unit mass, T is the absolute temperature, S is the entropy per unit mass and Z_i are the Cartesian components of the heat flux vector measured per unit area per unit time. The related local entropy production inequality (Clausius-Duhem) is given by

$$\rho T\dot{S} - \rho r + Z_{i,i} - Z_i(T_{,i}/T) \geq 0 \qquad (8.143)$$

With reference to (8.142), it is usually assumed that ε_{ij} and T are continuous in the interval $-\infty < t < \infty$ and that ε_{ij} tends to zero and T tends to T_0 as t tends to $-\infty$. Based on this assumption, the free energy can be expressed (Christensen, 1971), as a polynomial in a set of real, continuous linear functions of ε_{ij} and T as

$$\rho A = \rho A_0 + \int_{-\infty}^{t} D_{ij}(t-\tau)\frac{\partial \varepsilon_{ij}(\tau)}{\partial \tau}d\tau - \int_{-\infty}^{t} \beta(t-\tau)$$

$$\frac{\partial \theta(\tau)}{\partial \tau}d\tau + \frac{1}{2}\int_{-\infty}^{t}\int_{-\infty}^{t} R_{ijkl}(t-\tau, t-s)\frac{\partial \varepsilon_{ij}(\tau)}{\partial \tau}\frac{\partial \varepsilon_{kl}(s)}{\partial s}d\tau\,ds$$

$$- \int_{-\infty}^{t}\int_{-\infty}^{t} \phi_{ij}(t-\tau, t-s)\frac{\partial \varepsilon_{ij}(\tau)}{\partial \tau}\frac{\partial \theta(s)}{\partial s}d\tau\,ds$$

$$-\frac{1}{2}\int_{-\infty}^{t}\int_{-\infty}^{t} m(t-\tau, t-s)\frac{\partial \theta(\tau)}{\partial \tau}\frac{\partial \theta(s)}{\partial s}d\tau\,ds$$

(8.144

where $T=T_0+\theta$ and A_0 is the mean free energy. In (8.144), the integrating material function are assumed to be continuous for arguments $\tau_i \geq 0$, and vanish identically for $\tau_i < 0$, i.e.

$$\beta(\tau_1) = 0, \qquad D_{ij}(\tau_1) = 0, \qquad R_{ijkl}(\tau_1,\tau_2) = 0$$

$$\phi_{ij}(\tau_1,\tau_2) = 0 \qquad m(\tau_1,\tau_2) = 0 \qquad for \ \tau_1 < 0 \ and \ \tau_2 < 0$$

(8.145

For the proposed theory, these integrating functions are necessarily independent of strain an temperature. Now, if one combines equations (8.142) to (8.144), and, at the same time, carryin out the indicated differentiation with respect to time, one obtains, Christensen (1971),

$$\left\{ -D_{ij}(0) - \int_{-\infty}^{t} R_{ijkl}(t-\tau,0)\frac{\partial \varepsilon_{kl}(\tau)}{\partial \tau}\,d\tau \right.$$

$$\left. + \int_{-\infty}^{t} \phi_{ij}(0,t-\tau)\frac{\partial \theta(\tau)}{\partial \tau}\,d\tau + \sigma_{ij} \right\}$$

$$\dot{\varepsilon}_{ij}(t) + \left\{ \beta(0) + \int_{-\infty}^{t} m(t-\tau,0)\frac{\partial \theta(\tau)}{\partial \tau}\,d\tau + \int_{-\infty}^{t} \phi_{ij}(t-\tau, \right.$$

$$\left. \frac{\varepsilon_{ij}(\tau)}{\partial \tau}\,d\tau - \rho S \right\} \dot{\theta}(t) + \left\{ -\int_{-\infty}^{t} \frac{\partial}{\partial t} D_{ij}(t-\tau)\frac{\partial \varepsilon_{ij}(\tau)}{\partial \tau}\,d\tau \right.$$

$$\left. + \int_{-\infty}^{t} \frac{\partial}{\partial t}\beta(t-\tau)\frac{\partial \theta(\tau)}{\partial \tau}\,d\tau + \Lambda - Z_i\frac{\theta_{,i}}{T_0} \right\} \geq 0$$

(8.146)

where

$$\Lambda = -\frac{1}{2}\int_{-\infty}^{t}\int_{-\infty}^{t} \frac{\partial}{\partial t2} R_{ijkl}(t-\tau,t-s)\frac{\partial \varepsilon_{ij}(\tau)}{\partial \tau}\frac{\partial \varepsilon_{kl}(s)}{\partial s}\,d\tau\,ds$$

$$+\frac{1}{2}\int_{-\infty}^{t}\int_{-\infty}^{t} \frac{\partial}{\partial t}\phi_{ij}(t-\tau,t-s)\frac{\partial \varepsilon_{ij}(\tau)}{\partial \tau}\frac{\partial \theta(s)}{\partial s}\,d\tau\,ds \qquad (8.147)$$

$$+\frac{1}{2}\int_{-\infty}^{t}\int_{-\infty}^{t} \frac{\partial}{\partial t} m(t-\tau,t-s)\frac{\partial \theta(\tau)}{\partial \tau}\frac{\partial \theta(s)}{\partial s}\,d\tau\,ds$$

and the following symmetry properties are implied

$$R_{ijkl}(t-\tau, t-s) = R_{klij}(t-s, t-\tau) \tag{8.148}$$

The inequality (8.146), must hold for all arbitrary values of $\dot{\varepsilon}_{ij}(t)$ and $\theta(t)$, therefore, i is necessary that the coefficients of $\dot{\varepsilon}_{ij}(t)$ and $\theta(t)$ in (8.146) vanish. Hence

$$\sigma_{ij} = D_{ij}(0) + \int_{-\infty}^{t} R_{ijkl}(t-\tau, 0) \frac{\partial \varepsilon_{kl}(\tau)}{\partial \tau} d\tau$$

$$- \int_{-\infty}^{t} \phi_{ij}(0, t-\tau) \frac{\partial \theta(\tau)}{\partial \tau} d\tau \tag{8.149}$$

and

$$\rho S = \beta(0) + \int_{-\infty}^{t} \phi_{ij}(t-\tau, 0) \frac{\partial \varepsilon_{ij}(\tau)}{\partial \tau} d\tau + \int_{-\infty}^{t} m(t-\tau, 0) \frac{\partial \theta(\tau)}{\partial \tau} d\tau \tag{8.150}$$

Relations (8.149) and (8.150) are the constitutive relations for stress and entropy respectively. From these it is clear that $D_{ij}(0)$ is the initial stress and that $\beta(0)$ is the initia entropy, ρS_0. The integrating functions $R_{ijkl}(t-\tau, 0)$, $\phi_{ij}(0, t-\tau)$, $\phi_{ij}(t-\tau, 0)$ and $m(t-\tau, 0)$ ar appropriate relaxation function norms of the material properties. It is the relaxation functio $R_{ijkl}(t, 0)$ in this formulation which corresponds to the relaxation function $R_{ijkl}(t)$ in the isotherma theory.

8.9.3. REDUCTION TO THE ISOTROPIC THEORY

For isotropic materials, ϕ_{ij} must be taken as

$$\phi_{ij}(\tau, s) = \delta_{ij} \phi(t, s) \tag{8.151}$$

where δ_{ij} is the Kronecker delta. Using the definitions of deviatoric stress and strain, the fre energy for isotropic theory can be expressed (Christensen, 1971) with reference to (5.4:5) as:

$$\rho A = \frac{1}{2}\int\limits_{-\infty}^{t}\int\limits_{-\infty}^{t} G_1(t-\tau,t-s)\frac{\partial\varepsilon'_{ij}(\tau)}{\partial\tau}\frac{\partial\varepsilon'_{ij}(s)}{\partial s}\,d\tau\,ds$$

$$+\frac{1}{6}\int\limits_{-\infty}^{t}\int\limits_{-\infty}^{t} G_2(t-\tau,t-s)\frac{\partial\varepsilon_{kk}(\tau)}{\partial\tau}\frac{\partial\varepsilon_{jj}(s)}{\partial s}\,d\tau\,ds$$

$$-\int\limits_{-\infty}^{t}\int\limits_{-\infty}^{t}\phi(t-\tau,t-s)\frac{\partial\varepsilon_{kk}(\tau)}{\partial\tau}\frac{\partial\theta(s)}{\partial\tau}\,d\tau\,ds$$

$$-\frac{1}{2}\int\limits_{-\infty}^{t}\int\limits_{-\infty}^{t} m(t-\tau,t-s)\frac{\partial\theta(\tau)}{\partial\tau}\frac{\partial\theta(s)}{\partial s}\,d\tau\,ds$$

(8.152)

where the initial stress and initial entropy effects in (8.146) have been dropped.

Based on the form (8.152) for the free energy, it can be shown that the stress relaxation equations for isotropic materials are

$$\sigma_{ij}'(t) = \int\limits_{-\infty}^{t} R_1(t-\tau,0)\frac{\partial\varepsilon'_{ij}(\tau)}{\partial\tau}\,d\tau$$

(8.153a)

and

$$\sigma_{kk}(t) = \int\limits_{-\infty}^{t} R_2(t-\tau,0)\frac{\partial\varepsilon_{kk}(\tau)}{\partial\tau}\,d\tau - 3\int\limits_{-\infty}^{t}\phi(0,t-\tau)\frac{\partial\theta(\tau)}{\partial\tau}\,d\tau$$

(8.153b)

If the material functions appearing in the constitutive equations (8.149) and (8.152) are independent of temperature, which in the service life of the material might be true for only small temperature changes, or the temperature is time-wise constant, then, these constitutive equations will reduce to their counterparts in the isothermal theory. When these two conditions do not exist and it is desired to find experimentally the material functions without making any a priori assumptions about their temperature dependence, a large number of tests will be needed even for the uniaxial test situation particularly if the temperature varies in a cyclic or a discrete fashion with the time. In this case, one must subject the material specimen to the actual temperature history (Landel and Peng, 1986). The latter approach could prove to be quite impractical. There is, however, experimental evidence (see, e.g., Schapery, 1974) which implies that viscoelastic

characterization for transient temperature applications may be performed by using tests at a set o
different constant temperatures. Hence, the phenomenological viscoelastic response descriptioı
of a large class of polymeric materials and inorganic glasses under non-isothermal conditions iː
simplified by the adoption of the so-called *"temperature-time equivalence"*, also known as th
"thermorheologically simple hypothesis".

8.9.4. THERMORHEOLOGICALLY SIMPLE MATERIALS (TSM)

Thermorheologically simple materials are a special class of viscoelastic materials whosϵ
temperature dependence of mechanical properties are particularly responsive to analyticɑ
description.   This group of materials generally constitutes the simplest and most realistiϲ
viscoelastic constitutive equation for which response under constant temperatures can be used tϲ
predict response under transient temperatures. Two temperature states are studied here, i.e., th
constant temperature state and the nonconstant temperature one. For detailed description of th
temperature dependent properties of Thermorheologically Simple Materials, reference is made tϲ
Leadermann (1943), Schwarzl and Staverman (1952), Morland and Lee (1960), Ferry (1970) anɗ
Schapery (1974), among others.

Thermorheologically Simple Materials under Constant Temperature States
Following Schapery (1974), the uniaxial creep constitutive relation for Thermorheologicallˀ
Simple Materials can be expressed as

$$\varepsilon(t) = \int_0^t C(\xi - \xi') \frac{d\sigma}{d\xi'} \, d\xi' \tag{8.154}$$

where ε is the uniaxial strain due to stress only, i.e., the total strain less that due to thermɑ
expansion, $C(\xi-\xi')$ is the time-,temperature-dependent creep compliance. In this equation, ξ iː
called the *"reduced time parameter"* and defined by

$$\xi = \xi(t) = \int_0^t \frac{d\tau}{a_T} \tag{8.155}$$

and equivalently

$$\xi' = \xi(t') = \int_0^{t'} \frac{d\tau}{a_T} \tag{8.156}$$

where $a_T = a_T[T(\tau)]$ is the so-called *"temperature shift factor"*. The latter is dependent on the absolute temperature T within the time interval $\tau = \xi - \xi'$. In the constitutive Eqn. (8.154), it is assumed that $\sigma = \epsilon = 0$ when $t \leq 0$.

The inverse of (8.154), i.e., the relaxation constitutive equation, can be determined by using Laplace transform with respect to reduced time. In this context, it can be shown that

$$\sigma(t) = \int_0^t R(\xi - \xi') \frac{d\epsilon}{d\xi'} d\xi' \tag{8.157}$$

in which $R(\xi - \xi')$ is the time-, temperature-dependent relaxation modulus and both the reduced time parameters ξ, ξ' are as expressed above by (8.155) and (8.156), respectively.

The experimental bases for the constitutive equations (8.154) and (8.157) under constant temperature conditions may be treated by considering isothermal creep and isothermal relaxation tests. Namely, for the uniaxial creep test: $\sigma(t) = \sigma_0 H(t)$, where σ_0 is the constant stress input and $H(t)$ is the Heaviside step function, the creep constitutive equation (8.154) yields

$$\epsilon(\xi) = C(\xi)\sigma(t) \tag{8.158}$$

where ϵ is the resulting uniaxial strain due to the stress only. Similarly, for the relaxation test: $\epsilon(t) = \epsilon_0 H(t)$ where ϵ_0 is the constant strain input, the relaxation constitutive equation (8.157) yields

$$\sigma(\xi) = R(\xi)\epsilon(t) \tag{8.159}$$

with the understanding that for both types of isothermal tests

$$\xi = t/a_T \tag{8.160}$$

Equation (8.160) indicates that the effect of temperature on the mechanical properties $F(\xi)$ or $R(\xi)$ for a thermorheologically-simple material produces only horizontal translations when the property is plotted against log t. Conversely, if it is found that the constant temperature viscoelastic response curves (creep or relaxation) can be superposed so as to form a single curve (*master curve*) by means of only rigid, horizontal translations then the associated mechanical property (relaxation modulus or creep compliance) would depend only on time and temperature through the one parameter ξ. Such description is probably a more or less conventional definition of a thermorheologically simple material, (*see* Schapery, 1974 and Tobolosky and Catsiff, 1956)

An illustration of the time-temperature shift of the stress-relaxation curves at different temperatures to form a master curve associated with a particular reference temperature is given in Fig. 8.16. In Fig. 8.16(a), a series of relaxation moduli curves at different base temperature are plotted using experimental relaxation data on Bis-Phenol Polycarbonate (MW=40,000) from Mercier *et al.* (1965). For the purpose of constructing the master curve for these data, the relaxation curve corresponding to a reference temperature $T_R = 141°C$ is assigned as the reference curve. The other curves are then shifted along the log time scale until they superimpose. The relaxation moduli curves corresponding to temperatures above the reference temperature are shifted to the right while those corresponding to temperatures below the reference temperature are shifted to the left of the reference curve. The full master curve is consequently formed as shown in Fig. 8.16 (b). It is noticed that the master curve covers a much wider range of time as compared with the original time range covered by the individual relaxation curves. An analogous procedure is followed in Figures 8.17 (a) and (b) for constructing a master curve using experimental creep compliance data on hot setting epoxy resin from Theocaris (1962). The reference temperature for the data is chosen as $T_R=25°C$.

For a large class of polymeric systems near their transition temperature, the time temperature shift factor a_T is often expressed by the following WLF-equation (Williams, Lande and Ferry, 1955)

$$log\ a_T = \frac{-17.44(T-T_g)}{51.6 + (T-T_g)} \tag{8.161}$$

where T_g is the glass transition temperature, assumed as the reference temperature for the particular polymer under conside- ration. Equation (8.161) is considered to be valid for a polymer in the temperature range T_g to $T_g + 100$ (see, e.g., Gittus, 1975). The numerical constants in this equation are established by experiment within the indicated temperature range. If a temperature other than T_g is chosen as the reference temperature, a form analogous to (8.161) may be used to determine a_T, but with new numerical constants corresponding to the chosen reference temperature.

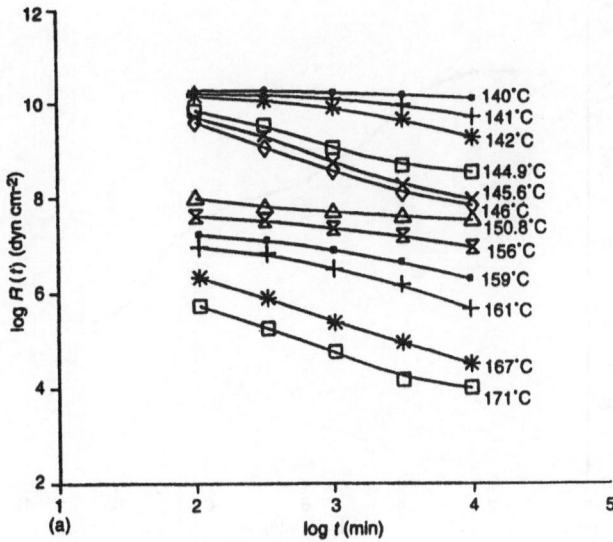

Figure 8.16(a). Variation of stress-relaxation moduli with time at different base temperatures for Bis-Phenol Polycarbonate (MW=40,000); experimental data from Mercier et al., (1965).

338

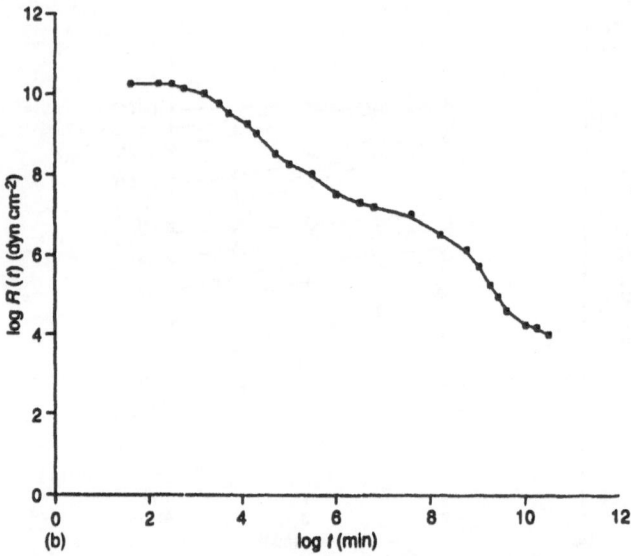

Figure 8.16(b). Master stress-relaxation curve based on stress-relaxation data presented in Fig. 8.16(a) for Bis-Phenol Polycarbonate (MW=40,000) with reference temperature $T_R = 141°C$.

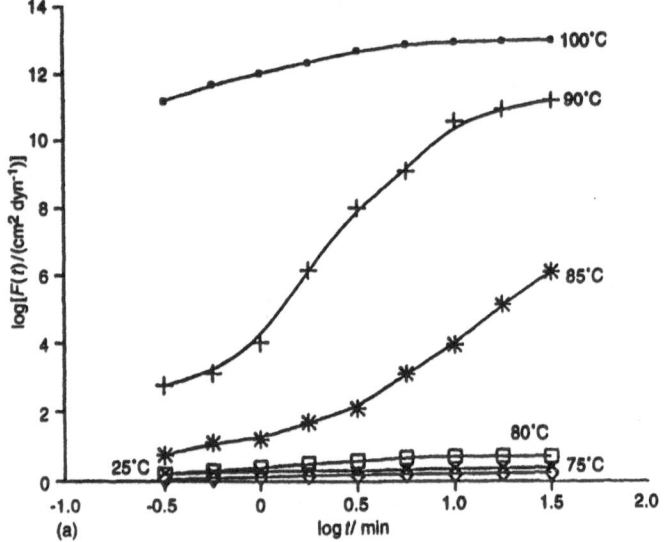

Figure 8.17(a). Variation of creep compliance with time at different temperatures for hot setting epoxy resin (Experimental data from Theocaris, 1962).

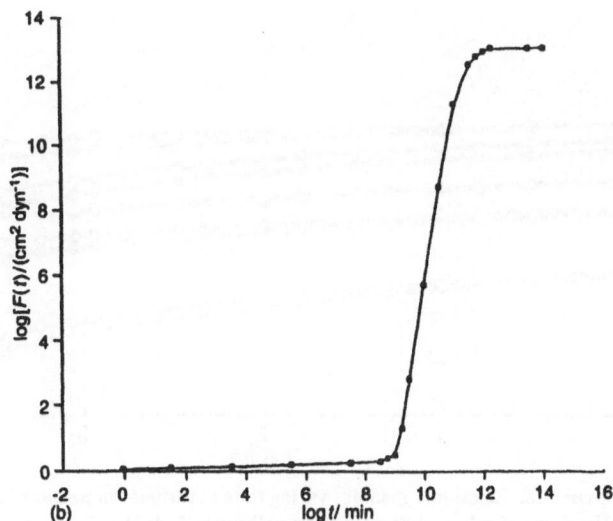

Figure 8.17(b). Master creep curve based on creep data, presented in Fig. 8.17(a), for hot setting epoxy resin with a reference temperature $T_R = 25°C$.

Although the method of time-temperature superposition has been shown to be useful in the characterization of the rheological properties of a large class of amorphous polymers over a wide range of time, it can only be applied to a much smaller range of time for many crystalline polymers (see, e.g., Onogi *et al.*, 1962 and Ferry, 1970). This is primarily due to the predominant nonlinear viscoelastic response of the latter polymers. In this, Onogi *et al.* (1962), for instance, investigated the applicability of the method of time-temperature superposition to the stress relaxation of PVA (Polyvinyl Alcohol) and Nylon 6 films.

In the case of PVA films, Fig. 8.18, two heat-treated specimens (with degree of crystallinity of 36.0 and 47.3%) were tested at temperatures varying between 20 to 100°C at 0% R.H. Fig. 8.18 shows the curves of relaxation modulus against log time for the tested PVA films. As can be seen from this figure, the time-temperature superposition cannot be applied to these relaxation curves. In other words, when the shown relaxation curves are shifted vertically along the relaxation modulus-axis, together with horizontal shifts along the log t-axis, the relaxation curves of PVA films cannot be superimposed to form a smooth master curve.

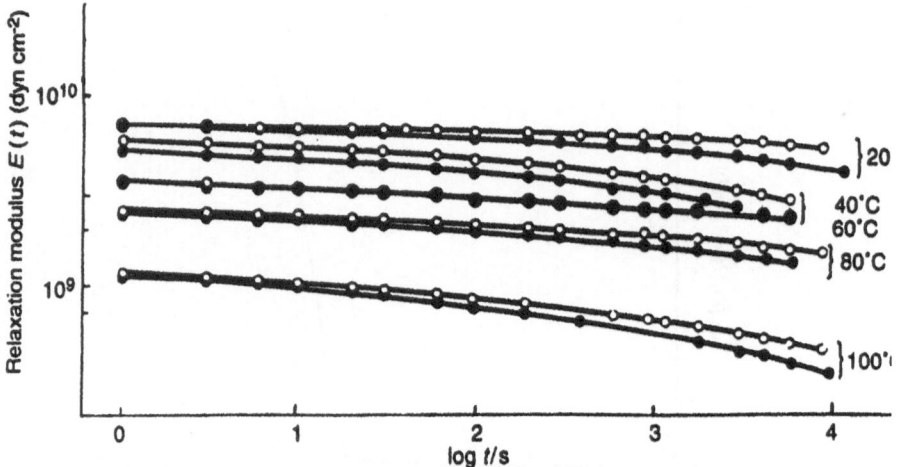

Figure 8.18. Relaxation modulus vs. log time at various temperatures for two PVA films of different degrees of crystallinity (0% R.H.). From: Onogi, S., Sasaguri, K., Adachi, T. and Ogihara, S. (1962). Time-Humidity Superposition in Some Crystalline Polymers, Journal of Polymer Science, Vol.58, 1-17 (John Wiley & Sons Inc. copyright © 1962). Reprinted by permission of John Wiley & Sons Inc.

In the case of Nylon 6 films, Fig. 8.19, the time-dependence of relaxation respon: examined for temperature range between 25 to 77°C. The tests were performed at 0% R. shown in Fig. 8.19, the curves at temperatures higher than 50°C can be superposed to 1 master curve for this temperature range, while those curves corresponding to lower temper than 50°C cannot be superposed. According to Onogi *et al.* (1962), the temperature conforms closely with the transition temperature of Nylon 6 film. That is, only the rela curves in the transition region can be superposed satisfactorily. The master curve obtain Onogi *et al.* (1962), with a reference temperature of 50°C is shown in Fig. 8.20. The shift log a_T versus the reciprocal absolute temperature is shown in Fig. 8.21. The tempe dependence of the shift factor can be represented well by the following form of the WLF-eq (Williams, Landel and Ferry, 1955), i.e.,

$$log\ a_T = [25.6(\theta - 50)]/(85.2 + \theta - 50)$$

where θ is the temperature in °C.

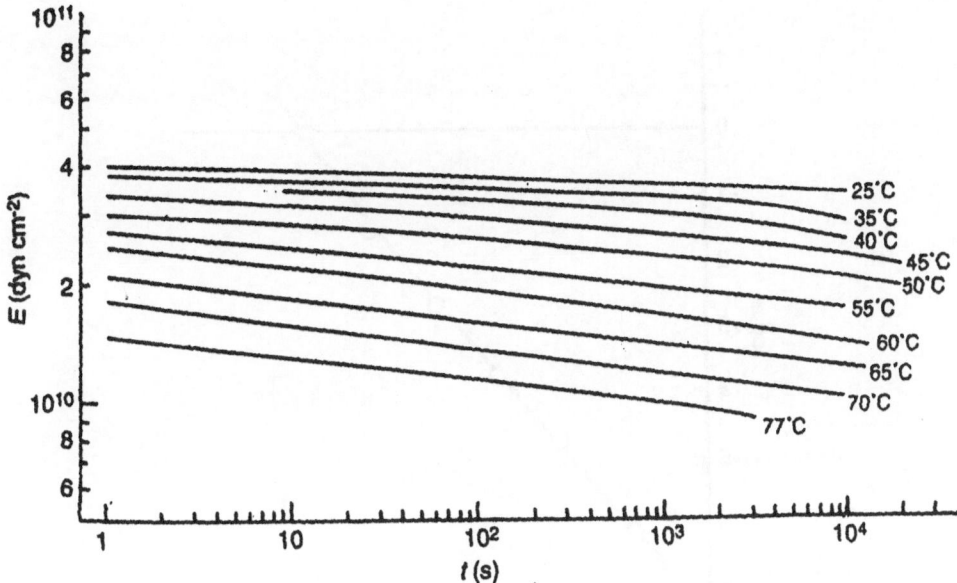

Figure 8.19. Relaxation modulus vs. log time at various temperatures for heat treated nylon 6 film (0% R.H.). From: Onogi, S., Sasaguri, K., Adachi, T. and Ogihara, S. (1962). Time-Humidity Superposition in Some Crystalline Polymers, Journal of Polymer Science, Vol.58, 1-17 (John Wiley & Sons Inc. copyright 1962). Reprinted by permission of John Wiley & Sons Inc.

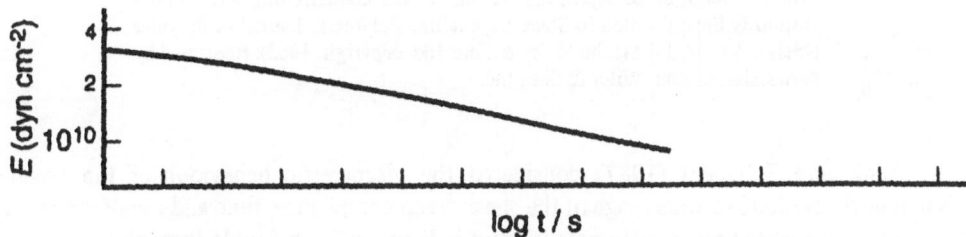

Figure 8.20. Master relaxation curve for heat-treated nylon 6 film (0% R.H.) as obtained from relaxation moduli curves of Fig. 8.19. From: Onogi, S., Sasaguri, K., Adachi, T. and Ogihara, S. (1962). Time-Humidity Superposition in Some Crys-talline Polymers, Journal of Polymer Science, Vol.58, 1-17 (John Wiley & Sons Inc. copyright 1962). Reprinted by permission of John Wiley & Sons Inc.

342

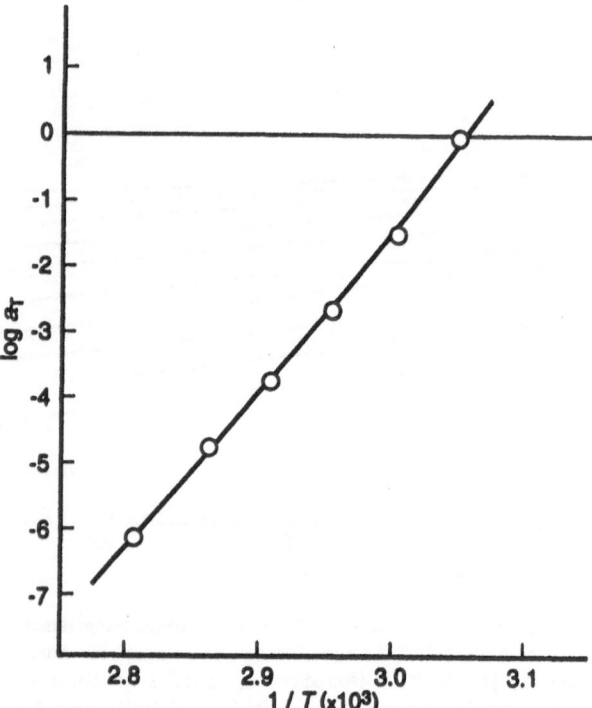

Figure 8.21. Shift factor a_T against reciprocal of absolute temperature (relaxation moduli data for heat-treated nylon 6 film at 0% R.H., Fig. 8.19). From: Onogi, S., Sasaguri, K., Adachi, T. and Ogihara, S. (1962). Time-Humidity Superposition in Some Crys-talline Polymers, Journal of Polymer Science, Vol.58, 1-17 (John Wiley & Sons Inc. copyright 1962). Reprinted by permission of John Wiley & Sons Inc.

Link and Schwarzl (1987) considered the viscoelastic behaviour of the technic: polystyrene PS N7000 in a wide range of the shear creep compliance, time and temperature. Th shear creep compliance versus creep time is given in Figure 8.22 in double-logarithmic scale. A seen in the figure, the compliance changes over seven orders in magnitude from 10^{-8} Pa^{-1} to 10 Pa^{-1}. The course of compliance is determined over more than seven decades in time. Meantim the temperature was varied between 95°C and 170°C. The creep compliance shows the wel known characteristic behaviour of this class of polymer with temperature and time. At the lowe temperatures, the transition region is seen with a steep rise of nearly constant slope. At th temperature of 100°C, the beginning of the rubbery plateau may first be seen at longer time: Increa-sing the temperature further, the rubbery plateau becomes shorter and the viscou contribution would start do dominate even at shorter times. In the flow region, the cree compliance increases with time with a slope which, as seen in the figure, approaches unity in double-logarithmic scale.

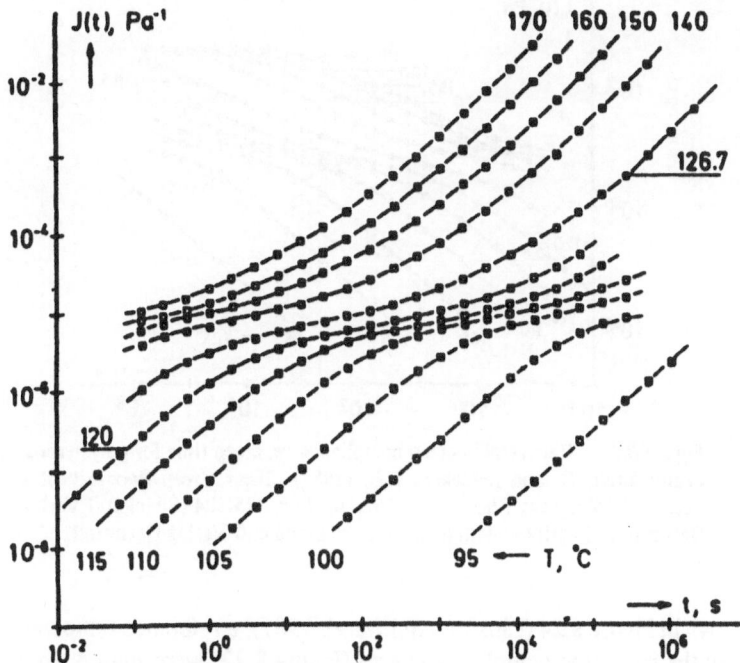

Figure 8.22. Double logarithmic plot of the creep compliance vs. time for polystyrene. From: Link, G. and Schwarzl, F.R. (1987). Shear Creep and Recovery of a Technical Polystyrene, Rheol. Acta, Vol.26, No.4, 375-384 (Steinkopff Verlag Darmstadt). Reprinted with permission of Steinkopff Verlag Darmstadt.

The recoverable creep compliance of polystyrene PS N7000 is shown at the same temperatures in Figure 8.23 due to link and Schwarzl (1987). As indicated by these authors, two transitions can be seen in the course of the recoverable creep compliance. In addition to the glass-rubber transition, a second pronounced transition becomes evident whereby the recoverable compliance rises from the rubbery level of about $5(10)^{-6}$ Pa^{-1} up to a long-time-limiting value of $4.7 (10^{-4})$ Pa^{-1} which is the steady recoverable compliance. This transition is often referred to as *"network transition"*.

344

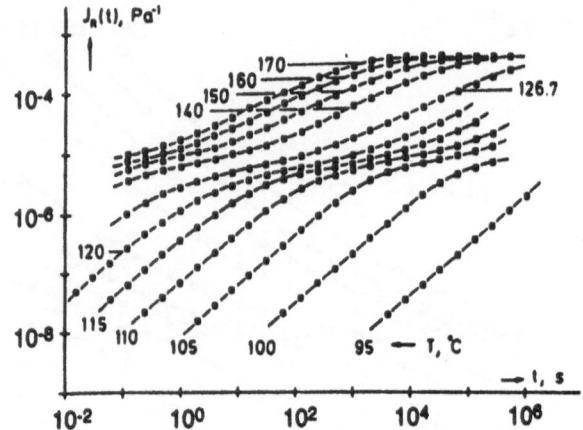

Figure 8.23. Recoverable creep compliance vs. creep time for polystyrene. From: Link, G. and Schwarzl, F.R. (1987). Shear Creep Recovery of a Technical Polystyrent, Rheol. Acta, Vol.26, No.4, 375-384 (Steinkopff Verlag Darmstadt). Reprinted with permission of Steinkopff Verlag Darmstadt.

As shown in Figure 8.24 (Link and Schwarzl, 1987), a reference temperature was chose as 126.7°C. All the measured compliance curves (Figure 8.22) were shifted along the time scal with the same time-temperature shift law. The latter was derived by shifting the creep complianc curves from 140°C to 170°C to coincide with the creep compliance curve at the referenc temperature of 126.7°C in the flow region at a compliance level of $4(10^{-3})$ Pa^{-1}. The correspondin master curve for the recoverable creep compliance is shown in Figure 8.25 at the same referenc temperature of 126.7°C.

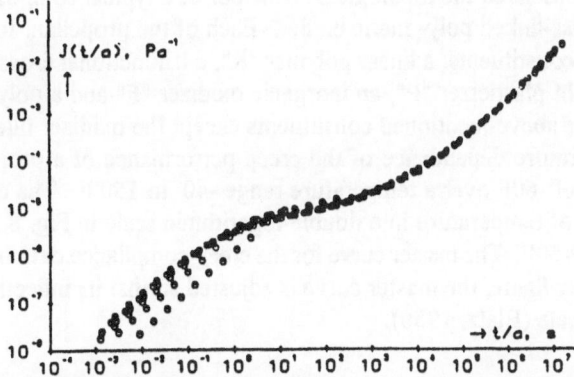

Figure 8.24. Creep compliance vs. reduced time t/a for polystyrene, reference temperature 126.7°C; applied shift function determined in the flow region. From: Link, G. and Schwarzl, F.R. (1987). Shear Creep and Recovery of a Technical Polystyrene, Rheol. Acta, Vol.26, No.4, 375-384 (Steinkopff Verlag Darmstadt). Reprinted with permission of Steinkopff Verlag Darmstadt.

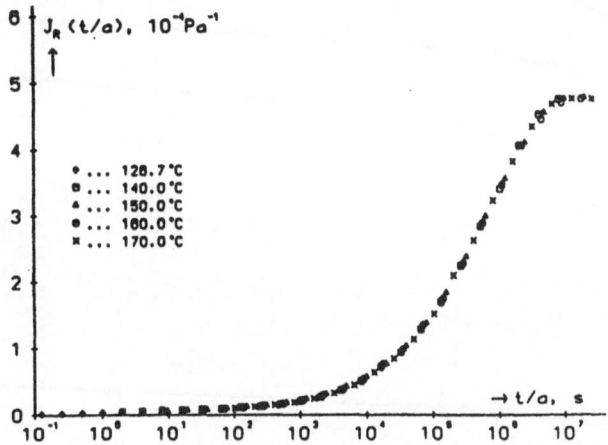

Figure 8.25. Semi-logarithmic plot of the recoverable creep compliance vs. reduced creep time t/a for polystyrene at various temperatures. From: Link, G. and Schwarzl, F.R. (1987). Shear Creep and Recovery of a Technical Polystyrene, Rheol. Acta, Vol.26, No.4, 375-384 (Steinkopff Verlag Darmstadt). Reprinted with permission of Steinkopff Verlag Darmstadt.

346

Blatz (1956) considered the rheological behaviour of a typical composite solid propellant that is based on a cross-linked poly- meric binder. Each of the propellant formulations studied include the following constituents: a linear polymer "R", a trifunctional cross-linking agent "X", a low molecular weight plasticizer "P", an inorganic oxidizer "F" and a polymeric binder. The latter com- prises all the above-mentioned constituents except the oxidizer filler "F". Figure 8.26 represents the temperature dependence of the creep performance of a propellant of weighted composition R-8X-0.6P-60F over a temperature range -40° to 150°F. The creep compliance is plotted as a function of temperature in a double-logarithmic scale in Fig. 8.27 for a propellant formulation R-6X-0.6P-60F. The master curve for the creep compliance data of Fig. 8.27 is given in Fig. 8.28. In the latter figure, the master curve is adjusted so that its inflection point is at unity on the reduced time scale (Blatz, 1956).

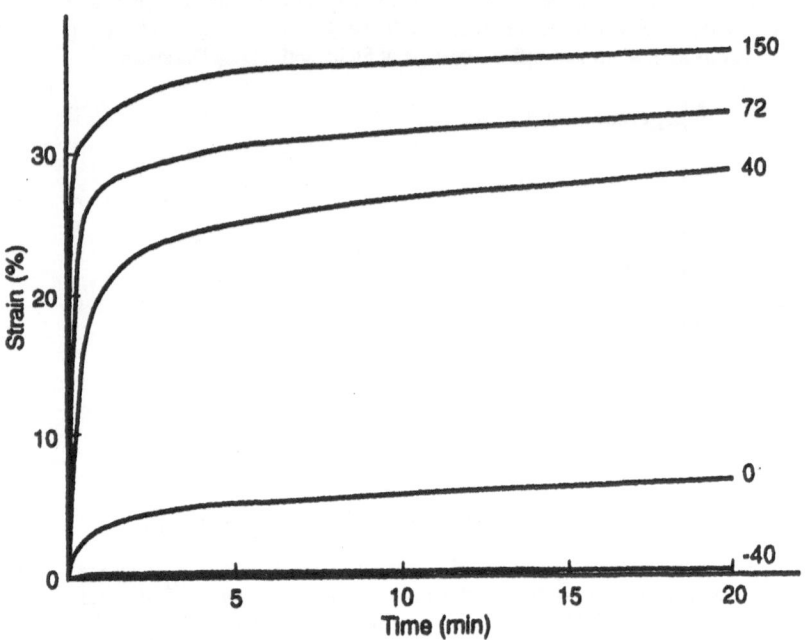

Figure 8.26. Effect of temperature (°F) on creep of a class of composite solid propellants. From: Blatz, P.J. (1956). Rheology of Composite Solid Propellants, J. Industr. & Engg. Chem., Vol.48, No.4, 327-329 (American Chemical Society). Reprinted with permission of American Chemical Society.

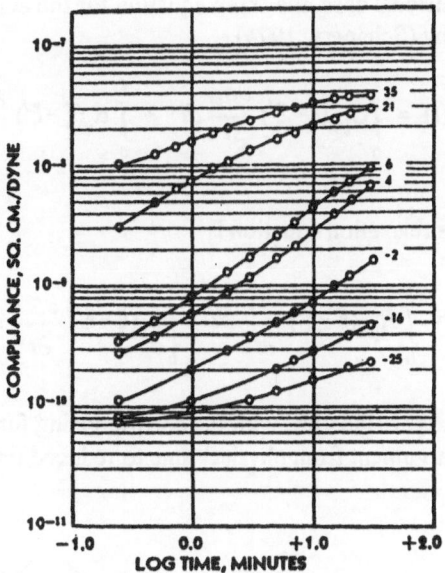

Figure 8.27. Flexural creep of a class of composite solid propellants as function of temperature. From: Blatz, P.J. (1956). Rheology of Composite Solid Propellants, J. Industr. & Engg. Chem., Vol.48, No.4, 727-729 (American Chemical Society). Reprinted with permission of American Chemical Society.

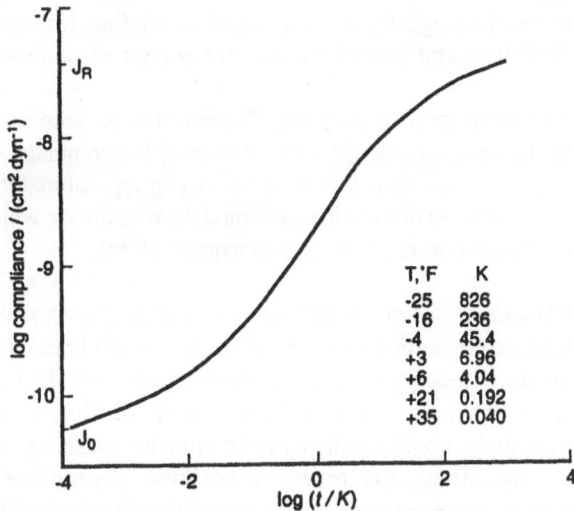

Figure 8.28. Master curve for creep of a class of composite solid propellants (Fig. 8.27). From: Blatz, P.J. (1956). Rheology of Composite Solid Propellants, J. Industr. & Engg. Chem., Vol.48, No.4, 727-729 (American Chemical Society). Reprinted with permission of American Chemical Society.

In the three-dimensional case, the creep equation for the anisotropic thermorheologicall simple material is given by (Schapery, 1974).

$$\varepsilon_{ij}(t) = \int_0^t C_{ijkl}(\xi - \xi') \frac{\partial \sigma_{kl}}{\partial t'} dt' + \int_0^t \alpha_{ij}(\xi - \xi') \frac{\partial \Delta T}{\partial t'} dt \tag{8.162}$$

The corresponding stress relaxation equation is

$$\sigma_{ij}(t) = \int_0^t R_{ijkl}(\xi - \xi') \frac{\partial \varepsilon_{kl}}{\partial t'} dt' - \int_0^t \beta_{ij}(\xi - \xi') \frac{\partial \Delta T}{\partial t'} dt' \tag{8.163}$$

where the material functions are identical with the corresponding functions in the isothermal cas except for the change in argument from physical time to reduced time where the latter is define by

$$\xi = \xi(t) = \int_0^t \frac{d\tau}{a_T}, \qquad \xi' = \xi(t') = \int_0^{t'} \frac{d\tau}{a_T} \tag{8.164}$$

as represented earlier by equations (8.155) and (8.156) respectively. The second order tensors α and β_{ij} in (8.162) and (8.163), respectively, are associated with the thermal expansio characteristics of the thermorheologically simple material and define, respectively, thermal strain in the absence of applied stress and thermal stresses in a completely constrained body.

The constitutive equations. (8.162) and (8.163) may be seen as results of the linea hereditary theory where the input variables (stresses or strains) are combined with the temperatur change which are not applied until time t=0 on a non-aging type of material of a reduced tim scale ξ. These equations have been derived from thermodynamics theory which predicts complet symmetry of the material functions involved (see Schapery, 1974).

Thermorheologically Simple Materials under Non-constant Temperature States
The effects to be studied here are outside the scope of the first order linear theory. Consequently a coupled thermoviscoelastic theory which includes the temperature dependence of mechanica properties is necessarily nonlinear. Guided by the work of Crochet and Naghdi (1969) Christensen (1971) presented a nonlinear theory of thermoviscoelasticity which, upon the usua linearization of stress and strain, still retains a nonlinear dependence upon temperature Concerning this, Christensen attempted to derive the special results appropriate to the stress-strai

constitutive relation without consideration of the other field variables such as energy, entropy, and the heat flux vector which necessarily are involved in the general theory. For this purpose, Christensen extended the uncoupled theory of linear thermoviscoelasticity to account for the temperature dependence of the relevant mechanical properties. The non-constant, non-uniform temperature history is considered to be known.

In Christensen's work (1971), the starting point is the statement of a general nonlinear function which expresses the dependence of the current value of strain upon the histories of stress and temperature, together with their current values. That is

$$E(t) = \underset{\tau=0}{\overset{\infty}{\psi}} \left\{ \Sigma(t-\tau), \ T(t-\tau), \ \Sigma(t), \ T(t) \right\} \tag{8.165}$$

In (8.165), $E(t)$ is the non-linear strain measure at time t, Σ is the Piola-Kirchhoff stress tensor and T is the absolute temperature.

For the material being in a stress free state, but with non-constant temperature history, Eqn. (8.165) may be expressed as a separate functional of temperatures only, i.e.,

$$E(t)|_{\Sigma} = \underset{\tau=0}{\overset{\infty}{\psi'}} \left\{ T(t-\tau), \ T(t) \right\} \tag{8.166}$$

The functional (8.166) could be restricted (Crochet and Naghdi, 1969 and 1979) to express strain at zero stress as a function of current temperature, i.e., (8.166) reduces to

$$\underset{\tau=0}{\overset{\infty}{\psi'}} \left\{ T(t-\tau), \ T(\tau) \right\} = \alpha T(t) \tag{8.167}$$

where α is the coefficient of linear thermal expansion. Equation (8.167) may be considered as a special case of the type of behaviour allowed in the infinitesimal theory. Thus, by decomposing $E(t)$ of (8.165) into two parts such that

$$\overset{\infty}{\underset{\tau=0}{\psi}} \{\Sigma(t-\tau),\ T(t-\tau),\ \Sigma(t),\ T(t)\} = \overset{\infty}{\underset{\tau=0}{\psi'}}\ \{T(t-\tau),\ T(t)\}$$

$$+ \overset{\infty}{\underset{\tau=0}{\psi''}} \{\Sigma(t-\tau),\ T(t-\tau);\ \Sigma(t),\ T(t)\}$$

(8.168)

where, by adopting (8.166), one must take

$$\overset{\infty}{\underset{\tau=0}{\psi''}} \{0, T(t-\tau),\ 0,\ T(t)\} = 0 \qquad (8.169)$$

Thus, using (8.166) and (8.167) in (8.168), gives the form

$$E(t) - \alpha T(t) = \overset{\infty}{\underset{\tau=0}{\psi''}} \{\Sigma(t-\tau),\ T(t-\tau),\ \Sigma(t),\ T(t)\} \qquad (8.170)$$

with the understanding from previous definitions, that $E(t)$ is the total strain including both the stress and temperature effects and $\alpha T(t)$ is the thermal strain in the absence of stress.

In a similar manner, on the assumption that the inverse of (8.165) exists, the stress relaxation equation can be written as

$$\Sigma(t) = \overset{\infty}{\underset{\tau=0}{Y}}\ \{E(t-\tau),\ T(t-\tau),\ E(\tau),\ T(t)\} \qquad (8.171)$$

Now, if one assumes, following Christensen (1971), that the non-isothermal stress constitutive relation is determined by the corresponding isothermal function with E replaced by $E-\alpha$ and with a modified time scale ξ_τ to account for the history of the temperature, the non isothermal stress relaxation may be expressed as

$$\Sigma(t) = \overset{\infty}{\underset{\tau=0}{Y}} \{[E(t-\xi_\tau) - \alpha(t-\xi_\tau)],\ E(t) - \alpha(t)\} \qquad (8.172)$$

In Eqn. (8.172), the modified time scale ξ_τ is given by

$$\xi_\tau = \mathop{\underset{s=0}{\overset{\infty}{\gamma}}} \{T(t-s), \tau\} \tag{8.173}$$

where $\mathop{\underset{s=0}{\overset{\infty}{\gamma}}} (\cdot)$ has properties such as

$$\xi_\tau |_{\tau=0} = 0 \tag{8.174a}$$

and

$$\frac{\partial \xi_\tau}{\partial \tau} \geq 0 \tag{8.174b}$$

and

$$\xi_\tau |_{T=T_0} = \tau \tag{8.174c}$$

with the understanding that T_0 designates some constant base temperature.

It has been shown by Christensen (1971) that the appropriate infinitesimal theory form of (8.172) is

$$\sigma_{ij}(t) = R_{ijkl}(0)\left[\varepsilon_{kl}(t) - \alpha_{kl}(t)\right] + \int_0^\infty \left[\varepsilon_{kl}(t-\xi_\tau)\right.$$

$$\left. - \alpha_{kl}(t-\xi_\tau)\right] \frac{\partial R_{ijkl}(\tau)d\tau}{\partial \tau} \tag{8.175}$$

In (8.175), the usual infinitesimal theory definitions of stress and strain are taken, however, the general nonlinear dependence on temperature is retained.

8.9.5. THERMORHEOLOGICALLY COMPLEX MATERIALS (TCM)

This is the class of viscoelastic materials whose temperature dependence of mechanical properties are not particularly responsive to analytical description through the time-temperature shift phenomenon. Following Schapery (1974), two classifications of such materials are defined namely, TCM-1 and TCM-2.

TCM-1: An example of this class of materials would be a composite material system consisting of two or more TSM (thermorheologically simple material) phases. The mechanical behaviour of different composite systems belonging to the class TCM-1 has been considered by different researchers under isothermal conditions at different temperatures. In this context, Halpin (1969) dealt with a composite system consisting of two types of elastomers with dif- ferent glass transition temperatures (-29 and -75°C). Fesko and Tscheogl (1971) considered the behaviour of two-phase block copolymers at various constant temperatures. The case of TCM-1 under transient temperatures is particularly complicated and further research effort is needed in this area (*see*, e.g., Schapery, 1966-1974).

TCM-2: This class of viscoelastic material is defined (Schapery, 1974) by the following uniaxial constitutive relationship for the cases of constant or transient temperatures

$$\varepsilon(t,T) = C_0(T)\sigma_0 + \int_0^t \Delta C(\xi - \xi') \frac{d}{dt'}\left(\frac{\sigma}{a_G(T)}\right) dt' \qquad (8.176)$$

where $C_0(T)$ is the initial value of creep compliance, $a_G(T)$ is a new shift factor for the class of material considered and ξ, ξ' are reduced time parameters introduced earlier by (8.164). Applying (8.176) to an isothermal creep test, the creep compliance, $C_T = \varepsilon/\sigma$, is expected to be of the form

$$C_T = C_0(T) + \Delta C(\xi)/a_G(T) \qquad (8.177)$$

where $\xi = t/a_T$. This is with the understanding that since $C_0(T)$ is the initial compliance, $\Delta C(0) = 0$. However, in order to relate the two shift factors a_T and a_G to experimental data, one writes (8.177) in the following logarithmic form

$$log(C_T - C_0(T)) = log \, \Delta C(\xi) - log \, a_G(T) \qquad (8.178)$$

One, also, recalls from (8.164) that

$$log \, \xi = log \, t - log \, a_T \qquad (8.179)$$

Equations (8.178) and (8.179) above imply that a plot of log $[C_T-C_0(T)]$ vs. log t at a test temperature T will be identical to that at an arbitrary selected reference temperature T_R, apart from rigid horizontal and vertical translations of $|\log t|$ and $|\log a_T|$ respectively.

A case of particular interest of the constitutive equation (8.176) is when

$$C_0 = C_0(T_R)/a_G(T) \tag{8.180}$$

In this particular case, the creep compliance (8.177) reduces to

$$C_T = C(\xi)/a_G(T) \tag{8.181}$$

in which

$$C(\xi) + C_0(T_R)/\Delta C(\xi) \tag{8.182}$$

On the other hand, assuming as a normalization case that $a_G=a_T=1$ at T_R, one can write, with reference to (8.177), the following equation

$$\Delta C(t) = C_T(t,T_R) - C_0(T_R) \tag{8.183}$$

Meantime, by using (8.183) with the argument t replaced by ξ, it can be shown that the constitutive equation (8.176) becomes

$$\varepsilon(t,T) = \int_0^t C(\xi-\xi')\frac{d}{dt'}\left(\frac{\sigma}{a_G}\right)dt' \tag{8.184}$$

The inversion of (8.184) may be expressed as

$$\sigma = a_G\int_0^t R(t-t')\frac{d\varepsilon}{dt'}dt' \tag{8.185}$$

Accordingly, the relaxation modulus, R_T, for a constant strain input, is

$$R_T = a_G(T)R(\xi) \tag{8.186}$$

Equations (8.181) and (8.186), within their applicability, indicate that master curves of $C=C(\xi)$ and $R=R(\xi)$ can be plotted by making horizontal ($\log a_T$) and vertical ($\log a$) shifts of the experimentally derived creep compliance C_T and relaxation modulus R_T.

For the experimental verification of the constitutive equations (8.176) and (8.184) reference is made to Schapery *et al.* (1973, 1974) and Watkins (1973). The reader is also referred to McCrum and Pogany (1970) for different procedures by which experimental data can be represented by master curves. At this point, it should be mentioned that the strain response expressed by (8.184) may be converted to total strain by using the procedure presented earlier for the case of thermorheologically simple materials (Section 8.10.4)

For a three-dimensional representation of thermorheologically complex materials (TCM) we restrict our analysis, following Schapery (1974), to the isotropic case. For this purpose, two independent material parameters are used. In the creep case, one can use both the uniaxial creep compliance as presented earlier by (8.177), i.e.,

$$C_T = C_0(T) + \Delta C(\xi)/a_G(T) \tag{8.187a}$$

together with the shear creep compliance

$$J_T = J_0(T) + \Delta J(\xi)/a_G(T) \tag{8.187b}$$

Thus, with reference to the one-dimensional constitutive creep equation (8.176), the components for the three-dimensional (isotropic) case are

$$\varepsilon_{11}(t,T) = C_1\sigma_{11} - \left(\frac{J_0}{2} - C_0\right)(\sigma_{22} + \sigma_{33}) + \int_0^t \Delta C(\xi - \xi')\frac{\partial}{\partial t'}\left(\frac{\sigma_{11}}{a_G}\right)dt'$$

$$- \int_0^t \left[\frac{\Delta J(\xi - \xi')}{2} - \Delta C(\xi - \xi')\right]\frac{\partial}{\partial t'}\left(\frac{\sigma_{22} + \sigma_{33}}{a_G}\right)dt' \tag{8.188a}$$

and two additional equations for ϵ_{22} and ϵ_{33}. Similarly for the shear components of the strain tensor, one has

$$2\epsilon_{23}(t,T) = J_0\sigma_{23} + \int_0^t \Delta C(\xi - \xi') \frac{\partial}{\partial t'}\left(\frac{\sigma_{23}}{a_G}\right) dt' \qquad (8.188b)$$

and two additional equations for ϵ_{12} and ϵ_{13}. From (8.188), it can be shown that the bulk creep compliance $K_T(t,T) = 3\epsilon_{ii}/\sigma_{ii}$, where σ_{ii} is constant, is expressed by

$$K_T(t,T) = 3[3C_T - J_T] \qquad (8.189)$$

Meantime, the isothermal value of Poisson's ratio is determined at each temperature by

$$\nu = (J_T/2C_T) - 1 \qquad (8.190)$$

The constitutive equations for thermorheologically simple materials can be immediately deduced from (8.188) by setting $a_G=1$ and assuming that C_1, J_1 and α are constants.

8.10. Problems

10. Elaborate briefly on the meaning of the terms: temperature-time equivalence, reduced-time parameter, and temperature shift factor.

11. What is meant by a *"master curve"*? Illustrate schematically how could one obtain a viscoelastic *"master curve"* in creep.

12. Elaborate briefly on the physical difference between *"thermorheologically simple"* and *"thermorheologically complex"* material. How would such difference reflect itself in the corresponding constitutive equation under constant temperature states?

13. Attempt the second part of Problem 12 above for the case of non-constant temperature states.

8.11. References

Bernal, J.D. (1958) Structure arrangements of macromolecules, *Discussions of Faraday Soc.* **25**, 7-18.
Biot, M.A. (1958) Linear thermodynamics and the mechanics of solids, *Proc. 3rd U.S. Nat. Congr. Appl. Mech.*, pp. 1-18.
Biot, M.A. (1973) Nonlinear thermoeleasticity, irreversible thermodynamics and elastic instability, *Indi. Math. J.*

356

23, 309-35.

Bland, D.R. (1960) *The Theory of Linear Viscoelasticity*, Pergamon, New York.

Blatz, P. J. (1956) Rheology of composite solid propellants, *J. Industr. Engg. Chem.* **48** (4), 727-9.

Boltzmann, L. (1874) Zür theorie der elastichen nachwirkung, Sitzungsber, Kaiserl, *Akad. Wiss. Wien, Mat. Naturkl* **70**(2), 275-306.

Boltzmann, L. (1877) Zür theorie der elastischen nachwirkung, *ibid* **76**, 815-42.

Boltzmann, L. (1878) Zür theorie der elastischen nach-wirkung, *Ann. d. Phys. u. Chem.*, N.F. Bd.V., 430-2.

Breuer, S. (1969) Lower bounds on work in linear viscoelasticity, *Quart. Appl. Math.* **27**(2), 139-46.

Breuer, S. and Onat, E.T. (1964) On the determination of free energy in linear viscoelastic solids, *Z. Angew Mat. Phys.* **15**, 184-91.

Christensen, R.M. (1971) *Theory of Viscoelasticity*, Academic Press, New York.

Christensen, R.M. and Naghdi, P.M. (1967) Linear non-isothermal viscoelastic solids, *Acta Mechanica* **3**, 1-1.

Coleman, B.D. (1964a) Thermodynamics of materials with memory, *Arch. Ration. Mech. Anal.* **17**, 1-46.

Coleman, B.D. (1964b) On thermodynamics, strain impulses, and viscoelasticity, *Arch. Ration. Mech. Anal.* **1** 230-54.

Coleman, B.D. and Mizel, V.J. (1963) Thermodynamics and departures from Fourier's law of heat conductio *Arch. Ration. Mech. Anal.* **13**, 245-61.

Coleman, B.D. and Mizel, V.J. (1964) Existence of coloric equations of state in thermodynamics, *J. Chem. Phy* **40**, 1116-25.

Coleman, B.D. and Noll, W. (1963) The thermodynamics of elastic materials with heat conduction and viscosit *Arch. Ration. Mech. Anal.* **13**, 167-78.

Crochet, M.J. (1975) A Non-isothermal theory of viscoelastic materials, in *Theoretical Rheology*, Eds. J.F. Hutto J.R.A. Pearson and K. Walters, Applied Science Publishers, London, pp. 111-22.

Crochet, M.J. and Naghdi, P.M. (1969) A Class of simple solids with fading memory, *Int. J. Eng. Sci.* **7**, 1173-9

Crochet, M.J. and Naghdi, P.M. (1974) On a restricted non-isothermal theory of simple materials, *J. de Mécaniqu* **13**, 97-114.

Day, W.A. (1970) Reversibility, recoverable work and free energy in linear viscoelasticy, *Quart. J. Mech. App Math.* **23**(1), 1-15.

Doi, M. and Edwards, S.F. (1986) *The Theory of Polymer Dynamics*, Clarendon Press, Oxford.

Eirich, F.R. (1956) *Rheology Theory and Applications*, Academic Press, New York.

Ferry, J.D. (1970) *Viscoelastic Properties of Polymers*, 2nd Edn., Wiley, New York.

Fesco, D.G. and Tscheogl, N.W. (1971) Time-temperature superposition in thermo-rheologically comple materials, *J. Polym. Sci.* **35**(C), 51-69.

Flügge, W. (1975) *Viscoelasticity*, Springer-Verlag, New York.

Fried, N. (1970) In: *Mechanics of Composite Materials*, Eds. F.W. Wendt, H. Liebowitz and N. Perron Pergamon, Oxford, pp.813-37.

Gittus, J. (1975) *Creep, Viscoelasticity and Creep Fracture in Solids*, John Wiley & Sons, New York.

Gross, B. (1953) *Mathematical Structure of the Theory of Viscoelasticity*, Hermann, Paris.

Haddad, Y. M. (1995) *Viscoelasticity of Engineering Materials*, Kluwer, Dordrecht.

Halpin, J.C. (1969) *Characterization of Orthotropic (Fiber-Reinforced) Polymeric Solids*, Doctoral Dissertatio Univ. of Akron, Ohio, U.S.A.

Hill, R.L. (1967) The creep behaviour of individual pulp fibres under tensile stress, *Tappi* **50**(8), 432-40.

Kauman, W.G. (1966) On the deformation and setting of the wood cell wall, *Holz als Roh und Wekstoff* **24**(11 551-6.

Landel, R.F. and Peng, S.T.T. (1986) Equations of state and constitutive equations, *J. Rheology* **30**(4), 741-6

Leaderman, H. (1943a) *Elastic and Creep Properties of Filamentous Materials and Other Polymers*, Texti Foundation, Washington, D.C., pp. 175-85.

Leaderman, H. (1943b) *Elastic and Creep Properties of Filamentous Materials and Other High Polymers*, Textile Foundation, Washington, D.C

Leaderman, H. (1958) *Viscoelasticity Phenomena in Amorphous High Polymeric Systems*, Ed. F. Eirich, Academ Press, New York, Vol.II, pp. 1-6.

Lethersich, W. (1950) The rheological properties of dielectric polymers, *Br. J. Appl. Phys.* **1**, 294-301.

Link, G. and Schwarzl, F.R. (1987) Shear creep and recovery of a technical polystyrene, *Rheol. Acta* **26**, 375-84.

Lockett, F. J. (1972) *Nonlinear Viscoelastic Solids*, Academic Press, New York.

Love, A. E. H. (1944) *A Treatise on the Mathematical Theory of Elasticity*, 4th Edition, Dover Publications, New York, pp. 1-31.

Mark, H. and Tobolosky, A. V. (1950) *Physical Chemistry of High Polymeric Materials*, Interscience Publishers, New York.

Markovitz, H. (1977) Boltzmann and the beginnings of linear viscoelasticity, *Trans. Soc. Rheol.* **21:3**, 381-98.

McCrum, N.G. and Pogany, G.A. (1970) Time-temperature superposition in the α-region of an epoxy resin, *J. Macromol Sci. Phys.* **B4(1)**, 109-25.

Meixner, J. (1969) Processes in simple thermodynamic materials, *Arch. Ration. Mech. Anal.* **33**, 33-53.

Mercier, J.P., Aklonis, M. Litt and Tobolsky, A.V. (1965) Viscoelastic behaviour of the Polycarbonate of Bisphenol A, *Journal of Applied Polymer Science* **9**, 447-59.

Meredith, R. (1954) Relaxation of stress in stretched cellulose fibers, *J. Textile Ins.* **45**, T438-T460.

Morland, L.W. and Lee, E.H. (1960) Stress analysis for linear viscoelastic materials with temperature variation, *Trans. Soc. Rheol.* **4**, 233-63.

Müller, I. (1967) On the entropy inequality, *Arch. Ration. Mech. Anal.* **26**, 118-41.

Onogi, S., Sasaguri, K., Adachi, T. and Ogihara, S. (1962) Time-humidity superposition in some crystalline polymers, *J. Polymer Science* **58**, 1-17.

Peters, L. (1955) A note on nonlinear viscoelasticity, *Textile Research Journal*, March 1955, 262-65.

Rivlin, R.S. (1975) The Thermodynamics of materials with fading memory, in *Theoretical Rheology*, Eds. J.F. Hutton, J.R.A. Pearson & K. Walters, Applied Science Publishers, London, pp. 83-103.

Schapery, R.A. (1964) Application of thermodynamics to thermo-mechanical, fracture, and birefringent phenomena in viscoelastic media, *J. Appl. Physics* **35(5)**, 1451-65.

Schapery, R.A. (1966) A theory of nonlinear thermoviscoelasticity based on irreversible thermodynamics, *Proc. 5th U.S. Nat. Congress of Appl. Mech.*, ASME, pp. 511-30.

Schapery, R.A. (1968) On a thermodynamic constitutive theory and its application to various nonlinear materials, *Proc. IUTAM Symp.*, East Kilbride, pp. 259-85.

Schapery, R.A. (1969) On a thermodynamic constitutive theory and its application to various nonlinear materials, *Proc. IUTAM Symp. on Thermoinelasticity*, Springer, Berlin.

Schapery, R.A., Beckwith, S.W. and Conrad, N. (1973) *Studies on the Viscoelastic Behaviour of Fiber-Reinforced Plastic*, Mech. Mat. Res. Center Rep. MM 2702-73-3 (AFML-TR-73-179), Texas A & M University.

Schapery, R.A. (1974) Viscoelastic behaviour and analysis of composite materials, in *Mechanics of Composite Materials*, G. Sandeskj (Ed.), Vol.2, Academic Press, New York, pp. 86-168,

Schwarzl, F. and Staverman, A.J. (1952) Time-temperature dependence of linear viscoelastic behaviour, *J. Appl. Phys.* **23(8)**, 838-43.

Staverman, A.J. and Schwarzl, F. (1956) Linear deformation behaviour of high polymers, in *Die Physik der Hochpolymeren*, Vol. IV, H.A. Stuart, ed., Springer, Berlin.

Steel, D.J. (1965) The creep and stress-rupture of reinforced plastics, *Trans. J. Plast. Inst.* **33**, 161-67.

Takehiro, S. and San-Ichiro, M. (1955) On the helical configuration of a polymer chain, *The Journal of Chemical Physics* **23(4)**, 707-11.

Theokaris, P.S. (1962) Viscoelastic properties of epoxy resins derived from creep and relaxation tests at different temperatures, *Rheologica Acta*, Band 2, Heft 2, 92-6.

Tobolosky, A.F. and Catsiff, E. (1956) Elastoviscous properties of polyisobutylene (and other amorphous polymers) from stress-relaxation studies, IX. A summary of results, *J. Polym. Sci.* **19**, 111-21.

Tobolosky, A. and Eyring, H.J. (1943) Mechanical properties of polymeric materials, *J. Chem. Phys.* **11**, 125-34.

Tschoegl, N.W. (1989). *The Phenomenological Theory of Linear Viscoelastic Behaviour, An Introduction*, Springer-Verlag, New York.

Truesdell, C. and Toupin, R.A. (1960) Classical field theories, In : *Handbuch der Physik*, Ed. S. Flügge, Vol. III/1. Springer, Berlin, 226-790.

Tsai, S.W. (1970) In: *Mechanics of Composite Materials*, Eds. F.W. Wendt, H. Liebowitz and N. Peronne,

358

Pergamon, Oxford, pp. 749-67.

Volterra, V. (1913) *Fonctions de Lignes*, Gauthier-Villard, Paris.

Volterra, V. and Peres, J. (1936) *Théorie générale des fonctionnelles*, Gauthier-Villard, Paris.

Ward, J.M. (1983) *Mechanical Properties of Solid Polymers*, Second Edition, John Wiley & Sons, New York

Watkins, L.A. (1973) *Creep of an Epoxy Resin under Transient Temperatures*, M.S. Thesis, Civil Eng., Texas *
& M. University.

Williams, M.L., Landel, R.F. and Ferry, J.D. (1955) The temperature dependence of relaxation mechanisms i
amorphous polymers and other flass-forming liquids, *J. Am. Chem. Soc.* **77**, (July 20, 1955), 3701-7

8.12. Further Reading

Aklonis, J. J. (1972) *Introduction to Polymer Viscoelasticity*, Wiley, New York.

Alexander, R. L. (1964) *Limits of Linear Viscoelastic Behaviour of an Asphalt Concrete in Tension an
Compression*, Institute of Transportation and Traffic Engineering, University of California, Berkeley, CA

Alfrey, T. (1948) *Mechanical Behaviour of High Polymers*, Interscience Publishers, New York.

Alfrey, T. and Doty, P.M. (1945) Methods of specifying the properties of viscoelastic materials, *J. Appl. Phys.* **1$**
700-13.

Andrade, E. N. (1910) The viscous flow in metals and allied phenomena, *Proc. of Royal Society, Series A* **84**, 1
12.

Andrews, R. D. (1952) Correlation of dynamic and static measurements on rubber-like materials, *Industr. Eng$*
Chem. **44**, 707-15.

Axelrad, D. R. (1970) Mechanical models of relaxation phenomena, *Advan. Mol. Relaxation Processes* **2**, 41-6$

Barker, L. M. and Hollenbach, R. E. (1964) System for measuring the dynamic properties of materials, *Rev. Sc$*
Inst. **35**(6), 742-46.

Bataille, J. and Kestin, J. (1979) Irreversible processes and physical interpretation of rational thermodynamics, $
Non Equil. Thermodynamics **4**, 229-58.

Bazant, Z.P. (1975) Theory of creep and shrinkage in concrete structures: Aprécis of recent developments, *Mech*
Today **2**, 1-93.

Benbow, J. J. (1956) The determination of dynamic moduli and internal friction of high polymers from cree$
measurements, *Proc. Phys. Soc.* **69** (9-B), 885-92.

Berry, D.S. and Hunter, S.C. (1956) The propagation of dynamic stresses in viscoelastic rods, *J. Mech. Phys. Solid*
4, 72-95.

Biot, M.A. (1954) Theory of stress-strain relationship in anisotropic viscoelasticity and relaxation phenomena, $
Applied Physics **(25)**11, 1385-91.

Bland, D. R. and Lee, E. H. (1956) On the determination of a viscoelastic model for stress analysis of plastics, $
Appl. Mech. **23**, 416-20.

Breuer, S. (1969) Lower bounds on work in linear viscoelasticity, *Quarterly of Applied Mathematics* **XXVII**, Jul
1969, No.2, 139-46.

Carson, J.R. (1926) *Electrical Circuit Theory and Operational Calculus*, McGraw-Hill, New York.

Christensen, R.M. (1972) Restrictions upon viscoelastic relaxation functions and complex moduli, *Trans. So$*
Rheol. **16**, 603-14.

Coleman, B.D. (1963) The thermodynamics of elastic materials with heat conduction and viscosity, *Arch. Ration*
Mech. Anal. **13**, 167-78.

Coleman, B.D. and Gurtin, M.E. (1967a) Thermodynamics with internal state variables, *J. Chem. Phys.* **47**, 59$
613.

Coleman, B.D. and Gurtin, M.E. (1967b) Equipresence and constitutive equations for rigid heat conductors, $
Angew Math. Phys. **18**, 199-208.

Coleman, B.D. and Mizel, V.J.(1967) A general theory of dissipation in materials with memory, *Arch. Ration*
Mech. Anal. **27**, 255-74.

Coleman, B.D. and Mizel, V.J. (1968) On the general theory of fading memory, *Arch. Ration. Mech. Anal.* **29**, 18-31.

Coleman, B.D. and Noll, W. (1960) An approximation theorem for functional with applications in continuum mechanics, *Arch. Ration. Mech. Anal.* **6**, 355-70.

Coleman, B.D. and Noll, W. (1961) Foundations of linear viscoelasticity, *Rev. Modern Phys.* **33**, 239-49.

Coleman, B.D. and Noll, W. (1964) Simple fluids with fading memory, *Proc. Int. Sympos., Second Order Effects, Heifa,* 1962, McMillan, New York, pp. 530-52.

Coleman, B.D. and Owen, D.R. (1970) On the thermodynamics of materials with memory, *Arch. Ration. Mech. Anal.* **36**, 245-69.

Creus, G. J. (1986) *Viscoelasticity; Basic Theory and Applications to Concrete Structures,* Springer-Verlag, Berlin.

Crochet, M.J. and Naghdi, P.M. (1979) On thermo-rheologically simple solids, *Proc. IUTAM Symp., Thermoelasticity,* Springer, New York, 59-86.

Davis, J.L. (1987) *Introduction to Dynamics of Continuous Media,* Macmillan, New York..

Day, M.A. (1970) Some results on the least work needed to produce a given strain in a given time in a viscoelastic material and a uniqueness theorem for dynamic viscoelasticity, *Q. J. Mech. Appl. Math.* **23**, 469-79.

Day, W.A. (1972) *The Thermodynamics of Simple Materials with Fading Memory,* Springer-Verlag, New York.

Duhamel, J. M. C. (1833) Sur la méthode générale relative au mouvement de la chaleur dans les corps solides plongés dans des milieux dont la température varie avec le temps, *J. Ec. Polytech.* (Paris) **14**, Cah. 22, 20-77.

Edelstein, W. S. and Gurtin, M. E. (1964) Uniqueness theorems in the linear dynamic theory of anisotropic viscoelastic solids, *Arch. Rat. Mech. Anal.* **17**, 47-60.

Eringen, A.C. (1960) Irreversible thermodynamics and continuum mechanics, *Phys. Rev.* **117**, 1174-83.

Findley, W.N. (1944) Creep characteristics of plastics, Symposium on Plastics, *ASTM,* pp. 118-34.

Findley, W.N. and Khosla, G. (1955) Application of the superposition principle and theories of mechanical equation of state, strain and time hardening to creep of plastics under changing loads, *J. Appl. Phys.* **26**(7), 821-32.

Finnie, I. and Heller, W. R. (1959) *Creep of Engineering Materials,* McGraw-Hill, New York.

Freudenthal, A.M. (1954) Effect of rheological behaviour on thermal stresses, *J. Appl. Phys.* **25**(9), 1110-17.

Froehlich, H. (1949) *Theory of Dielectrics,* Clarendon Press, Oxford.

Fung, Y.C. (1965) *Foundations of Solid Mechanics,* Prentice-Hall, Englewood Cliffs, N.J.

Gibson, R.F., Hwang, S. J. and Sheppard, C. H. (1990) Characterization of creep in polymer composites by the use of frequency-time transformations, *J. Composite Materials* **24**, 441-53.

Glauz, R.D. and Lee, E.H. (1954) Transient wave analysis in linear time-dependent material, *J. Appl. Phys.* **25**, 947-53.

Green, A.E. and Laws, N. (1967) On the formulation of constitutive equations in thermomechanical theories of continua, *Quart. J. Mech. Appl. Math.* **20**, 265-75.

Green, A.E. and Naghdi, P.M. (1965) A general theory of an elastic-plastic continuum, *Arch. Ration. Mech. Anal.* **18**, 251-81.

Green, A.E., Rivlin, R.S. and Spencer, A.J.M. (1959) The mechanics of nonlinear materials with memory, Part II, *Arch. Ration. Mech. Anal.* **3**, 82-90.

Gross, B. (1947) On creep and relaxation, *J. Applied Physics* **18**, Feb. 1947, 212-21.

Gross, B. and Pelzer, H. (1951) On creep and relaxation III, *J. Appl. Phys.* **22**, 1035-39.

Gurtin, M.E. (1965) Thermodynamics and the possibility of spatial interaction in elastic materials, *Arch. Ration. Mech. Anal.* **19**, 339-52.

Gurtin, M.E. and Herrara, R.I. (1965) On dissipation inequalities and linear viscoelasticity, *Quart. Appl. Math.* **23**, 235-45.

Gurtin, M.E. and Sternberg, E. (1962) On the linear theory of viscoelasticity, *Arch. Rational Mech. Anal.* **11**, 291-356.

Gurtin, M.E. and Williams, W.O. (1966) On the inclusion of the complete symmetry group in unimodular group, *Arch. Ration. Mech. Anal.* **23**, 163-72.

360

Haddad, Y.M (1988) On the theory of the viscoelastic solid, *Res Mechanica* **25**, 225-59.

Halpin, J.C. (1968) Introduction to viscoelasticity, in *Composite Materials Workshop*, Eds. S.W. Tsai, J.C. Halpi and N.J. Pagano, Technomic, Stamford, Conn., pp. 87-152.

Hilton, H. H. (1964) Viscoelastic analysis, in *Engineering Design for Plastics*, Ed. E. Baer, Reinhold, New Yorl

Hopkins, I. L. and Hamming, R. W. (1957) On creep and relaxation, *J. Appl. Phys.* **28(8)**, 906-9.

Hopkinson, J. (1876) The residual charge of the Leyden Jar, *Philos. Trans.* **166**, 489-94.

Hunter, S. C. (1960) Viscoelastic waves, in *Progress in Solid Mechanics*, Eds. I. N. Sneddon and R. Hill, Vol. : Ch.1, North Holland Publ. Co., Amsterdam, pp. 1-57.

Hunter, S.C. (1961) Tentative equations for the propagation of stress, strain and temperature fields in viscoelasti solids, *J. Mech. Phys. Solids* **9**, 39-51.

Kestin, J. (1966) *A Course in Thermodynamics*, Vol. 1, Blaisdell, Waltham, Mass.

Koh, S.L. and Eringen, A.C. (1963) On the foundation of nonlinear thermoviscoelasticity, *Int. J. Engg. Sci.* **1**, 199 229.

Kolsky, H. (1960) Viscoelastic waves, in: *Proc. Int. Symposium on Stress Wave Propagation in Materials*, Ed. N Davids, Interscience Publishers, New York, pp. 59-90.

Kolsky, H. (1965) Experimental studies of the mechanical behaviour of linear viscoelastic solids, *Proc. of the 4 Symposium on Naval Structural Mechanics*, Pergamon Press, London, pp. 381-442.

König H. and Meixner, J. (1958) Linear systeme und lineare transformationen, *Math. Nachr* **19**, 256-322.

Laws, N. (1967) On the thermodynamics of certain materials with memory, *Int. J. Eng. Science* **5**, 427-34.

Leaderman, H. (1954a) Approximations in linear viscoelasticity theory: Delta function approximation, *J. App Phys.* **25**, 294-6.

Leaderman, H. (1954b) Rheology of polyosobutylene, IV. Calculation of the retardation time function an dynamic response from creep Data, *Proc. 2nd Int. Congr. on Rheology*, Butterworth, London, pp. 203-1:

Leaderman, H. (1958) *Viscoelasticity Phenomena in Amorphous High Polymeric Systems*, Ed. F. Eirich, Academi Press, New York, Vol. II, pp. 1-6.

Lee, E.H. (1955) Stress analysis in viscoelastic bodies, *Quart. Appl. Math.* **13(2)**, 183-90.

Lee, E. H. (1956) Special issues on rheology of polymers, *J. Appl. Phys.* **27**, 665-72.

Lee, E. H. (1960a) *Viscoelasticity: Phenomenological Aspects*, Ed. J.T. Bergen, Academic Press, New York.

Lee, E. H. (1960b) Viscoelastic stress analysis, in: *First Symposium on Naval Structural Mechanics*, Pergamo Press, New York, 456-82.

Lee, E. H. (1962) Viscoelasticity, in *Handbook of Engineering Mechanics*, Ed. W. Flügge, McGraw-Hill, Nev York, pp.53/1-53/22.

Leitman, M. J. and Fisher, G. M. C. (1973) The linear theory of viscoelasticity, *Handbuch der Physik*, Ed. (Flügge VI a/3, Springer-Verlag, Berlin, pp. 1-123.

Lubliner, J. and Salkman, J.L. (1967) On uniqueness in general linear viscoelasticity, *Q. Appl. Math.* **25**, 129-3(

Macey, B. (1948) On the application of Laplace Pairs to the analysis of relaxation Curves, *J. Sci. Instrum.* **25**, 251 3.

Marvin, R.S. (1952) A New approximate conversion method for relating stress relaxation and dynamic modulu: *Phys. Rev.* **86**, 644-5.

Mazilu, P. (1973) On the constitutive law of Boltzmann-Volterra, *Revue Roumaine de Mathématiques Pures (Appliquées* **18**, 1067-9.

McHenry, D. (1943) A new aspect of creep in concrete and its application to design, *Proc. American Society fc Testing on Materials* **43**, 1064-86.

Meixner, J. (1965) Linear passive systems, *Proc. Int. Symp. Stat. Mech. Thermodyn.*, 1964, North-Holland Pub Co., Amsterdam, 52-68.

Noll, W. (1958) A mathematical theory of the mechanical behaviour of continuous media, *Arch. Ration. Mecl Anal.* **2**, 197-226.

Odeh, F. and Tadjbakhsh, I. (1965) Uniqueness in the linear theory of viscoelasticity, *Arch. Ration. Mech. Ana* **18**, 244-50.

Owen, D.R. (1968) Thermodynamics of materials with elastic range, *Arch. Ration. Mech. Anal.* **31**, 91-112.

Owen, D.R. (1970) A mechanical theory of materials with elastic range, *Arch. Ration. Mech. Anal.* **37**, 85-110.

Pindera, J.T. and Straka, P. (1974) On physical measures of rheological responses of some materials in wide ranges of temperature and spectral frequency, *Rheol. Acta* **13**, 338-51.

Pipkin, A.C. (1972) *Lectures on Viscoelasticity Theory*, Springer, New York.

Prager, W. (1956) Thermal stresses in viscoelastic structures, *J. Appl. Math. Phys.* **7**, 230-8.

Rivlin, R.S. (1972) On the principles of equipresence and unification, *Quart. App. Math.* **30**, 227-8.

Roesler, F.C. (1955) Some applications of Fourier Series in the numerical treatment of linear behaviour, *Proc. Phys. Soc. (B)* **68**, 89-96

Roesler, F. C. and Pearson, J. R. A. (1954) Determination of relaxation spectra from damping measurements, *Proc. Phys. Soc. (B)* **67**, 338-47.

Roesler, F.C. and Twyman, W.A. (1955) An iteration method for the determination of relaxation spectra, *Proc. Phys. Soc..***68** **(2-B)**, 97-105.

Roscoe, R. (1950) Mechanical models for the representation of viscoelastic properties, Brit. *J. Appl. Phys.* **1**, 171-3.

Roy, M. (1966) *Milieux continus*, Dunod, Paris.

Schapery, R.A. (1974) Viscoelastic behaviour and analysis of composite materials, in *Mechanics of Composite Materials*, Ed. G. Sendeckj, Vol. 2, 86-168, Academic Press, New York.

Schausberger, A., Knoglinger, H. and Janeschitz-Kriegl, H. (1987) The role of short chain molecules for the rheology of polystyrene melts, II. Linear viscoelastic properties, *Rheol. Acta* **26**, 468-73.

Schwarzl, F. (1951) Näherungsmethoden in der theorie des viscoelastischen verhaltens I, *Physica* **XVII**, 830-40.

Schwarzl, F. and Staverman, A. J. (1952) Time-temperature dependence of linear viscoelastic behaviour, *J. Appl. Physics* **23(8)**, 838-43.

Scott-Blair, G.W. (1949) *Survey of General and Applied Rheology*, Pitman and Son, London.

Struik, L.C.E. (1987) The accuracy of some formulae for the interconversion of creep and relaxation data, *Rheol. Acta* **26**, 7-13.

Stuart, H.A. (1956) *Die Physik der Hochpolymeren*, Vol. 4, Springer, Berlin.

Tapsell, H. J. and Johnson, A. E. (1940) Creep under combined tension and torsion, *Engineering* **150**, 24-28.

Truesdell, C. (1951) A new definition of a fluid. II. The Maxwelliam fluid, *J. Math. Pures Appl.* **30**, 111-58.

Wilson, A.H. (1957) *Thermodynamics and Statistical Mechanics*, Cambridge Univ. Press, Cambridge.

Wolosewick, R.M. and Gratch, S. (1965) Transient response in a viscoelastic material with temperature-dependent properties and thermomechanical coupling, *J. Appl. Mech., Transactions of the ASME*, September 1965, 620-2.

APPENDIX A

CURVILINEAR TENSORS

A.1. Introduction

Following our presentation of Cartesian tensors in Chapter 1, we introduce, briefly, in this appendix, the more general subject of curvilinear components. Here we consider the general formulation of curvilinear tensors which the reader might find helpful for reading much of the advanced literature on the mechanics of deformation of engineering materials.

A.2. Preliminary Material

A.2.1. THE "e-SYSTEM" AND GENERALIZED KRONECKER DELTA

Complete Symmetry
A system of quantities, say $A^{i_1 i_2 \cdots i_n}$ (or $A_{i_1 i_2 \cdots i_n}$), depending upon n indices is said to be completely symmetric if the numerical value of $A^{i_1 i_2 \cdots i_n}$ is unaltered by any permutation of the indices; e.g. $A^{ijkl} = A^{ikjl} = A^{kilj} = \text{etc.}$

Complete skew- or anti-symmetry
A system of quantities, say $A^{i_1 i_2 \cdots i_n}$ (or $A_{i_1 i_2 \cdots i_n}$), depending on n indices is said to be completely skew- or anti-symmetric if the numerical value of $A^{i_1 i_2 \cdots i_n}$ is unchanged by an even permutation of the indices and merely changes sign on an odd permutation of the indices; $A_{ijkl} = - A_{ikjl} = A_{kjli} = \text{etc.}$

It immediately follows that any term of a skew -symmetric system having two like indices has the value of zero.
A system of quantities depending upon n≥3 indices may be symmetric or skew symmetric in any two or more of its indices. The notation $A^{(i_1 i_2) \cdots i_n}$ and $A^{[i_1 i_2] \cdots i_n}$ designates the two indices which are symmetric and skew-symmetric respectively. For example:

Completely skew-symmetric:

$$
A^{(\alpha\beta)} = \begin{pmatrix} A^{11} & A^{12} \\ A^{12} = A^{21} & A^{22} \end{pmatrix} \quad ; \quad A^{[\alpha\beta]} = \begin{pmatrix} 0 & A^{12} \\ - A^{12} = A^{21} & 0 \end{pmatrix}
$$

The e-System

The e-system $e^{i_1 i_2 \cdots i_n}$ or $e_{i_1 i_2 \cdots i_n}$ depending on n indices is defined as:

$$\begin{cases} = 1 & \text{if } i_1 \ldots i_n \text{ are distinct and are an even permutation of } 123\ldots n, \\ = -1 & \text{if } i_1 \ldots i_n \text{ are distinct and are an odd permutation of } 123\ldots n, \\ = 0 & \text{if } i_1 \ldots i_n \text{ are indistinct.} \end{cases}$$

Note that if $A^{i_1 i_2 \cdots i_n}$ is completely skew-symmetric and the range of each index equals the order of the system of quantities, in this case n, then:

$$A^{i_1 i_2 \cdots i_n} = e^{i_1 \cdots i_n} A^{12 \ldots n}$$

For example,

$$A^{ijk} = e^{ijk} A^{123}$$

$$= A^{123} \begin{pmatrix} 0\ 0\ 0 & 0\ 0\ -1 & 0\ 1\ 0 \\ 0\ 0\ 1 & 0\ 0\ 0 & -1\ 0\ 0 \\ 0\ -1\ 0 & 1\ 0\ 0 & 0\ 0\ 0 \end{pmatrix}$$

Generalized Kronecker Delta

The generalized Kronecker delta $\delta^{i_1 i_2 \cdots i_k}_{j_1 j_2 \cdots j_k}$ depending on 2k indices, $k \leq n$, is defined as

$$\delta^{i_1 i_2 \cdots i_k}_{j_1 j_2 \cdots j_k} = \begin{cases} 1 & \text{if } i_1 \ldots i_k \text{ are distinct integers from the range n, } k \leq n \text{ and} \\ & \text{if } j_1 \ldots j_k \text{ is an even permutation of } i_1 \ldots i_k. \\ -1 & \text{if } i_1 \ldots i_k \text{ are distinct integers from the range n, } k \leq n \text{ and} \\ & \text{if } j_1 \ldots j_k \text{ is an odd permutation of } i_1 \ldots i_k. \\ 0 & \text{in all other cases.} \end{cases}$$

Combining the e-systems with the generalized Kronecker delta, it follows that:

i) $\qquad \delta^{i_1 \cdots i_n}_{j_1 \cdots j_n} = e^{i_1 \cdots i_n} e_{j_1 \cdots j_n}$

ii) $\qquad \delta^{i_1 \cdots i_n}_{12 \ldots n} = e^{i_1 \cdots i_n}$

iii) $\qquad \delta^{12 \ldots n}_{j_1 \cdots j_n} = e_{j_1 \cdots j_n}$

The student should attempt to verify the validity of the above expressions. These expressions are considered as definitions of the ϵ - δ relationship.

An important property of generalized Kronecker delta is expressed by:

$$\delta_{j_1 \cdots j_r}^{i_1 \cdots i_r} = \frac{(n-k)!}{(n-r)!} \, \delta_{j_1 \cdots j_r, \, i_{r+1} \cdots i_k}^{i_1 \cdots i_r, \, i_{r+1} \cdots i_k} \quad ; \quad r \le k \le n,$$

where the summation convention applies, in contrast to the Cartesian tensor convention, over repeated raised (*contravariant*) and lowered (covariant) indices.

EXAMPLES

The student should be able to verify the following relations:

i)
$$e^{i_1 \cdots i_r \, i_{r+1} \cdots i_n} \, e_{j_1 \cdots j_r \, i_{r+1} \cdots i_n} = \delta_{j_1 \cdots j_r \, i_{r+1} \cdots i_n}^{i_1 \cdots i_r \, i_{r+1} \cdots i_n}$$
$$= (n-r)! \, \delta_{j_1 \cdots j_r}^{i_1 \cdots i_r}$$

ii)
$$\delta_{i_1 \cdots i_r}^{i_1 \cdots i_r} = \frac{n!}{(n-r)!}$$

iii) If $n = 3$

a) $\delta_{lmn}^{ijk} = e^{ijk} \, e_{lmn}$

b) $\delta_{lm}^{ij} = \dfrac{(3-3)!}{(3-2)!} \, \delta_{lmk}^{ijk} = \delta_{lmk}^{ijk}$

c) $\delta_{l}^{i} = \dfrac{(3-3)!}{(3-1)!} \, \delta_{ljk}^{ijk} = \dfrac{1}{2} \, \delta_{ljk}^{ijk}$

d) $\delta_{ij}^{ij} = \dfrac{3!}{(3-2)!} = 6$

e) $\delta_{ijk}^{ijk} = \dfrac{3!}{(3-3)!} = 6$

and   f) $\delta_{i}^{i} = \dfrac{3!}{(3-1)!} = 3$

It has now become apparent that δ_j^i corresponds to the familiar Cartesian Kronecker delta and has the familiar subtitution properties, e. g.,

$$\delta_s^r \, A_{.ij}^s \;=\; A_{.ij}^r$$
$$\text{or:} \qquad \delta_s^r \, A_{rij} \;=\; A_{sij}$$

We shall return back to this point at a later stage.

A.2.2. APPLICATION TO DETERMINANTS

As an instructive and useful exercise in the manipulation of e-systems and generalized Kronecker deltas, let us apply them to simple operations between determinants of order n. The nth order determinant:

$$a = |\, a_j^{\; i}\, | = \begin{vmatrix} a_1^{\; 1} & \cdots & a_n^{\; 1} \\ \vdots & & \\ a_1^{\; n} & \cdots & a_n^{\; n} \end{vmatrix}$$

may be shown by an inductive argument to be equivalent to either:

$$a = e^{\,i_1 \cdots i_n}\; a_{i_1}^{\; 1}\; a_{i_2}^{\; 2} \cdots a_{i_n}^{\; n}$$
$$\text{or} \qquad a = e_{\,i_1 \cdots i_n}\; a_1^{\; i_1}\; a_2^{\; i_2} \cdots a_n^{\; i_n}$$

For example, if n = 2 then:

$$\text{i)} \quad a = |\, a_\beta^\alpha\, | = \begin{vmatrix} a_1^{\; 1} & a_2^{\; 1} \\ a_1^{\; 2} & a_2^{\; 2} \end{vmatrix} = a_1^{\; 1}\, a_2^{\; 2} - a_2^{\; 1}\, a_1^{\; 2}$$

$$\text{ii)} \quad a = e^{\,\alpha\beta}\, a_\alpha^{\; 1}\, a_\beta^{\; 2} = a_1^{\; 1}\, a_2^{\; 2} - a_2^{\; 1}\, a_1^{\; 2}$$

$$\text{iii)} \quad a = e_{\,\alpha\beta}\, a_1^{\; \alpha}\, a_2^{\; \beta} = a_1^{\; 1}\, a_2^{\; 2} - a_1^{\; 2}\, a_2^{\; 1}$$

A further useful identity, which is proved in essentially the same manner as that illustrated for

Cartesian tensors, is either,

$$e^{i_1 \cdots i_n} a = e^{j_1 \cdots j_n} a_{j_1}^{i_1} a_{j_2}^{i_2} \cdots a_{j_n}^{i_n}$$

or:
$$e_{i_1 \cdots i_n} a = e_{j_1 \cdots j_n} a_{i_1}^{j_1} a_{i_2}^{j_2} \cdots a_{i_n}^{j_n}$$

i) Determinant Multiplication

$$ab = a\, e^{i_1 \cdots i_n} b_{i_1}^{1} \cdots b_{i_n}^{n}$$
$$= e^{j_1 \cdots j_n} a_{j_1}^{i_1} b_{i_1}^{1} \cdots a_{j_n}^{i_n} b_{i_n}^{n}$$
$$= e^{j_1 \cdots j_n} c_{j_1}^{1} \cdots c_{j_n}^{n} = c$$

where

$$c_{j_1}^{1} = b_{i_1}^{1} a_{j_1}^{i_1} \quad \text{etc.}$$

ii) Cofactors

Let A_i^j be the cofactor of the element a_j^i (note order of indices) in the determinant $a = |\, a_j^i \,|$. Carrying through the analysis for one element then:

$$a = e_{i_1 \cdots i_n} a_1^{i_1} a_2^{i_2} \cdots a_n^{i_n}$$
$$= a_1^{i_1} e_{i_1 \cdots i_n} a_2^{i_2} \cdots a_n^{i_n}$$
$$= a_1^{i_1} A_{i_1}^{1} \qquad\qquad \text{where,} \qquad\qquad A_{i_1}^{1} = e_{i_1 \cdots i_n} a_2^{i_2} \cdots a_n^{i_n}$$

Now,
$$\delta_{1 j_2 \cdots j_n}^{1 j_2 \cdots j_n}$$

$$A_{i_1}^{1} = e_{i_1 \cdots i_n} a_2^{i_2} \cdots a_n^{i_n}$$
$$A_{i_1}^{1} = e^{1 j_2 \cdots j_n} e_{1 j_2 \cdots j_n} A_{i_1}^{1}$$
$$= e^{1 j_2 \cdots j_n} e_{i_1 i_2 \cdots i_n} a_{j_2}^{i_2} a_{j_3}^{i_3} \cdots a_{j_n}^{i_n}$$

Since we can consider $A_{i_1}^{1}$ to be an n-1 order determinant, then

$$(n-1)! \; A_{i_1}^1 = \delta_{i_1 i_1 \cdots i_n}^{1 j_2 \cdots j_n} \; a_{j_2}^{i_2} \; a_{j_3}^{i_3} \; \cdots \; a_{j_n}^{i_n}$$

which, when considering an arbitrary element in **a**, generalizes to

$$(n-1)! \; A_{i_1}^{j_1} = \delta_{i_1 i_1 \cdots i_n}^{1 j_2 \cdots j_n} \; a_{j_2}^{i_2} \; a_{j_3}^{i_3} \; \cdots \; a_{j_n}^{i_n}$$

iii) Show that $a_k^m A_m^{\ell} = a \, \delta_k^{\ell}$

$$
\begin{aligned}
a_k^m A_m^{\ell} &= a_k^m \frac{1}{2!} \delta_{mrs}^{\ell pq} a_p^r a_q^s \\
&= \frac{1}{2} e^{\ell pq} e_{mrs} a_k^m a_p^r a_q^s \\
&= \frac{1}{2} e^{\ell pq} e_{kpq} \, a \\
&= \frac{1}{2} \delta_{kpq}^{\ell pq} \, a \\
&= \frac{1}{2} \cdot 2 \delta_k^{\ell} \, a \qquad\qquad = \quad a \, \delta_k^{\ell}
\end{aligned}
$$

In a similar manner it can be shown that:

$$a_m^k A_{\ell}^m = a \, \delta_{\ell}^k$$

Now, if we have other forms of tensor, for example, a completely covariant or a completely contravariant one, then, we may define, in a manner similar to above:

i)   if A_{ji} is the cofactor of the term a_{ij} in **a** then $a_{km} A_{m\ell} = a \, \delta_{k\ell}$

ii)   if A^{ji} is the cofactor of the term a_{ij} in **a** then $a_{km} A^{m\ell} = a \, \delta_k^{\ell}$

iii)   if A^{ij} is the cofactor of the term a_{ij} in **a** then $a_{km} A^{\ell m} = a \, \delta_k^{\ell}$

As is well known, the concept of the cofactor is of importance in solving linear equations. For example, suppose we have three equations in the unknowns x^i in the form:

$$a_{ij} \, x^j = b_i$$

then, defining the cofactor in the form of (ii) above we may solve these three equations by noting that:

thus,
$$A^{ki} a_{ij} x^j = A^{ki} b_i$$
$$a \, \delta_j^k x^j = A^{ki} b^i$$

i.e.,
$$x^k = \frac{A^{ki}}{a} b_i$$

which is referred to as the *"Cramer's rule"*.

As another example, let us transform the variables x^i into a new set of variables y^i by the linear transformation:

$$y^i = \alpha_j^i x^j$$

If $\alpha = |\, \alpha_j^i \,|$ is the determinant of the transformation, then, it is singular if $\alpha = 0$, non-singular if $\alpha \neq 0$. Hence:

$$x^k = \frac{A_i^k}{\alpha} y^i$$

where A_i^k is the cofactor of the term α_k^i in α.

iii)   General Transformations

Rather than developing the entire subject, we just rather quote certain well known results from vector analysis which will be of use to us at a later date.

Consider the general transformation:

$$y^{i_1} = y^{i_1}(x^{j_1}) \quad ; \quad i_1, j_1 = 1, \ldots, n,$$

then, the Jocobian of the transformation is naturally given by:

$$J = \left| \frac{\partial y^{i_1}}{\partial x^{j_1}} \right| = \begin{vmatrix} \dfrac{\partial y^1}{\partial x^1} & \cdots & \dfrac{\partial y^1}{\partial x^n} \\ \vdots & & \\ \dfrac{\partial y^n}{\partial x^1} & \cdots & \dfrac{\partial y^n}{\partial x^n} \end{vmatrix}$$

a)	A necessary and sufficient condition that the y^{i_1} be independent is for $J \neq 0$.

b)	A necessary and sufficient condition that to the set y^{i_1} there corresponds a unique set x^{i_1} is for $J \neq 0$.

c)	If $y^{i_1} = y^{i_1}(x^{j_1})$ and $z^{k_1} = z^{k_1}(y^{i_1})$ then:

$$\left| \frac{\partial z^{k_1}}{\partial x^{j_1}} \right| = \left| \frac{\partial z^{k_1}}{\partial y^{i_1}} \right| \left| \frac{\partial y^{i_1}}{\partial x^{j_1}} \right|$$

A.2.3. CONTRAVARIANT, COVARIANT AND MIXED COMPONENTS OF A TENSOR

Definition.
A set of quantities U^{i_1}, associated with a point P in n-dimensional space is said to be the *contravariant* components of a vector U if they transform, on a change of coordinates, according to the transformation law:

$$U^{i_1'} = \frac{\partial x^{i_1'}}{\partial x^{i_2}} U^{i_2} \qquad\qquad i_1, i_2 = 1, \dots, n.$$

Note that the index on x is almost invariably a *contravariant* one. The set of quantities $U^{i_1 i_2}$, again associated with the point P, are said to be *contravariant* components of a 2ⁿᵈ order tensor if they transform according to:

$$U^{i_1' i_2'} = \frac{\partial x^{i_1'}}{\partial x^{j_1}} \frac{\partial x^{i_2'}}{\partial x^{j_2}} U^{j_1 j_2}$$

The appropriate definitions of tensors of third, fourth or higher orders are immediately suggested by the above. Naturally, a scalar or invariant transforms according to the law:

$$\vartheta' = \vartheta$$

Definition
A set of quantities U_{j_1}, again associated with the point P in n-dimensional space, is said to be the covariant components of a vector U if they transform, on a change of coordinates, according to the transformation law:

$$U_{j_1}' = \frac{\partial x^{j_2}}{\partial x^{j_1}} \, U_{j_2} \, , \qquad j_1, j_2 = 1, \dots, n.$$

where the partial derivatives are evaluated at P. The set of quantities $U_{j_1 j_2}$ are said to form the *covariant* components of a 2nd order tensor if they transform according to the equation:

$$U_{j_1 j_2}' = \frac{\partial x^{i_1}}{\partial x^{j_1}} \frac{\partial x^{i_2}}{\partial x^{j_2}} \, U_{i_1 i_2}$$

Higher order covariant tensor definitions are immediately suggested.

> *It is a well established convention that indices indicating contravariant character are placed as superscripts while those indicating covariant character are placed as subscripts.*

Definition

Having set down the definitions of contravariant and covariant tensors, definitions of mixed tensors immediately follow. For example, suppose a set of quantities $U_{j_1 j_2}^{i_1}$ transform according to:

$$U_{j_1 j_2}^{i_1} = \frac{\partial x^{i_1}}{\partial x^{k_1}} \frac{\partial x^{\ell_1}}{\partial x^{j_1}} \frac{\partial x^{\ell_2}}{\partial x^{j_2}} \, U_{\ell_1 \ell_2}^{k_1}$$

we would say that the quantities are the components of a mixed tensor of third order with one contravariant and two covariant indices. The Kronecker delta, δ_s^r, is an example of a mixed second order tensor since:

$$\delta_s^r = \frac{\partial x^r}{\partial x^s} = \frac{\partial x^r}{\partial x^k} \frac{\partial x^k}{\partial x^s} = \frac{\partial x^r}{\partial x^k} \frac{\partial x^\ell}{\partial x^s} \, \delta_\ell^k$$

A.2.4. ADDITION, MULTIPLICATION AND CONTRACTION OF TENSORS

Definition

Two tensors of the same kind and order may be <u>added</u> together to form a third tensor of the same kind and order. For example.

$$A_{jk}^{i} + B_{jk}^{i} = C_{jk}^{i}$$

From any second or higher order tensor one can construct two tensors, one of which is symmetric while the other is skew-symmetric in two of the indices, e.g.: let $A_{k\ell}^{ij}$ be a given tensor, then:

$$A_{k\ell}^{(ij)} = \frac{1}{2}\left(A_{k\ell}^{ij} + A_{k\ell}^{ji}\right) , \quad \text{which is unique,}$$

while

$$A_{k\ell}^{[ij]} = \frac{1}{2}\left(A_{k\ell}^{ij} - A_{k\ell}^{ji}\right) , \quad \text{which is again unique.}$$

Hence:

$$A_{k\ell}^{ij} = A_{k\ell}^{(ij)} + A_{k\ell}^{[ij]}$$

Definition
Multiplication. The multiplication of two tensors of the same range but of any order is also a tensor. For example, if A_{ij} and $B_{\ell m}^{k}$ are tensors then so is:

$$C_{ij\ell m}^{k} = A_{ij} B_{\ell m}^{k}$$

Definition
Contraction. Given any mixed tensor, a process of contraction is accomplished by identifying a contravariant index with a covariant one. For example, one may obtain:

$$C_{ik\ \ell m}^{\quad k} = A_{ik} B_{\ell m}^{k} = D_{i\ell m}$$

by contracting on the indices j and k of the previous example. Note that not only does $D_{i\ell m}$ obey a third order covariant tensor transformation low but also that contraction on a contravariant and covariant component reduces its order by two.

Definition
Inner Product. The inner product is obtained by applying the process of contraction to any multiplication of two tensors, e.g., the $D_{j\ell m}$ as defined above could alternatively have been

defined by $D'_{j\ell m} = C_{kj}{}^{k}{}_{\ell m} = A_{kj}\, B_{\ell m}^{k}.$

A.2.5. AUXILIARY TESTS FOR TENSOR CHARACTER

In a manner similar to that indicated in the discussion of Cartesian tensors, a series of tests, frequently visual, exist for testing the tensor character of a system. For example, if X^i, Y^i and B^i are first order contravariant tensors, φ is a scalar and $A_{\bar{i}}$, $A_{\bar{i}\bar{j}}$, $A_{\bar{i}}^j$ and $A_{(\bar{i}\bar{j})}$ are quantities whose tensor characters are in doubt, then they are tensors if:

$$i) \quad A_{\bar{i}}\, X^i = \varphi,$$
$$ii) \quad A_{\bar{i}\bar{j}}\, X^i Y^j = \varphi,$$
$$iii) \quad A_{\bar{i}}^j\, X^i = B^j,$$
and
$$iv) \quad A_{(\bar{i}\bar{j})}\, X^i X^j = \varphi.$$

There are many more tests for tensor character which may be conjectured by extending the above concepts.

A.3. Differential Geometry

In this section, curvilinear coordinates are introduced along with an illustration of base vectors and metric tensors. We will then discuss associative and relative tensors before bringing the section to a close with a description of physical components. There are various notations for the quantities about to be introduced but we will use a notation which is extremely common in the discussion of continuum mechanics, e.g., C. Truesdell and R. Toupin (1960).

A.3.1. THE METRIC TENSOR

Let z^i, $I = 1, 2, 3$, be the set of rectangular Cartesian coordinates which describes the position of an arbitrary point in a continuum and let x^i be the set of curvilinear coordinates which also describes the same point. Hence,

$$x^i = x^i(z^1, z^2, z^3)$$

and upon assuming that $J = \left| \dfrac{\partial x^i}{\partial z^j} \right| \neq 0,$ then $z^i = z^i(x^1, x^2, x^3).$

In the Cartesian coordinate space let us now consider two neighbouring points, P and Q with the coordinates:

$$P = \{z^i\} \quad ; \quad Q = \{z^i + \Delta z^i\}$$

Hence,
$$\vec{PQ} = \{\Delta z^i\}$$

and as $Q \rightarrow P$ the square of the linear element, namely:

$$|\vec{PQ}|^2 = \Delta z^i \Delta z^i$$

becomes, in the limit,

$$ds^2 = dz^i dz^i$$

Turning our attention to the curvilinear coordinate system x^i, the same square of the linear element relates to the above expression through the relation:

$$ds^2 = \frac{\partial z^i}{\partial x^k} \frac{\partial z^i}{\partial x^l} dx^k dx^l$$

then,
$$ds^2 = g_{kl} dx^k dx^l$$

where
$$g_{kl} = \frac{\partial z^i}{\partial x^k} \frac{\partial z^i}{\partial x^l}$$

In as much as the square of the linear element is a scalar we can immediately conclude from the above that the set of point functions, g_{kl}, represents a symmetric second order covariant tensor. This tensor, as described in the above expression, is called the <u>Metric Tensor</u> since, as will be shown, all essential metric properties of the space are completely determined by this tensor. At this point, it is worth noting that $g_{ij} \neq \delta_{ij}$ exept in the case where x^i represents another Cartesian coordinate frame.

A.3.2. THE RECIPROCAL METRIC TENSOR

If we define:

i) $g = |g_{ij}|$,

ii) G^{ij} to be the cofactor of the term g_{ij} in g,

iii) $g^{ij} = \dfrac{G^{ij}}{g}$

then g^{ij} is known as the *"reciprocal metric tensor"*. From the previous chapter's deliberations we note the following important relationships between the metric tensor and the reciprocal metric tensor:

i) $\qquad g_{im} G^{mj} = g \delta_i^j,$

ii) $\qquad g_{im} g^{mj} = g_{mi} g^{jm} g_{im} g^{jm} = g_{mi} g^{mj} = \delta_i^j$

Any of the relations in (ii) above may be taken as a definition of the reciprocal metric tensor. Furthermore, by simple observations, g^{ij} is a second order symmetric contravariant tensor.

A.3.3. GEOMETRIC FOUNDATIONS

In the alternative coordinate systems z^i and x^i, the position vector P may be represented by either:

$$P = z^i e_i = x^i g_i$$

where the g_i, also termed "base vectors" but *not "unit base vectors"*, are position dependent. This basic fact is the sole reason why curvilinear tensor analysis is far more complicated than Cartesian tensor analysis. These base vectors are defined by:

$$g_i = \frac{\partial z^j}{\partial x^i} e_j \equiv \begin{cases} g_1 = \cos x^2 e_1 + \sin x^2 e_2 \\ g_2 = -x^1 \sin x^2 e_1 + x^1 \cos x^2 e_2 \\ g_3 = e_3 \end{cases}$$

Naturally, these curvilinear base vectors can be related to the metric tensor of 1.9.1. Let us consider

$$d P = \frac{\partial P}{\partial x^i} d x^i = d x^i g_i$$

with this interpretation, the square of the of the linear element used in defining the metric tensor becomes:

$$d s^2 = d P . d P$$
$$g_{kl} d x^k d x^l = g_k . g_l d x^k d x^l$$
$$\therefore g_{kl} = g_k . g_l$$

Furthermore, *"reciprocal base vectors"* are defined solely through the relation:

$$g^i \cdot g_j = \delta^i_j$$

Note, however, that since it has already been shown that:

$$g^{im} g_{mj} = g^{im} g^m \cdot g_j = \delta^j_i$$

We will return to this property of the metric tensor to "raise" or to "lower" indices at a later date. In the meantime it follows that:

$$\begin{aligned}
g^i \cdot g^j &= g^{im} g_m \cdot g^{jn} g_n \\
&= g^{im} g^{jn} g_{mn} \\
&= g^{im} \delta^j_m \\
&= g^{ij}
\end{aligned}$$

and in this way, the reciprocal base vectors are related to the reciprocal metric tensor has already been stated that the base and hence reciprocal base vectors are not unit base vectors.

Their magnitudes are given by:

$$\|g_i\| = \left(g_{\bar{i}\bar{i}}\right)^{1/2} = \left(g_{\bar{i}} \cdot g_{\bar{i}}\right)^{1/2} ; i = 1, 2, 3$$

and

$$\|g^j\| = \left(g^{\bar{j}\bar{j}}\right)^{1/2} = \left(g^{\bar{j}} \cdot g^{\bar{j}}\right)^{1/2} ; j = 1, 2, 3$$

where the bar "−" over repeated indices implies the negation of the summation convention rather than doubt of tensor character.

A.3.4. ASSOCIATED TENSORS

By this time it will have been noticed that any arbitrary vector, say **U**, may be expressed in terms of either the base vector system g_i or the reciprocal base vector system g^i, depending on the requirements, by:

$$U = U^i \, g_i = U_i \, g^i$$

where the contravariant components of **U**, namely U^i, have different values than the covariant components U_i due to the different magnitudes of g_i and g^i. Luckily, however, there is a simple relation between the two since:

$$U^i g_i = U_j g^j$$
$$= U_j g^{ij} g_i$$
and, then,
$$U^i = g^{ij} U_j$$

In a similar manner it may be shown that:

$$U_i = g_{ij} U^j$$

Hence, we have raised and lowered the indices of U, respectively. Hence U_i and U^j are called associated vectors. Naturally, this property of space metrics may be generalized to define associated tensors of any order.

For example,

$$A^{\cdot i}_{jk} = g^{im} g_{jn} g_{kl} A^{\cdot nl}_{m}$$

A.3.5. RELATIVE TENSORS

The concept of relative tensors is perhaps best introduced by considering some important examples before discussing relative tensors in general.

How are the determinants of the metric tensors of two different curvilinear coordinate systems related?

$$g^i = \left| g'_{ij} \right| = \left| \frac{\partial x^h}{\partial x^{i'}} \frac{\partial x^l}{\partial x^{j'}} g_{kl} \right|$$

$$= \left| \frac{\partial x^k}{\partial x^{i'}} \right| \left| \frac{\partial x^l}{\partial x^{j'}} \right| |g_{kl}|$$

$$g^1 = J^2 g$$

How are the permutation symbols pertaining to two different coordinate systems related?

a) $$J = \left| \frac{\partial x^k}{\partial x^{i'}} \right| = e_{ijk} \frac{\partial x^i}{\partial x^{1'}} \frac{\partial x^j}{\partial x^{2'}} \frac{\partial x^k}{\partial x^{3'}}$$

then, $\quad e_{rst}\, J = e_{ijk}\, \dfrac{\partial x^i}{\partial x^{r'}}\, \dfrac{\partial x^j}{\partial x^{s'}}\, \dfrac{\partial x^k}{\partial x^{t'}}$

thus, $\quad e_{rst} = J^{-1}\, e_{ijk}\, \dfrac{\partial x^i}{\partial x^{r'}}\, \dfrac{\partial x^j}{\partial x^{s'}}\, \dfrac{\partial x^k}{\partial x^{t'}}$

b) $\quad J = \left| \dfrac{\partial x^k}{\partial x^{i'}} \right| = e^{ijk'}\, \dfrac{\partial x^1}{\partial x^{i'}}\, \dfrac{\partial x^2}{\partial x^{j'}}\, \dfrac{\partial x^3}{\partial x^{k'}}$

then, $\quad e^{rst}\, J = e^{ijk'}\, \dfrac{\partial x^r}{\partial x^{i'}}\, \dfrac{\partial x^s}{\partial x^{j'}}\, \dfrac{\partial x^t}{\partial x^{k'}}$

thus, $\quad \dfrac{\partial x^{\ell}}{\partial x^r}\, \dfrac{\partial x^m}{\partial x^s}\, \dfrac{\partial x^n}{\partial x^t}\, e^{rst}\, J = e^{ijk'}\, \delta^{\ell}_i\, \delta^m_j\, \delta^n_k$

$\quad\quad\therefore \quad e^{\ell mn'} = J\, e^{rst}\, \dfrac{\partial x^{\ell}}{\partial x^r}\, \dfrac{\partial x^m}{\partial x^s}\, \dfrac{\partial x^r}{\partial x^t}$

Generalizing these two examples we say that a quantity τ is a *"relative invariant of weight M"* if it transforms according to the law:

$$\tau' = J^M \tau \;\; ; \;\; J = \left| \dfrac{\partial x^{\ell}}{\partial xj} \right|$$

A relative scalar of weight zero is called an *"absolute scalar"*. With reference to the above examples, **g** is a relative invariant of weight 2. Note in passing that since J=1 for transformations between two Cartesian frames this problem did not arise in the analysis of Cartesian tensors.

We secondly define a set of quantities $\tau^{i_1 \dots i_n}_{j_1 \dots j_m}$ as a *relative tensor of weight M* if it transforms according to the law:

$$\tau^{i_1 \dots i_n'}_{j_1 \dots j_m} = J^M \, \tau^{k_1 \dots k_n'}_{\ell_1 \dots \ell_m}\, \dfrac{\partial x^{i_1}}{\partial x^{k_1}} \cdots \dfrac{\partial x^{i_n'}}{\partial x^{kn}}\, \dfrac{\partial x^{\ell_1}}{\partial x^{j_1'}} \cdots \dfrac{\partial x^{\ell m}}{\partial x^{jm'}} \;\; ; \;\; J = \left| \dfrac{\partial x^{i_1}}{\partial x^{j_1'}} \right|$$

A relative tensor $\tau^{\cdots}$ of weight zero is an *"absolute tensor"*. Again with reference to our earlier presentation, e_{rst} is a relative tensor of weight -1, while $e^{\ell mn}$ is a relative tensor of weight +1. Furthermore, all Kronecker deltas are absolute tensors.

A.3.6. THE "ε-SYSTEM"

The above relative nature of the e-systems is, as can be imagined, rather annoying since absolute tensors are far more tractable. Noting that g is a relative scalar of weight 2 we introduce absolute ε-systems as follows:

$$\epsilon^{ijk} = \frac{1}{\sqrt{g}} e^{ijk} \quad ; \quad \epsilon_{ijk} = \sqrt{g} \, e_{ijk}$$

A.4. Physical Components

Let us again consider our arbitrary vector U referred to the base vectors g_k and the reciprocal base vectors g^k through:

$$U = U^k g_k = U_k g^k$$

Furthermore, it will be recalled that the base vectors have magnitudes given by:

$$\|g_k\| = \left(g_{\bar{k}\bar{k}}\right)^{1/2} ; \|g^k\| = \left(g^{\bar{k}\bar{k}}\right)^{1/2} \quad ; \text{ no sum on k}$$

Hence, it is possible to define unit base vectors in terms of the base vectors and their magnitudes via:

$$\hat{g}_k = \frac{1}{\left(g_{\bar{k}\bar{k}}\right)^{1/2}} g_k \quad ; \quad \hat{g}^k = \frac{1}{\left(g^{\bar{k}\bar{k}}\right)^{1/2}} g^k \quad ; \text{ no sum on k}$$

Corresponding to these unit vectors, the contravariant physical components of U may be derived as follows:

$$U = U^k g_k = U^k \left(g_{\bar{k}\bar{k}}\right)^{1/2} \hat{g}^k = \hat{U}^k \hat{g}_k$$

where,

$$\hat{U}^k = U^k \left(g_{\bar{k}\bar{k}}\right)^{1/2}$$

while the covariant physical components are

$$U = U_k \mathbf{g}^k = U_k \left(g^{\bar{k}\bar{k}}\right)^{1/2} \hat{\mathbf{g}}^k = \hat{U}_k \hat{\mathbf{g}}^k$$

where,

$$\hat{U}_k = U_k \left(g^{\bar{k}\bar{k}}\right)^{1/2}$$

It is to a certain extent unfortunate, from the point of view of confusion, that since:

$$g^{\bar{k}\bar{k}} = \frac{1}{g_{\bar{k}\bar{k}}}$$

the physical components are often defined by:

$$\hat{U}_k = \frac{1}{\sqrt{g_{\bar{k}\bar{k}}}} \, U_k \quad ; \quad \hat{U}^k = \frac{1}{\sqrt{g^{\bar{k}\bar{k}}}} \, U^k$$

The physical components of second and higher order tensors may be obtained by generalizing the above concepts. For example, the mixed second order tensor $A_t^{\;k}$ is related to its physical components $A_t^{\;k}$ by the relation:

$$\hat{A}^k{}_\ell = \left(g_{\bar{k}\bar{k}} \; g^{\bar{i}\bar{i}}\right)^{1/2} A^k{}_\ell$$

A.5. Tensors Calculus

In this section, we give an account of tensor analysis which includes the concepts of covariant differentiation, differential operators, gradients, divergences and curls.

A.5.1. THE CHRISTOFFEL SYMBOLS; COVARIANT DIFFERENTIATION

Upon expressing our arbitrary vector U in terms of the base vectors $\mathbf{g}_k$ we see that:

$$U = U^k \mathbf{g}_k$$

and hence,

$$\frac{\partial U}{\partial x^i} = \frac{\partial U^k}{\partial x^i} \mathbf{g}_k + U^k \frac{\partial \mathbf{g}_k}{\partial x^i}$$

The new quantity in this expression is: $\dfrac{\partial \mathbf{g}_k}{\partial x^i}$. Let us analyze it. We already know that

$$g_k = \frac{\partial z^l}{\partial x^k}\, e_l \quad ; \quad e_m = \frac{\partial x^n}{\partial z^m}\, g_n$$

and hence,

$$\frac{\partial g_k}{\partial x^i} = \frac{\partial^2 z^l}{\partial x^i \partial x^k}\, \frac{\partial x^n}{\partial z^l}\, g_n$$

The motivation for replacing the e_l in the above expression is simply because it is nice to express the derivative of a base vector in terms of the existing base vectors. The notation:

$$\left\{ \begin{matrix} n \\ i\ k \end{matrix} \right\} = \frac{\partial^2 z^l}{\partial x^i \partial x^k}\, \frac{\partial x^n}{\partial z^l}$$

is known as the *"Christoffel symbol of the second kind"*. It is symmetric in the indices i and k and vanishes for Rectangular Cartesian Components. Continuing the above those equations results in:

$$\frac{\partial U}{\partial x^i} = \left(\frac{\partial U^n}{\partial x^i} + U^k \left\{ \begin{matrix} n \\ i\ k \end{matrix} \right\} \right) g_n \quad ; \quad \frac{\partial g_k}{\partial x^i} = \left\{ \begin{matrix} n \\ i\ k \end{matrix} \right\} g_n$$

Upon defining the *"covariant partial derivative of a contravariant component"* of a vector as:

$$U^n_{;i} = U^n_{,i} + \left\{ \begin{matrix} n \\ i\ k \end{matrix} \right\} U^k$$

then:
$$\frac{\partial U}{\partial x^i} = U^n_{;i}\, g_n$$

Before proceeding with a detailed discussion of these quantities, the corresponding relations when $U = U_k\, g^k$ turn out to be:

Again examining $\dfrac{\partial g^k}{\partial x^i}$ we first recall that

$$g^i \cdot g_j = \delta^i_j$$

then,
$$\frac{\partial g^i}{\partial x^k} \cdot g_j + \frac{\partial g_j}{\partial x^k} \cdot g^i = 0$$

i.e.,

$$\frac{\partial g^i}{\partial x^k} \cdot g_j = - \begin{Bmatrix} n \\ j\ k \end{Bmatrix} g_n \cdot g^i = - \begin{Bmatrix} i \\ j\ k \end{Bmatrix}$$

This relation suggests that we set

$$\frac{\partial g^i}{\partial x^k} = - \Gamma^i_{kn} g^n$$

where Γ^i_{kn} is to be evaluated. Dotting both sides with g_j we immediately see that from ( ):

$$- \begin{Bmatrix} i \\ j\ k \end{Bmatrix} = - \Gamma^i_{kn} g^n \cdot g_j = - \Gamma^i_{kj}$$

Hence,
$$\frac{\partial g^i}{\partial x^k} = \begin{Bmatrix} i \\ k\ n \end{Bmatrix} g^n$$

These concepts may again be generalized to cover higher order tensors. For example, for the <u>absolute</u> tensors:

$$\tau^{k\ell}_{;m} = \tau^{k\ell}_{,m} + \begin{Bmatrix} k \\ i\ m \end{Bmatrix} \tau^{i\ell} + \begin{Bmatrix} \ell \\ i\ m \end{Bmatrix} \tau^{ki}$$

$$\tau^k_{\ell;m} = \tau^k_{\ell,m} + \begin{Bmatrix} k \\ i\ m \end{Bmatrix} \tau^i_\ell - \begin{Bmatrix} i \\ \ell\ m \end{Bmatrix} \tau^k_i$$

$$\tau_{k\ell;m} = \tau_{k\ell,m} + \begin{Bmatrix} i \\ k\ m \end{Bmatrix} \tau_{i\ell} - \begin{Bmatrix} i \\ \ell\ m \end{Bmatrix} \tau_{kj}$$

etc.

It is left to the student to write down the required laws for relative tensors of weight M.

So far in our investigation we have always expressed the derivative of a base vector in terms of base vectors and the derivative of a reciprocal base vector in terms of reciprocal base vectors. Mixing this up a little we now define *"Christoffel symbols of the first kind"* by expressing the partial derivative of g_k in terms of the reciprocal base vectors g^k, namely

$$\frac{\partial g_k}{\partial x^i} = [ki, \ell] g^\ell$$

From our knowledge of the index lowering properties of $g_{j\ell}$ and upon comparison with the expansion already noted in terms of the base vectors, we see that:

$$[ki, \ell] = \left\{ {}_k{}^j{}_i \right\} g_{j\ell}$$

and again, $[k i, \ell]$ is symmetric in the indices k and i.

A.5.2. FAMILIAR VECTOR OPERATIONS

In order to indicate the usefulness of the Christoffel symbols and covariant differentiation, let us examine familiar vector operations in curvilinear coordinates.

The Gradient of a Scalar
Suppose that φ is an arbitrary scalar function of position. We may define the covariant vector, grad φ, by:

$$\text{grad } \varphi = \nabla \varphi = g^k \frac{\partial}{\partial x^k} \varphi = \varphi_{,k} \, g^k$$

The Divergence of a Vector
Suppose that our arbitrary vector U is a function of position. The invariant, div. U, is given by:

i)
$$\text{div } U = \nabla \cdot U = g^k \frac{\partial}{\partial x^k} \cdot U^\ell g_\ell$$
$$= g^k \frac{\partial}{\partial x^k} \left(U^\ell g_\ell \right)$$
$$= g^k \cdot U^\ell {}_{;k} g_\ell$$
$$= \delta^k_\ell \, U^\ell {}_{;k}$$
$$= U^k {}_{;k}$$

ii)
$$= g^k \frac{\partial}{\partial x^k} \cdot U_\ell g^\ell$$
$$= g^k \cdot \frac{\partial}{\partial x^k} \left(U_\ell g^\ell \right)$$
$$= g^k \cdot U_{\ell;k} g^\ell$$
$$= g^{k\ell} U_{\ell;k}$$

The Curl of a vector.

Again considering **U**, then the vector, curl **U**, may become, among other forms:

$$\text{Curl } \mathbf{U} = \nabla \times \mathbf{U} = \mathbf{g}^k \frac{\partial}{\partial x^k} \times U_\ell \mathbf{g}^\ell$$

$$= \mathbf{g}^k \times \mathbf{g}^\ell \; U_{\ell;k} = \epsilon^{k\ell m} U_{\ell;k} \mathbf{g}_m$$

The Laplacian Operator.

Returning to the arbitrary scalar function φ, then the scalar, Laplacian φ, is given by:

$$\nabla^2 \varphi = \text{div } (\text{grad } \varphi) = \nabla \cdot \nabla \varphi$$

$$= \mathbf{g}^k \frac{\partial}{\partial x^k} \cdot \varphi_{,\ell} \mathbf{g}^\ell$$

$$= \mathbf{g}^k \cdot \mathbf{g}^\ell (\varphi_{,\ell})_{;k} = g^{k\ell} (\varphi_{,\ell})_{;k}$$

It is now apparent that these expressions in their present form are of little use to us except for a convenient way of writing them. We therefore reexamine them in order to carry their development further.

Now, since we have the identity:

$$\frac{1}{2g} \frac{\partial g}{\partial x^i} = \frac{1}{2g} \frac{\partial g}{\partial \sqrt{g}} \frac{\partial \sqrt{g}}{\partial x^i} = \frac{2\sqrt{g}}{2g} \frac{\partial \sqrt{g}}{\partial x^i} = \frac{1}{\sqrt{g}} \frac{\partial \sqrt{g}}{\partial x^i}$$

then we also have the identity:

$$\frac{1}{\sqrt{g}} \frac{\partial}{\partial x^k} \left(\sqrt{g} \, U^k \right) = U^k{}_{,k} + \frac{1}{\sqrt{g}} \frac{\partial \sqrt{g}}{\partial x^k} U^k = U^k{}_{,k} + \frac{1}{2g} \frac{\partial g}{\partial x^\ell} U^\ell$$

Hence:

$$\text{div } \mathbf{U} = \frac{1}{\sqrt{g}} \frac{\partial}{\partial x^k} \left(\sqrt{g} \, U^k \right)$$

$$= \frac{1}{\sqrt{g}} \sum_{k=1}^{3} \frac{\partial}{\partial x^k} \left(\frac{\sqrt{g}}{\sqrt{g_{kk}}} \, \hat{U}^k \right)$$

$$\text{div } \mathbf{U} = \frac{1}{\sqrt{g}} \left[\frac{\partial}{\partial x^1} \left(\frac{\sqrt{g}}{\sqrt{g_{11}}} \, \hat{U}^1 \right) + \frac{\partial}{\partial x^2} \left(\frac{\sqrt{g}}{\sqrt{g_{22}}} \, \hat{U}^2 \right) + \frac{\partial}{\partial x^3} \left(\frac{\sqrt{g}}{\sqrt{g_{33}}} \, \hat{U}^3 \right) \right]$$

$$\text{Curl } \mathbf{U} = \varepsilon^{k\ell m} U_{\ell;k} \, \mathbf{g_m}$$

$$= \varepsilon^{k\ell m} U_{\ell;k} \, \mathbf{g_m} - \varepsilon^{k\ell m} \overset{\text{S. AS.}}{\underset{0}{\begin{Bmatrix} n \\ k \ell \end{Bmatrix}}} U_n \, \mathbf{g_m}$$

$$= \frac{e^{k\ell m}}{\sqrt{g}} U_{\ell,k} \, \mathbf{g_m} = \sum_{k\ell m = 1}^{3} e^{k\ell m} \frac{1}{\sqrt{g}} \frac{\partial}{\partial x^k} \left(\sqrt{g_{\bar{\imath}\bar{\imath}}} \, \hat{U}_\ell \right)$$

$$\sqrt{g_{\bar{m}\bar{m}}} \, \hat{\mathbf{g}}_m$$

then, $\text{Curl } \mathbf{U} \dfrac{1}{\sqrt{g}} \begin{vmatrix} \mathbf{g_1} & \mathbf{g_2} & \mathbf{g_3} \\ \dfrac{\partial}{\partial x^1} & \dfrac{\partial}{\partial x^2} & \dfrac{\partial}{\partial x^3} \\ U_1 & U_2 & U_3 \end{vmatrix}$

$$= \frac{1}{\sqrt{g}} \begin{vmatrix} \sqrt{g_{11}} \, \hat{\mathbf{g}}_1 & \sqrt{g_{22}} \, \hat{\mathbf{g}}_2 & \sqrt{g_{33}} \, \hat{\mathbf{g}}_3 \\ \dfrac{\partial}{\partial x^1} & \dfrac{\partial}{\partial x^2} & \dfrac{\partial}{\partial x^3} \\ \sqrt{g_{11}} \, \hat{U}_1 & \sqrt{g_{22}} \, \hat{U}_2 & \sqrt{g_{33}} \, \hat{U}_3 \end{vmatrix}$$

A.6. Problems

1. Show that if $A_{\bar{\imath}\bar{\jmath}} X^i Y^j$ is an invariant then $A_{\bar{\imath}\bar{\jmath}}$ is a second order covariant tensor if X^i and Y^j arbitrary contravariant vectors.

2. If numbered Latin indices have the range n, show that:

   (i) $\quad | \delta_{j_1}^{i_1} | = 1$

   (ii) $\quad a_{j_1}^{i_1} A_{j_1}^{i_1} = n \, a$

   (iii) $\quad e^{i_1 \cdots i_n} A_{i_1}^{j_1} = e^{j_1 \cdots j_n} a_{j_2}^{i_2} \cdots a_{j_n}^{i_n}$

   (iv) $\quad a = \dfrac{1}{n!} \delta_{j_1 j_2 \cdots j_n}^{i_1 i_2 \cdots i_n} a_{i_1}^{j_1} \cdots a_{i_n}^{j_n}$

   where $A_{j_1}^{i_1}$ is the cofactor of the $a_{i_1}^{j_1}$ term in a.

3. If in rectangular Cartesian coordinates, $z^i = (x, y, z)$ while in spherical coordinates, $x^i = (r, \theta, \varphi)$ and the transformation between the two takes the form:

$$z^1 = x^1 \sin x^2 \cos x^3 \; ; \; z^2 = x^1 \sin x^2 \cos x^3 \; ; \; z^3 = x^1 \cos x^2$$

determine:

      a) the metric g_{ij}

      b) the reciprocal metric g^{ij}

      c) g_i in terms of e_i

      d) g^i in terms of e_i

      e) the validity of i) $g^i \cdot g_j = \delta^i_j$ ii) $g^i \cdot g^j = g^{ji}$

and   f) the magnitudes of g_i and g^i.

4.   If A^{ji} is the cofactor of the a_{ij} term in $a = |\, a_{ij}\, |$, show that:

      i) $\dfrac{\partial a_{mn}}{\partial a_{ij}} = \delta^i_m \delta^j_n$

      ii) $\dfrac{\partial a}{\partial a_{mn}} = A^{nm}$

      iii) $\dfrac{\partial a}{\partial x^k} = A^{ij} \dfrac{\partial a_{ji}}{\partial x^k}$

5.   With reference to problem 3 above, find in sperical coordinates:

      i)   grad φ

      ii)   div U

      iii)   curl U

      iv) $\nabla^2 \varphi$

6.   Prove that the Christoffel symbols $\left\{ {i \atop j\ k} \right\}$ and $[\, ij,\, k]$ are tensors.

7. Show that:

i) $$\frac{\partial g}{\partial x^i} = 2g\begin{Bmatrix} m \\ i\ m \end{Bmatrix}$$

ii) $$\frac{\partial \log \sqrt{g}}{\partial x^i} = \begin{Bmatrix} m \\ i\ m \end{Bmatrix}$$

8. Prove that:

$$[ki,\ell] = \frac{1}{2}\left(\frac{\partial g_{k\ell}}{\partial x^i} + \frac{\partial g_{i\ell}}{\partial x^k} - \frac{\partial g_{ki}}{\partial x^\ell} \right)$$

9. Prove Ricci's theorems:

i) $g_{ij;k} = 0$

ii) $\delta^i_{j;k} = 0$

iii) $g^{ij}_{;k} = 0$

10. Prove that $g_{m\ell,j} = [mj,\ell] + [\ell j,m]$

A.7. Further Reading

Aris, R. (1962) *Vectors, Tensors and the Basic Equations of Fluid Mechanics*, Prentice-Hall, Englewood Cliffs, NJ.

Bishop, R.L. and Goldberg. S.I. (1968) *Tensor Analysis and Manifolds*, Dover Publications, New York.

Borg, S.F. (1963) *Matrix-Tensor Methods in Continuum Mechanics*, Van Nostrand, New York.

Brillouin, L. (1946) *Les Tenseurs en Mécanique et en Élasticité*, Dover Publications, New York.

Coburn, N. (1955) *Vector and Tensor Analysis*, Macmillan, New York.

Cotter, B.A. and Rivlin, R.S. (1955) *Tensors Associated with Time-Dependent Stress, Quart. Appl. Math.* 13, 177-82

Eisenhardt, L.P. (1926) *An Introduction to Differential Geometry*, Princeton University Press.

Eisenhart, L.P. (1949) *Riemannian Geometry*, Princeton University Press. Princeton, NJ.

Ericksen, J.L. (1960) Tensor fields, in *Encyclopedia of Physics*, Vol. 3/1 (ed. S. Flügge), Springer, Berlin, pp. 794-858.

Eringen, A.C. (1962) *Nonlinear Mechanics of Continua*, McGraw-Hill, New York.

Eringen, A. C. (1967) *Mechanics of Continua*, Wiley, New York.

Eringen, A. C. (1971) *Continuum Physics*, Vol. 1, Academic Press, New York.

Fung, Y.C. (1965) *Foundations of Solid Mechanics*, Prentice Hall, Englewood Cliffs, NJ.

Hay, G.E. (1953) *Vector and Tensor Analysis*, Dover Publications, New York.

Jefferys, H. (1931) *Cartesian Tensors*, Cambridge University Press, Cambridge.

Landau, L. and Lifshitz, E. (1951) *The Classical Theory of Fields* (translated from Russian by H. Hammermesh), Addison-Wesley, Reading, MA.

Lass, H. (1950) *Vector and Tensor Analysis*, McGraw-Hill, New York.

Levi-Civita, T. (1927) *The Absolute Differential Calculus* (translated from Italian by M. Long), Blackie, London.

Lodge, A.S. (1951) *On the Use of Convected Coordiante Systems in the Mechanics of Contineous Media*, *Proc. Cambridge Phil. Soc.* **47**, 575-84.

Malvern, L.E. (1969) *Introduction to the Mechanics of a Contineous Medium*, Prentice Hall, Englewood Cliffs, NJ.

McConnell, A.J. (1946) *Applications of the Absolute Differential Calculus*, Blackie, London.

Michal, A.D. (1947) *Matrix and Tensor Calculus with Applications to Mechanics, Elasticity and Aeronautics*, Wiley, New York.

Naghdi, P.M. and Wainwright, W.L. (1961) On the Time Derivative of Tensors in Mechanics of Continna, *Quart. Appl. Math.* **19**, Number 1, 95-119.

Nakada, O. (1960) Theory of Nonlinear responses, *J. Phys. Soc. Jpn*, **15**, 2280-8.

Ricci, G. and Levi-Civita, T. (1901) *Methodes du Calcul différentiel absolu et leurs applications*, *Math. Ann.* **54**, 125-201.

Schouten, J. (1951) *Tensor Analysis for Physicists*, Oxford University Press, New York.

Sokolnikoff, I. (1964) *Tensor Calculus*, Wiley, New York.

Spain, B. (1953) *Tensor Calculus*, Interscience, New York.

Spiegel, M.R. (1959) *Theory and Problems of Vector Analysis and an Introduction to Tensor Analysis*, Schaum, New York.

Synge, J. and Schild, A. (1949) *Tensor Calculus*, University of Toronto Press, Toronto.

Thomas, T.Y. (1955) Kinematically preferred coordinate system, *Proc. Natl. Acad. Sci.* **41**, 762-70.

Truesdell, C. (1953) The Physical Components of Vectors and Tensors, *Zeis angew. Math. U. Mech.* **33**, 345-56.

Truesdell, C. and Noll, W. (1965) The Nonlinear field theories of mechanics, in *Encyclopedia of Physics*, Vol. III/3 (ed. S. Flügge), Springer, Berlin.

Truesdell, C. and Toupin, R.A. (1960) Classical field theories, in *Handbuch der Physic*, Vol. III/1 (ed. S. Flügge), Springer, Berlin, pp. 226-90.

Wills, A.P. (1938) *Vector Analysis with an Introduction to Tensor Analysis*, Prentice-Hall, Englewood Cliffs, NJ.

APPENDIX B

DELTA AND STEP FUNCTIONS

B.1. The Delta Function $\delta(t)$

The delta function is defined by

$$\delta(t) = \lim_{a \to 0} \delta(t;a) \tag{1}$$

where a is a parameter of arbitrary positive value and the function $\delta(t;a)$ is given by

$$\delta(t;a) = \frac{1}{\pi}\int_0^\pi \exp(-apt)\cos pt \; dp$$
$$= \frac{a}{\pi(a^2+t^2)} \tag{2}$$

Thus, by combining (1) and (2), the delta function may be defined as

$$\delta(t) = \frac{1}{\pi} \lim \frac{a}{a^2+t^2}$$
$$= \lim_{a \to 0} \frac{1}{\pi}\int_0^\infty \exp(-ap)\cos pt \; dp \tag{3}$$

Following the above, it can be shown that the delta function $\delta(t)$ has the properties

$$\delta(t) = \begin{cases} \infty \; ; \; t=0 \\ 0 \; ; \; t\neq 0 \end{cases} \tag{4}$$

and

$$\int_{-\infty}^{\infty} \delta(t)dt = 1 \tag{5}$$

388

The dimension of the delta function is the reciprocal of the dimension of its argument. The delta function is considered as an even function. Further, an important property of the delta function is the so-called "*shifting property*". That is

$$\int_{-\infty}^{\infty} f(t)\ \delta(t)dt\ =\ f(0) \tag{6}$$

i.e., the operation of integrating over $f(t)\delta(t)$ shifts the function $f(t)$ to $f(0)$. Equation (6) would remain valid, if one changes the interval of integration from $-\infty \le t \le \infty$ to $0 \le u \le t$. That is

$$\int_{0}^{t} f(u)\delta(u)du\ =\ d(0) \tag{7}$$

Further, through a change of variable in (5), it follows that

$$\delta(t/c)\ =\ c\delta(t) \tag{8}$$

with the dimension c/t. The function $c\ \delta(t)$ appearing in eqn. (8) above, is often referred to as an impulse of strength c. The definition of the delta function can be extended to include the argument $(t-t')$. Thus, with reference to (4), one can write

$$\int_{-\infty}^{\infty} \delta(t-t')dt\ =\ 1 \tag{9}$$

with, in view of (5),

$$\delta(t-t')\ =\ \begin{cases} \infty;\ t=t' \\ 0;\ t \neq t' \end{cases} \tag{10}$$

The function $\delta(t-t')$ is known as the "*shifted*" delta function. In view of Eqn. (6), the function $\delta(t-t')$ is treated as an even function. That is

$$\delta(t-t')\ =\ \delta(t'-t) \tag{11}$$

The shifting property (7) can be also applied to $\delta(t-t')$, i.e.,

$$\int_{-\infty}^{\infty} f(t)\delta(t-t')dt = \int_{-\infty}^{\infty} f(t-t')\delta(t)dt$$

$$= f(t') \tag{12}$$

Further, in analogy to (7), one can write

$$\int_{0}^{t} f(u)\,\delta(t-u)du = \int_{0}^{t} f(t-u)\delta u\, du = f(t) \tag{13}$$

B.2. The Step "Heaviside" Function H(t)

Integrating the function $\delta(t;a)$ of (2) leads to a function $H(\infty;a)=1$ such that

$$H(t;a) = \frac{1}{2} + \frac{1}{\pi}\arctan\frac{t}{a} \tag{14}$$

In (14), as the value of the parameter a decreases, the value of the function $H(t;a)$ approaches a straight line at $H(t; a)=1$ for $0 \le t \le \infty$. The resulting function at the limit as $a \to 0$ is known as the "*unit step function*" and is given the notation $H(t)$. The latter is often referred to in the literature as the "*Heaviside*" unit step function.

Combining (3) and (14), it can be shown that

$$H(t) = \frac{1}{2} + \frac{1}{\pi}\lim_{a\to0}\int_{0}^{\infty} \exp(-ap)\frac{\sin pt}{p}dp \tag{15}$$

from which it is apparent that

$$\delta(t) = \frac{d\,H(t)}{dt} \tag{16}$$

i.e., the delta function $\delta(t)$ is the derivative of the unit step function $H(t)$. For t=0, the unit step function $H(t)$ is undefined unless one distinguishes between $t=0^-$ and $t=0^+$ as the last point of negative time and the first point of positive time, respectively (Flügge, 1975).

One can also demonstrate that the Heaviside unit step function H(t) is an odd function of its argument, i.e.,

$$H(t) = -H(-t) \tag{17}$$

By its definition, the unit step function H(t) can be used as a restrictive device to limit the values of a given function to its values for a particular range of the argument t.

The definition of the Heaviside function may be extended to include the argument (t-t') such that

$$H(t-t') = \begin{cases} 0; & t < t' \\ 1; & t > t' \end{cases} \tag{18}$$

B.3. References

Carslaw, H.S. and Jaeger, J.C. (1941) *Operational Methods in Applied Mathematics*, Oxford University Press, London.

Churchill, R.V. (1958) *Operational Methods*, McGraw-Hill, New York.

Flügge, W. (1975) *Viscoelasticity*, Springer-Verlag, New York.

Goldman, S. (1949) *Transformation Calculus and Electrical Transients*, Constable and Co., Ltd., London.

Tschoegl, N.W. (1989) *The Phenomenological Theory of Linear Viscoelastic Behaviour. An Introduction*, Springer-Verlag, New York.

APPENDIX C

INTEGRAL TRANSFORMS

C.1. Introduction

In this appendix, we introduce briefly the basic concepts and properties concerning the operation of integral transformation of a function F (t), in "*t-space*", into another function in "*s-space*". If we denote the integration operator by the letter I, then, the integral transform is expressed by

$$I\{F(t)\} = \int_a^b \Gamma(t,s) \, F(t) \; dt \; = \; f(s) \tag{1}$$

where Γ (t, s) denote some prescribed function of the variable t and a parameter s.

The function f (s) is called the "*integral transform*" of F(t). The class of function F(t) and the range of the parameter s are to be specified in a manner such that the integral (1) exists. Hence, the transformation I{F(t)} applies to all integral transform functions whereby the function f(s) may be interpreted as the "*image*" of the original function F(t) under the referred-to transformation.

In the above example, an inverse transformation exists in the sense that when the image function f(s) is given, a corresponding function F(t) exists which has this image. The inverse transformation of (1) is expressed as

$$F(t) = I^{-1}\{f(s)\}$$

$$\tag{2}$$

An integral transformation I{F(t)} is described as linear if for every pair of functions $F_1(t)$ and $F_2(t)$ and for each pair of constants a and b, the following relation is satisfied.

$$I\left\{a\,F_1(t) \, + \, b\,F_2(t)\right\} \; = \; a\,I\{F_1(t)\} \; + \; b\,I\{F_2(t)\}$$

$$\tag{3}$$

That is, in the case of linear transformation, the integral transform of a linear combination of two functions is the same linear combination of the transforms of these functions. Expression (3) above is referred to as the "*linearity property*".

Integral transforms have many physical applications. Linear integral transformations are particularly useful in solving problems in differential equations. In this context, with certain kernels $\Gamma(t, s)$, the transformation (1) when applied to prescribed linear differential forms in F(t), changes those forms into algebraic expressions in f(s) that would involve certain bounding values of the object function F(t). Hence, classes of problems in ordinary differential equations transform into much simpler algebraic problems to solve. Accordingly, if an inverse transformation is possible, the solution of the original problem can be determined.

Within the realm of the theory of linear integral transformation, two special classes of integral transforms are of particular importance; i.e., those of the operational mathematics of Laplace and Fourier transformations. In our presentation below, we deal first with the definition and basic properties of Laplace transform (Section C.2). This is followed, in Section C.3, by those pertaining to Fourier transformation.

C.2. Laplace Transform

Laplace transformation is a form of operational mathematics that is of significant importance in the treatment of problems concerning differential equations. Laplace (1749-1827) and Cauchy (1789-1857) were two of the earlier contributors to the development of Laplace transform and the pertaining operational calculus. For a comprehensive review of the subject matter, reference is made to Churchill (1958), Doetch (1974), among others. A list of references is given at the end of this Appendix.

If a function $F(t)$, defined for all positive values of the variable t, is multiplied by a kernel function e^{-st} of the variable t and a parameter s, and integrated with respect to t, an image function $\bar{f}(s)$ is expressed as

$$\int_0^\infty e^{-st} F(t)dt = \bar{f}(s)$$

$$\mathcal{L}\{F(t)\} = \int_0^\infty e^{-st} F(t)dt = \bar{f}(s) \tag{4}$$

We note in (4) that the integration is carried out over the infinite interval from $t = 0$ to $t = \infty$.

Expression (4) is known as the Laplace transformation of the original function F(t) into the image function $\bar{f}(s)$. This transformation is given here the notation $\mathcal{L}\{F(t)\}$. Hence, with reference to (4), the image function $\bar{f}(s)$ is referred to as the Laplace transform of the object function $F(t)$. Although the transform operator may be considered to be real, it generally assumes complex values.

Example C.1

Determine the Laplace transform of $F(t) = 1$ for $t > 0$.

$$\{F(t)\} = \mathcal{L}\{1\} = \int_0^\infty e^{-st} dt = -\frac{1}{s} e^{-st} \Big|_0^\infty$$

Thus, for $s > 0$;

$$\mathcal{L}\{1\} = \frac{1}{s}$$

Example C.2

Determine the Laplace transform of $F(t) = e^{bt}$ for $t > 0$; where b is a constant.

$$\mathcal{L}\{F(t)\} = \mathcal{L}\{\exp\{bt\}\} = \int_0^\infty e^{bt} e^{-st} dt$$

$$= \frac{1}{b-s} e^{-(s-b)t} \Big|_0^\infty$$

Thus, for $s > b$,

$$\mathcal{L}\{e^{bt}\} = \frac{1}{s-b}$$

Example C.3

Consider the Laplace transform of the unit step function.

$$H(t_1,t) = 0 \quad \text{for} \quad 0 < t < t_1$$
$$= 1 \quad \text{for} \quad t > t_1$$

$$\mathcal{L}\{H(t_1,t)\} = \int_t^\infty e^{-st} dt = -\frac{1}{s} e^{-st} \Big|_{t_1}^\infty$$

Thus, for $s > 0$;

$$\mathcal{L}\{H(t_1,t)\} = \frac{e^{-t_1 s}}{s}$$

Example C.4

The Laplace transforms of many other functions can be determined. For instance,

$$\mathcal{L}\{t^2\} = \frac{2}{s^3}$$

$$\mathcal{L}\{\sin at\} = \frac{a}{s^2 + b^2}$$

$$\mathcal{L}\{\cos bt\} = \frac{s}{s^2 + b^2}$$

C.2.1. EXISTENCE OF LAPLACE TRANSFORMATION

<u>Two</u> conditions need to be satisfied in order for the Laplace transform $\bar{f}(s)$ of $F(t)$ to exist:

(i) $F(t)$ be '*sectionally continuous*' in every finite interval for the variable $t \geq 0$, *and*

(ii) $F(t)$ be of '*exponential order*' as t tends to ∞.

The above conditions for the existence of the Laplace transform of a function are sufficient rather than necessary conditions. However, they are convenient for the majority of applications (e.g., Churchill, 1958).

A function $F(t)$ is considered to be sectionally continuous on a finite interval $0 \le t \le t_1$, if it is such that the interval $(0, t_1)$ can be subdivided into a finite number of intervals, in each of which $F(t)$ is continuous and has finite limits as t approaches either of such limits of the sub-interval from inside. The integral of every function of this class over the interval $(0, t_1)$ exists; it is the sum of the integrals of the pertaining continuous functions over the sub-intervals.

An example of a sectionally continuous function is the unit step function $H(\tau,t)$, where

$$H(\tau,t) = 0 \quad \text{for} \quad 0 < t < \tau$$
$$= 1 \quad \text{for} \quad t > \tau$$

It is a sectionally continuous function in the interval $0 \le t \le t_1$ for every positive number t_1.

The second condition for the existence of Laplace transform $\bar{f}(s)$ for a function $F(t)$ is that $F(t)$ be of exponential order as the variable t tends to infinity. In other words, $F(t)$ must not grow at a greater rate than that of exponential as $t \to \infty$. This can be expressed as

$$\lim_{t \to \infty} e^{-\alpha t} F(t) = 0 \tag{5}$$

under the condition that a constant α exists.

As an example of a function of exponential order is the function e^{2t}. It is of the order of $e^{\alpha t}$ as $t \to \infty$ for $\alpha \ge 2$. The unit step function $H(t_1, t)$, mentioned above, as well as the function t^n are also of the order $e^{\alpha t}$ as $t \to \infty$ for any positive α. It is further emphasized that not every function of s is a transform. The class of functions $\bar{f}(s)$ that are transforms is limited by several conditions concerning the continuity of $\bar{f}(s)$. Under these conditions, $\bar{f}(s)$ is continuous when $s > \alpha$ and that $\bar{f}(s)$ vanishes as s tends to infinity (e. g., Churchill, 1958).

C.2.2. TRANSFORM OF THE DERIVATIVE

A fundamental property of Laplace transformation is that concerned with the Laplace transformation of derivatives. This property enables us to replace the operation of differentiation of order n by a single algebraic operation on the transform. As a first step, we consider the Laplace transform of the first derivative of the original function $F(t)$. Let $F(t)$

be continuous with a sectionally continuous derivative F'(t), in every finite interval $0 \leq t \leq t_1$. The function F(t) is, further, assumed to be of exponential order as $t \to \infty$. Then,

$$\mathscr{L}\{F'(t)\} = \int_0^{\infty} e^{-st} F'(t)dt = e^{-st} F(t)\Big|_0^{\infty} + s\int_0^{\infty} e^{-st}F(t)dt$$

$$= e^{-st}F(t)\Big]_0^{\infty} + s\bar{f}(s)$$

Since $F(t)$ is of exponential order $e^{\alpha t}$, then, for s greater than α, the first derivative on the right hand side of the above expression becomes equal to $-F(0)$ and accordingly

$$\mathscr{L}\{F'(t)\} = s\,\bar{f}(s) - F(0) \tag{6}$$

To obtain the transformation of the derivative of the second order, consider $F'(t)$ be continuous and $F''(t)$ to be sectionally continuous in each finite interval. Also, let $F(t)$ and $F'(t)$ be of exponential order as t tends to ∞. Thus, for $s > \alpha$, it can be shown that

$$\mathscr{L}\{F''(t)\} = s^2\,\bar{f}(s) - sF(0) - F'(0) \tag{7}$$

The same procedure above can be applied to obtain the Laplace transformation of the nth derivative of $F(t)$. Let the function $F(t)$ and its first (n - 1) derivatives be continuous. Also, consider $F^n(t)$ to be sectionally continuous in every finite interval $0 \leq t \leq t_1$ and $F(t)$, $F'(t)$,..., $F^{n-1}(t)$ be of exponential order $e^{\alpha t}$ as the variable t tends to ∞. Accordingly, it can be shown, by mathematical induction, that the Laplace transfer of the nth derivative of F(t), for $s > \alpha$, is given by

$$\mathscr{L}\{F^n(t)\} = s^n f(s) - s^{n-1}F^1(0)$$
$$- ... - sF^{n-2}(0) - F^{(n-1)}(0) \tag{8a}$$

where

$$F^k(0) = \frac{d^k F(t)}{d\,t^k}\Big|_{t=0} \tag{8b}$$

Example C.5

Consider $\mathcal{L}\{t\}$.

The function $F(t) = t$ and its first derivative $F'(t) = 1$ are continuous and of exponential order $e^{\alpha t}$ for any positive α. Hence, with reference to (6),

$$\mathcal{L}\{F'(t)\} = s\,\mathcal{L}\{F(t)\} - F(0)$$
$$\mathcal{L}\{1\} = s\,\mathcal{L}\{t\}$$

But $\mathcal{L}\{1\} = 1/s$ (see Example 1), then,

$$\mathcal{L}\{t\} = 1/s^2$$

Example C.6

Determine $\mathcal{L}\{\sin(at)\}$.

The function $F(t) = \sin(at)$ and its derivatives are all continuous and of exponential order $e^{\alpha t}$ for $\alpha > 0$. Thus, in view of (7),

$$\mathcal{L}\{F''(t)\} = s^2\mathcal{L}\{F(t)\} - sF(0) - F'(0)\,;\,(s > 0)$$

Substitution for $F(t) = \sin(at)$ in the above expression, it follows that

$$-a^2\,\mathcal{L}\{\sin at\} = s^2\mathcal{L}\{\sin at\} - a$$

i.e.,

$$\mathcal{L}\{\sin at\} = \frac{a}{s^2 + a^2}\,;\,(s > 0)$$

C.2.3. INITIAL VALUE THEOREM

"If F(t) has a Laplace transform, say $\bar{f}(s)$, then the behaviour of F(t) in the neighbour-hood of t = 0 corresponds to the behaviour of s $\bar{f}(s)$ in the neighbourhood of s = 0".

The initial value F(0) of F(t) can be obtained from the transform $\bar{f}(s)$ through the relation

$$\underset{t \to 0}{\text{Lim}}\ F(t) = \underset{s \to 0}{\text{Lim}}\ s\ \bar{f}(s) \qquad (9)$$

This result is of particular interest as it may be generalized to obtain *F(t)* for small values of the variable *t*; If $\bar{f}(s)$ can be expanded in a power series of terms involving $(1/_s)^n$, $n \geq 1$, then, in view of the linearity property (3), a term by term inversion could be applied.

C.2.4. THE INVERSE TRANSFORM

If $\mathcal{L}\{F(t)\} = \bar{f}(s)$, then, the inverse Laplace transform is defined as

$$F(t) = \mathcal{L}^{-1}\ \{\bar{f}(s)\} \qquad (10)$$

That is *F(t)* is the inverse Laplace transform of *$\bar{f}(s)$*.

In the strict sense of the concept of uniqueness of functions, the inverse Laplace transform is not unique. A theorem due to Lerch (e.g., Carslaw and Jaeger, 1941) concerning the uniqueness of the inverse transform is of interest here. It states that if two functions $F_1(t)$ and $F_2(t)$ have the same Laplace transform $\bar{f}(s)$, then

$$F_2(t) = F_1(t) + \phi(t) \qquad (11)$$

where $\phi(t)$ is a null function. The latter is expressed by

$$\int_0^{t_1} \phi(t)dt = 0 \qquad (12)$$

for every positive t_1.

Hence, a given transform function $\bar{f}(s)$ cannot have more than one inverse transform F(t) that is continuous for each positive t. On the other hand, it is possible that a function $\bar{f}(s)$ would not have a continuous inverse transform.

C.2.5. SHIFTING THEOREM

If the inverse transformation of $\bar{f}(s)$ is $F(t)$, then, the inverse transformation of $e^{-as}\, \bar{f}(s)$ is given by

$$\mathcal{L}^{-1}\, e^{-as}\, \bar{f}(s) \;=\; F(t-a)\, H(t-a) \tag{13}$$

where $H(t-a)$ is the unit step function and a is a constant.

C.2.6. BASIC PROPERTIES OF LAPLACE TRANSFORM

Linearity
An important property of Laplace transform and its inverse is Linearity. The latter follows from the definition of the transform. Thus, recalling (3)

$$
\begin{aligned}
\mathcal{L}\{aF_1(t) + bF_2(t)\} &= a\;\mathcal{L}\{F_1(t)\} + b\;\mathcal{L}\{F_2(t)\} \\
&= a\;\bar{f}_1(s) + b\;\bar{f}_2(s)
\end{aligned}
\tag{14}
$$

where a and b are constants. That is, the Laplace transform of a linear combination of two functions is the linear combination of the transforms of these functions. It can also be demonstrated that the linearity property can be extended to linear combination of more than two functions. Such functions again must be sectionally continuous and each be of exponential order for their Laplace transforms to exist.

The linearity property applies also to the inverse of the transform. Thus, the inverse of (14) can be written as

$$
\begin{aligned}
\mathcal{L}^{-1}\{a\bar{f}_1(s) + b\bar{f}_2(s)\} &= a\;F_1(t) + b\;F_2(t) \\
&= a\;\mathcal{L}^{-1}\{\bar{f}_1(s)\} + b\;\mathcal{L}^{-1}\{\bar{f}_2(s)\}
\end{aligned}
\tag{15}
$$

Substitution (or Shift of Origin)
Let the object function $F(t)$, of exponential order $e^{\alpha t}$, be such that its Laplace integral converges when $s > \alpha$.

Recalling the Laplace transform expression (4), one may replace the argument of the transform $\bar{f}(s)$ by $(s-a)$ where a is a constant, then

$$\bar{f}(s-a) = \int_0^\infty e^{-(s-a)t} F(t)dt$$

$$= \int_0^\infty e^{-st} e^{at} F(t)dt$$

then, for $(s - a) > \alpha$, one has

$$\bar{f}(s-a) = \mathcal{L}\{e^{at} F(t)\} \tag{16}$$

That is, the substitution of $(s - a)$, where a is a constant, for the parameter s of $\bar{f}(s)$ would translate into multiplying the original function $F(t)$ by the function e^{at} as expressed in (16).

Example C.7

Consider the Laplace transform.

$$\mathcal{L}\{t^m\} = \frac{m!}{s^{m+1}}$$
$$(m = 1,2,...; \ s > 0)$$

Replacing the argument s by $(s - a)$, leads to

$$\frac{m!}{(s-a)^{m+1}} = \mathcal{L}\{t^m e^{at}\}; \ (s > 0)$$

Example C.8

Consider the following Laplace transform.

$$\mathcal{L}\{\cos bt\} = \frac{s}{s^2 + b^2}; \ (s > 0)$$

Replacing the parameter s by (s + a), then

$$\frac{(s + a)}{(s + a)^2 + b^2} = \mathcal{L}\{e^{-at} \cos bt\};(s > -a)$$

Change of scale
One may also replace the argument of the object function F(t) from *t* to *(at)* where *a* is a real positive constant, then,

$$\mathcal{L}\{F(at) = \frac{1}{a} \bar{f}\left(\frac{s}{a}\right) \tag{17}$$

Translation
Consider translation of the argument of the object function F(t) from t to (t - τ) where both t and τ are variables and F(t) = 0 for t < o, then,

$$\mathcal{L}\{F(t-\tau)\} = e^{-s\tau} \bar{f}(s) \tag{18}$$

Differentiation of Transforms
It can be shown that

$$\frac{d^n}{dt^n} \bar{f}(s) = \mathcal{L}\{(-1)^n t^n F(t)\} \tag{19}$$

Integration of Transforms
The following property can be proved

$$\int_0^\infty \bar{f}(s)\, ds = \int_0^\infty \frac{F(t)}{t}\, dt \tag{20}$$

Transform of Integral
Consider the integral

$$\int_0^t F(\tau)\, d\tau,$$

then,

$$\mathcal{L}\{\int_0^t F(\tau)\ d\tau\} = \frac{1}{s}\ \bar{f}(s)$$

and

$$\mathcal{L}\{\underbrace{\int_0^t d\tau \int_0^t d\tau \ ... \int_0^t F(\tau)\ d\tau}_{n}\} = \frac{1}{s^n}\bar{f}(s) \tag{21}$$

Transform of convolution integral

Consider two functions $F(t)$ and $G(t)$, both are sectionally continuous and of exponential order. The convolution of $F(t)$ and t $G(t)$ is defined by $\int F(t)G(t - \tau)dt$. It is conventionally denoted by $F(t) * G(t)$.o

The following properties of the convolution of functions $F(t)$, $G(t)$ and $J(t)$ defined over $-\infty < t < \infty$ can be verified:

Commutativity. $F(t) * G(t) = G(t) * F(t)$

Associativity. $F(t) * \{G(t) * J(t)\} = \{F(t) * G(t)\} * J(t)$
                              $= F(t) * G(t) * J(t)$

Distributivity. $F(t) * \{G(t) + J(t)\} = F(t) * G(t) + F(t) * J(t)$

Titchmarsh Theorem. $F(t) * G(t) = 0$ implies that $F(t) = 0$ or $G(t) = 0$

The Laplace transform of a convolution integral is expressed as

$$\mathcal{L}\{F(t) * G(t)\} = \bar{f}(s)\ \bar{g}(s) \tag{22}$$

with the property:

$$\bar{f}(s)\ \bar{g}(s) = \bar{g}(s)\ \bar{f}(s)$$

That is, the Laplace transform of the convolution integral is also commutative.

404

C.3. Problems

1. Show the validity of the following transformations, where a, b, and c are constants.

a) $\mathcal{L}\{a + bt\} = \dfrac{as + b}{s^2}$

b) $\mathcal{L}\{\sinh ct\} = \dfrac{c}{s^2 - c^2}$

c) $\mathcal{L}\{e^{at}\} = \dfrac{1}{s - an!}$

d) $t^n (n = 1, 2, \dots) = \dfrac{}{s^{n+1}}$

e) $d^{at} - e^{bt} (a>b) = \dfrac{a-b}{(s-a)(s-b)}$

f) $\dfrac{1}{a}\sin(at) - \dfrac{1}{b} - \sin(bt) = \dfrac{b^2 - a^2}{(s^2 + a^2)(s^2 + b^2)}$

g) $\cos(at) - \cos(bt) = \dfrac{(b^2 - a^2)s}{(s^2 + a^2)(s^2 + b^2)}$

2. Find the Laplace transform of each of the following functions.

a) $\sin(t) + 2\cos(t)$

b) $\cos^2 t$

c) $\sin(t)\cos(t)$

3. Find

$$\mathcal{L}^{-1}\left\{\dfrac{s+1}{s^2+2s}\right\}$$

$$\mathcal{L}^{-1}\left\{\dfrac{a^2}{s(s+a)^2}\right\}$$

4. Obtain the following inverse transforms where a and b are constants.

$$\mathcal{L}^{-1}\left\{\frac{a^2}{s(s+a)^2}\right\}$$

$$\mathcal{L}^{-1}\left\{\frac{b}{s(s+b)}\right\}$$

$$\mathcal{L}^{-1}\left\{\frac{b^3}{s(s+b)^3}\right\}$$

C.4. Fourier Transform

Similar to Laplace transformation, *Section B.2 above*, Fourier transform is a linear integral transformation with operational properties under which differential functions are converted into algebraic forms involving boundary values.

The Fourier transformation of a sectionally continuous function $F(t)$, defined for all positive values of the variable t, is denoted hereby $S\{F(t)\}$ and expressed by

$$S\{F(t)\} = \int_{-\infty}^{\infty} e^{-i\omega t}F(t)dt = f(\omega) \tag{23}$$

With reference to (23), the Fourier transform is based on the kernel function $e^{-i\omega t}$ and, hence, on this kernel's real and imaginary parts, i.e., $\cos \omega t$ and $\sin \omega t$ where ω is a constant parameter. Since such kernel function is often used to describe the propagation of waves in different media, Fourier transform is used extensively in such studies, e.g., for extraction of information from different phases of waves.

In equation (23), it is understood that $f(\omega)$ is the Fourier transform, or the image, of its object function $F(t)$; and that $F(t)$ is the inverse transform of $f(\omega)$. If t represents the time variable, for instance, then, equation (23) and its inverse imply that $F(t)$ can be analyzed into an integral sum of harmonic oscillations over a continuous range of frequencies. If $F(t)$ exists only for the range $t>0$, then, a special simplification may be possible in terms of the finite Fourier sine transform "$f_s(\omega)$" and the finite Fourier cosine transform "$f_c(\omega)$". These forms possess a complete symmetry and could be used in a situation where the variable t represents the time and $F(t)$ implies some stimulus applied to a particular system from zero time onwards. We shall introduce the Fourier sine and cosine transforms later in this section.

C.4.1. RELATIONS BETWEEN FOURIER PAIRS

(1) Linearity:
If $f_1(\omega)$ and $f_2(\omega)$ are the Fourier transforms of $F_1(t)$ and $F_2(t)$, respectively, and a and b are two arbitrary constants, then,

$$
\begin{aligned}
S\{a\,F_1(t) + b\,F_2(t)\} &= a\,S\{F_1(t)\} + b\,S\{F_2(t)\} \\
&= a\,f_1(\omega) + b\,f_2(\omega)
\end{aligned}
\tag{24}
$$

(2) Scaling:
If a is a real constant, then,

$$
S\{F(at)\} = \frac{1}{|a|}\, f\!\left(\frac{\omega}{a}\right)
\tag{25}
$$

(3) Symmetry:
If $F(t)$ is an even function, then, its Fourier transform $f(\omega)$ is also even. Also, if the object function $F(t)$ is an odd function, then, $f(\omega)$ is also odd.

(4) Shifting:
If the function $F(t)$ is shifted by a constant a, then, its Fourier spectrum remains the same, but, its phase angle is adjusted by a linear term $a\omega$:

$$
\begin{aligned}
S\{F(t \pm a)\} &= f(\omega)\, e^{\pm i\omega t} \\
&= A(\omega)\, e^{i(\phi(\omega)\pm a\omega)}
\end{aligned}
\tag{26}
$$

With ω_0 as a real constant, the Fourier integral of $e^{i\omega t}$ is obtained by shifting $f(\omega)$ by ω_0. That is

$$
S\{e^{i\omega_0 t}\, F(t)\} = f(\omega - \omega_0)
\tag{27}
$$

(5) Differentiation:
If $S\{F(t)\} = f(\omega)$, then,

$$
S\!\left\{\frac{d^{\,n}F(t)}{dt^{\,n}}\right\} = (i\omega)^n\, f(\omega)
\tag{28}
$$

and

$$S\{(-it)^n F(t)\} = \frac{d^n f(\omega)}{d\omega^n} \tag{29}$$

C.4.2. FINITE FOURIER SINE TRANSFORM

Consider a function $F(x)$ that is sectionally continuous and defined over the interval between $x=0$ and $x=\pi$. The Fourier sine transformation of $F(x)$ on that interval is denoted here by $S_n\{F(x)\}$ and is expressed as

$$S_n\{F(x)\} = \int_0^\pi F(x) \sin(nx)\,dx = f_s(n) \tag{30}$$

$$(n = 1,2,\dots)$$

where $f_s(n)$ is the finite sine transform. This transformation sets up a correspondence between functions $F(x)$ defined within the internal $0 < x < \pi$ and sequences of numbers $f_s(n)$ $(n=1,2,\dots)$. For example,

The function $F(x) = 1$ has the transform

$$f_s(n) = \int_0^\pi \sin(nx)\,dx = \frac{1-(-1)^n}{n}$$

$$(n = 1,2,\dots)$$

Also, the function $F(x)\ (0 < x < \pi)$ has the transform

$$f_s(n) = \int_0^\pi x \sin(nx)\,dx = \pi\frac{(-1)^{n+1}}{n}$$

$$(n=1,2,\dots)$$

In order to obtain an inversion formula for the transformation (30), consider both the object function $F(x)$ and its first derivative $F'(x)$ to be sectionally continuous functions. Let, also, $F(x)$ be defined at each point x_0 of discontinuity $0 < x_0 < \pi$, by its mean value.

$$F(x_0) = \frac{1}{2}\Big[F(x_0 + 0) + F(x_0 - 0)\Big]$$

$$(0 < x_0 < \pi)$$

(31)

It follows, according to the classical theory of Fourier series, that the Fourier sine series for F(x) converges to the function

$$(x) = \frac{2}{\pi} \sum_{n=1}^{\infty} \sin(nx) \int_0^{\pi} F(\xi) \sin(n\xi)\, d\xi$$

$$(0 < x < \pi)$$

(32)

Thus, with reference to (30),

$$F(x) = \frac{2}{\pi} \sum_{n=1}^{\infty} f_s(n) \sin(nx)$$

$$(0 < x < \pi)$$

(33)

which is the inversion formula for the Fourier-sine transformation.

In the class of sectionally continuous functions with sectionally continuous derivatives of the first order, there is only one function with a given transform as demonstrated by (33). In other words, the inverse transformation is unique. It is apparent that both the transformation $S_n\{F(x)\}$ and its inverse are linear transformations.

Recalling (30), the Fourier sine transform of a function on an interval $0 < x < c$ is expressed in terms of the transform on the standard interval $(0 < x < \pi)$ by the substitution $x'=\pi x/c$ in (30), i.e.,

$$\int_0^c F(x) \sin\left(\frac{n\pi x}{c}\right) dx = \frac{c}{\pi} \int_0^{\pi} F\left(\frac{cx'}{\pi}\right) \sin(nx')\, dx'$$

$$= \frac{c}{\pi} S_n\left\{F\left(\frac{cx}{\pi}\right)\right\}$$

(34)

As an example, the function $F(x)$ $(0 < x < c)$ has the sine transform

$$S_n\{x\}\Big|_c = \frac{c}{\pi} S_n\left\{\frac{cx}{\pi}\right\} = \frac{c^2}{\pi} \frac{(-1)^{n+1}}{n}$$

$$(n=1,2,...)$$

An important property of $S_n\{F(x)\}$

If $F(x)$ and $F'(x)$ are continuous and $F''(x)$ is sectionally continuous, then,

$$\int_0^\pi F''(x) \sin(nx)\, dx = F'(x) \sin nx\Big|_0^\pi - n \int_0^\pi F'(x) \cos(nx)\, dx$$

$$= -n \cos(nx) F(x)\Big|_0^\pi - n^2 \int_0^\pi F(x) \sin(nx)\, dx$$

which can be written as

$$S_n\{F''(x)\} = -n^2 S_n\{F(x)\} + n\big[F(0) - (-1)^n F(\pi)\big] \tag{35}$$

That is, the finite Fourier sine transformation resolves the differential form $F''(x)$ into a linear algebraic form in the transform $f_s(n)$ and the boundary values $F(0)$ and $F(\pi)$ as expressed by (35) above. Formula for the transforms of other derivatives $F^{(2n)}(x)$ of even order may be determined in the same manner. This property is employed in the construction of tables of Fourier transforms.

Example C.9

Consider $F(x) = x^2$, then, $F''(x)=2$ and

$$S_n\{2\} = -n^2 S_n\{x^2\} - n(-1)^n \pi^2$$

Also, $S_n\{2\} = 2 S_n\{1\} = 2[1-(-1)^n]/n$

Thus,

$$S_n\{x^2\} = \frac{\pi^2}{n}(-1)^{n-1} - \frac{2}{n^3}\left[1 - (-1)^n\right]$$

C.4.3. FINITE FOURIER COSINE TRANSFORM

The finite Fourier cosine transformation of a function $F(x)$, $0 < x < \pi$, is denoted below by $C_n\{F(x)\}$ and expressed by

$$C_n\{F(x)\} = \int_0^\pi F(x) \cos nx\, dx = f_c(n)$$

(36)

$$(n=0,1,2,...)$$

where $f_c(n)$ is the resulting finite Fourier cosine transform.

Consider both the object function $F(x)$ and its first derivative to be sectionally continuous; and $F(x)$ is defined, at each point of discontinuity within the said interval, by its mean value, expressed by eqn. (31) above. Thus, the inverse transformation $F(x)$ is given by the Fourier cosine series

$$F(x) = \frac{1}{\pi} f_c(0) + \frac{2}{\pi} \sum_{n=1}^{\infty} f_c(n) \cos nx$$

(37)

$$0 < x < \pi$$

As in the case of Fourier-sine transform, the cosine transform of each sectionally continuous function $F(x)$ exists and it is unique.

In a corresponding manner to the sine transform, the cosine transformation resolves the differential form $F''(x)$ into an algebraic form in $f_c(n)$ and the boundary values $F'(0)$ and $F'(\pi)$, i.e.,

$$C_n\{F''(x)\} = -n^2 f_c(n) - F'(0) - (-1)^n F'(\pi)$$

(38)

$$(n=0,1,2,...)$$

This is again under the condition that $F(x)$ and $F'(x)$ are continuous and $F''(x)$ is sectionally continuous.

C.4.4. JOINT PROPERTIES OF $S_n\{F(x)\}$ AND $C_n\{F(x)\}$

Under the conditions that $F(x)$ is continuous and $F'(x)$ is sectionally continuous, it can be shown that

$$S_n\{F'(x)\} = -n\, C_n\{F(x)\}$$

$$(n=1,2,...)$$

(39)

and

$$C_n\{F'(x)\} = n\, S_n\{F(x)\} - F(0) + (-1)^n F(\pi)$$

$$(n=0,1,2,...)$$

(40)

Alternative formulations of the joint properties of S_n and C_n are

$$S_n\{G(x)\} = -nC_n \int_0^x \{G(x')dx'\}$$

$$(n=1,2,...)$$

(41)

and

$$C_n\{G(x) - \frac{1}{\pi}g_c(0)\} = n\, S_n\left\{\int_0^x G(x')dx' - \frac{x}{\pi}g_c(0)\right\}$$

$$(n=0,1,...)$$

(42)

where $G(x)$ is any sectionally continuous function and $g(n)$ is its Fourier transform.

C.4.5. FOURIER SINE AND COSINE TRANSFORMS OVER UNBOUNDED INTERVALS

Consider $F(x)$ to be a function defined on a specified unbounded interval, sectionally continuous on each finite subinterval and that the integral of $F(x)$ over the unbounded interval exists. If k denotes a real parameter, the Fourier-sine transformation of $F(x)$ is expressed as

$$S_k\{F(x)\} = \int_0^{\infty} F(x) \sin(kx)dx = f_s(k)$$

(43)

$$(x \geq 0, \ k \geq 0)$$

At the same time, the Fourier-cosine transformation is defined by

$$C_k\{F(x)\} = \int_0^{\infty} F(x) \cos(kx) \, dx = f_c(k)$$

(44)

$$(x \geq 0, \ k \geq 0)$$

When $F'(x)$ is also sectionally continuous on each finite subinterval $0 \leq x \leq x_1$, then, $F(x)$ may be represented by either the Fourier-sine or cosine integral formula.

That is

$$F(x) = \frac{2}{\pi} \int_0^{\infty} f_s(k) \sin kx \, dk = \frac{2}{\pi} S_x\{f_s(k)\}$$

(45)

and

$$F(x) = \frac{2}{\pi} \int_0^{\infty} f_c(k) \cos kx \, dk = \frac{2}{\pi} C_x\{f_c(k)\}$$

(46)

i.e., the inverse transforms are given by the transforms themselves.

When $F(x)$ is continuous, $F'(x)$ is sectionally continuous, and $F(\infty) = 0$, it can be shown that the Fourier sine and cosine transforms are interconnected by the following two relations

$$S_k\{F'(x)\} = -k \, C_k\{F(x)\}$$

(47)

$$C_k\{F'(x)\} = k \, S_k\{F(x)\} - F(0)$$

(48)

Further, when the function $F(x)$ is replaced by its first derivative $F'(x)$, the following relations can be written

$$S_k\{F''(x)\} = -k^2 f_s(k) + kF(0) \tag{49}$$

and

$$C_k\{F''(x)\} = -k^2 f_c(k) - F'(0) \tag{50}$$

In the derivation of (49) and (50), it is assumed that $F(x)$ and $F'(x)$ are both continuous and can be integrated, $F''(x)$ is sectionally continuous and that $F(\infty)=F'(\infty) = 0$. The same procedure can be used for obtaining the transform of $F^{(2n)}(x)$.

C.5. Problems

1. Show that

$$S_n^{-1}\left\{\frac{1-(-1)^n}{n^3}\right\} = \frac{x}{2}(\pi-x)$$

   and

$$S_n^{-1}\left\{\frac{1}{n^3}(-1)^{n+1}\right\} = \frac{x(\pi^2-x^2)}{6\pi}$$

2. If $F(x)$ and $F'(x)$ are continuous except that $F'(x)$ has a jump b at $x=c$, where $0 < c < \pi$, and if $F''(x)$ is sectionally continuous, show that

$$S_n\{F''(x)\} = -n^2 f_s(n) + n\left[F(0)-(-1)^n F(\pi)\right] - b \sin nc$$

3. Where a is a constant, prove that

   (a) $f_c(n+a) = C_n\{F(x)\cos ax\} - S_n\{F(x)\sin ax\}$

   (b) $f_s(n+a) = S_n\{F(x)\cos ax\} + C_n\{F(x)\sin ax\}$

(c) $2C_n\{F(x)\sin ax\} = f_s(n+a) - f_s(n-a)$

(d) $2S_n\{F(x)\cos ax\} = f_s(n-a) + f_s(n+a)$

(e) $S_r\{e^{ax}\} = \dfrac{r}{r^2 + a^2}$

4. Prove the following relations, where a is a constant,

(a) $2S_r\{F(x)\cos ax\} = f_s(r+a) + f_s(r-a)$

(b) $C_r\{e^{ax}\} = \dfrac{a}{r^2 + a^2}$

C.6. References

Carslaw, H.S. and Jaeger J.C. (1941) *Operational Methods in Applied Mathematics*, Oxford University Press, London.

Churchill, R.V. (1958) *Operational Mathematics*, McGraw-Hill Book Company, New York.

Doetsch, G. (1974) *Introduction to the Theory and Application of the Laplace Transformation*, Springer, Berlin.

C.7. Further Reading

Abramowitz, M. and Stegun, I.A., Eds. (1965) *Handbook of Mathematical Functions*, Dover Publications, Inc., New York.

Bochner, S. and Chandrasekhoran, L. (1947) *Fourier Transforms*, Princeton University Press, Princeton.

Churchill, R.V. (1941) *Fourier Series and Boundary Value Problems*, McGraw-Hill, New York.

Churchill, R.V. (1958) *Operational Mathematics*, Second Edition, McGraw-Hill Book Company, Inc., New York.

Cost, T. L. (1964) Approximate Laplace transform inversion in viscoelastic stress analysis, *AIAA*, **2**, 2157-66.

Donoghue, W.F. (1969) *Distributions and Fourier Transforms*, Academic Press, New York.

Dortsch, G. (1950, 1955, 1956) *Handbuch der Laplace-Transformation*, Birkhauser, Basel, Vol.1 (1950); Vol.2 (1955); Vol. 3 (1956).

Erdeli, A., Magnus, W., Oberhettinger, F. and Tricomi, F. (1954) *Tables of Integral Transforms*, Volumes 1 and 2, McGraw-Hill, New York.

Paley, R. and Wiener, N. (1934) *Fourier Transforms in the Complex Domain*, American Mathematical Society, Providence, R.I.

Schapery, R.A. (1962) Approximate methods of transform inversion for viscoelastic stress analysis, *Proc. 4th U.S. Nat. Congr. of Applied Mechanics*, pp. 1075-85.

Sneddon, I.N. (1951) *Fourier Transforms*, McGraw-Hill, New York.

Titchmarch, E.C. (1937) *Introduction to the Theory of Fourier Integrals*, Clarendon Press, Oxford.

Tranter, C.J. (1956) *Integral Transforms in Mathematical Physics*, Second Edn., Methuen and Company, London.

Widder, D.V. (1941) *The Laplace Transform*, Princeton University Press, Princeton.
Wiener, N. (1933) *The Fourier Integral*, Cambridge University Press, London.

SUBJECT INDEX

A

Admissibility, 141
Alloys
    Creep of, 244-257
    Stress-relaxation of, 257-265
Anelastic strain, 259

B

Boundary Value Problem
    Elastic, 171, 172, 171-203
    Plastic, 224-235

C

Cauchy's
    deformation tensor, 87
    first equation of motion, 53, 167
    first fundamental theorem, 94
    second equation of motion, 55, 157
    second fundamental theorem, 96
    stress (*see stress tensor*), 209
Christoffel symbol, 291-293
Clausius-Duhem inequality,127, 131, 141, 148, 149, 159, 160
Clausius inequality, 148
Clausius integral, 148
Cofactor, 278
Compatibility condition, 101, 102
Compliance,
    Complex, 295, 299, 310
    Loss, 295, 296, 299, 300
    Storage, 295, 299, 300
Conservation
    of energy, 125, 127
    of mass, 40
Constitutive equation, 137, 217, 218, 249-252

Continuity
    of mass, 41
    of momentum, 42, 55
Continuum, 38, 39, 150
Contravariant
    physical component, 378, 379
    tensor, 16, 369,370, 380
Covariant
    derivative, 379, 380
    physical component, 378, 379
    tensor, 16, 369
Creep
    recovery, 280, 289
    response, 244-257, 259, 274, 277, 279, 280, 284-286, 289, 290, 299, 301, 306, 317, 319
    of metals, 244-256
Curl of a vector, 30
Curvilinear tensor, 362-386

D

Deformation
    Definition, 85, 86, 88
    Elastic, 205
    Homogeneous, 98
    Inelastic, 205
    Isochoric, 97
    maps, 254-256
    rate, 108
    Rigid, 97
    Simple extension, 98
    Simple sheer, 99
Delta function, 388-391
Determinant, 16, 32, 365-368
Differential geometry, 284
Dilatation, 94, 97, 169
Divergence of a vector, 29

417